Underwater Science

Underwater Science

An Introduction to Experiments by Divers

Edited by

J. D. Woods and J. N. Lythgoe

London
Oxford University Press
New York Toronto
1971

Oxford University Press, Ely House, London W.1
GLASGOW NEW YORK TORONTO MELBOURNE WELLINGTON
CAPE TOWN SALISBURY IBADAN NAIROBI DAR ES SALAAM LUSAKA ADDIS ABABA
BOMBAY CALCUTTA MADRAS KARACHI LAHORE DACCA
KUALA LUMPUR SINGAPORE HONG KONG TOKYO

ISBN 0 19 217622 6

Printed by photo-lithography and made in Great Britain at the Pitman Press, Bath

Contents

Figures

Preface

One hundred and twenty-five years ago, Professor Henri Milne-Edwards of the University of Paris dived under the sea off Messina in primitive apparatus 'to pursue marine animals into their hidden retreats'. During the century that followed this pioneering work many scientists ventured under water extending the application of natural history techniques that had proved so successful on dry land. This period saw the invention and development of many tools used by the modern underwater scientist. The standard (helmet) diving apparatus was perfected by Siebe and a variety of self-contained systems was being tried. Underwater photography, pioneered in 1893 by Boutan, achieved a high technical standard despite the rather elementary apparatus then available, and twenty years later motion pictures were being made for both commercial and scientific purposes by Williamson.

This pioneering era of natural history in the sea was terminated by the Second World War. After the war, the new Cousteau-Gagnan aqualung revolutionized diving methods and created a popular sport. The simplicity of the new apparatus and the great mobility it offered attracted a growing number of marine scientists to the possibility of *in situ* observation under the sea. The great majority of these new recruits continued to apply, with increased efficiency, the earlier tradition of natural history in the sea. But a few began to explore the possibility of testing their hypotheses concerning the marine environment and its inhabitants by means of *in situ* experiments. Thus, thanks to the aqualung, the last quarter century has seen the first steps in the transition from natural history to underwater science, comprising the full range of activities from observation, through hypothesis, to experimental test.

The transition from passive observer and collector to active experimenter has revealed the limitations of the aqualung diver, so new methods and apparatus are being developed. Simultaneously, physiologists and psychologists are studying the performance of man in the sea, with particular regard to his ability to perform exacting tasks. The aim of this book is to illustrate these contemporary trends in underwater science in nine essays by active diving scientists. The coverage is selective rather than complete. For example, we have deliberately excluded any reference to the medical aspects of diving, since these have recently been covered in great detail by Bennett and Elliot (1969).[1] But we have included a strong section on recent advances in experimental psychology of divers. These chapters (2 and 3) contain examples of some of the most successful underwater experiments which have required great skill in planning and execution and have yielded invaluable information concerning the diver's performance. This knowledge has helped other underwater scientists when they set out to make measurements in their own experiments.

It soon becomes apparent that the successful underwater scientist is the one who can design his experimental apparatus and method to overcome the special human limitations of working in the sea. In Chapter 1 we identify a number of problems confronting all underwater scientists regardless of their particular discipline, and we have indicated the most practical solutions available at present. The more specialized equipment and methods are dealt with later in the appropriate chapters.

An important theme in Chapter 1 is the need for scientists to accept and make allowance for their lack of regular diving practice; each item of equipment and every technique must be examined with this in mind. The underwater scientist needs the performance of a professional diver without the hours of experience that this normally entails.

Originally we had planned to include a separate chapter on photography, since this is the most powerful tool available to the diving scientist. In general it provides the centrepiece of his experimental design and a carefully interpreted photograph or film can largely compensate for the diver's forgetfulness and poor judgement of time and distance while in the sea. The importance of photography cannot be over-emphasized in underwater research and the diving scientist must become skilled in its application. Happily there exists a large literature on the scientific use of photography and a variety of introductory books on photography under the sea, so we have been able to limit our remarks on the subject to a brief account of commercially available equipment and some recent applications to scientific work.

Our original aim in preparing this book was to provide the non-diving oceanographer and the lay diver with an account of what diving scientists have achieved during the past decade or so. The specialist chapters were designed to help the marine scientist to identify the aspect of his research that might profitably be studied under water. And, being

[1]Bennett, P. B., and D. H. Elliott (1969), *The physiology and medicine of diving*. London: Baillière, Tindall, and Cassell.

aware that many diving projects are not well planned, we hope the book will provide the necessary background to guide science administrators faced with judging proposals for such work. Finally, while no book can take the place of practical experience, we hope ours will help the newcomer to plan his work effectively.

J. D. W.

J. N. L.

1 Apparatus and Methods for the Diving Scientist

J. D. Woods and J. N. Lythgoe

The Inefficient Diver

As soon as the diver leaves the surface he begins to get cold and fuddled and stupid. If he is breathing air his depth is limited because oxygen becomes toxic at high pressures and nitrogen has a narcotic effect. If he is to avoid the bends his working time is limited by the need for slow decompression. His vision is commonly restricted to one metre or less and it almost never exceeds 40 m. He cannot communicate without special equipment and all his equipment must withstand high pressures. The conduct of each dive is ruled by physical and physiological limits which it is always dangerous and sometimes fatal to exceed. Compared to a scientist at his bench a diver is very inefficient indeed.

It is not surprising that the scientist's first reaction when he is faced with a problem requiring divers is to hire somebody else. This of course is only sense when there is an exactly defined task to do, but this is not often true in research. It is hard to think of a more talented team of scientist and engineer-diver than Isaac Newton and Mr. Halley. Yet the following quotation from Newton's *Opticks* serves as a classic example of the errors to which a scientist relying on another person's information is prone.

> Of this kind is an Experiment lately related to me by Mr. Halley, who, in diving deep into the Sea in a diving Vessel, found in a clear Sunshine Day, that when he was sunk many Fathoms deep into the Water the upper part of his Hand on which the Sun shone directly through the Water and through a small Glass Window in the Vessel appeared of a red Colour, like that of a Damask Rose, and the Water below and the under part of his Hand illuminated by Light reflected from the Water below look'd green. For thence it may be gather'd, that the Sea-Water reflects back the violet and blue-making Rays most easily, and lets the red-making Rays pass most freely and copiously to great Depths. For thereby the Sun's direct Light at all great Depths, by reason of the predominating red-making Rays, must appear red; and the greater the Depth is, the fuller and intenser must that red be. And at such Depths as the violet-making Rays scarce penetrate unto, the blue-making, green-making, and yellow-making Rays being reflected from below more copiously than the red-making ones, must compound a green.

Newton concluded that the light at the bottom of the sea is red—he could hardly have made this mistake had he accompanied Mr. Halley.

If possible it is better for the scientist to make his own underwater observations and the rest of this chapter broadly describes ways in which he can make himself more efficient, comfortable, and safe. Once he has defined his task so precisely that apparatus can be designed to do it for him he is probably better off sitting in a warm room interpreting the data the instrument has recorded. The chapters which follow this one contain accounts of research where the greater intelligence and adaptability of the diver over remotely-controlled apparatus has been important.

Training and Safety

Diving to normal depths is NOT dangerous, although it is classified quite rightly as a high risk activity. This apparent paradox is the cause of many administrative problems in connection with underwater research programmes such as rules and regulations, payment, and insurance. It is natural that the director of an institute is particularly conscious of the risks involved in diving activities, while the actual diving scientists are more conscious of the research opportunities. The laws of physics and physiology impose a number of basic limitations to man's activities under water and these are learnt by the novice diver during his training. To exceed these limitations is foolhardy and dangerous, but if diving regulations during and after training are framed in such a way that all operations are conducted within them, then diving can be very safe indeed. Most of the events and factors affecting each and every dive can be predicted and plans made for various contingencies. By contrast, driving a car in heavy traffic involves many more unknowns, and the forming of quite unwarranted assumptions about environmental factors (e.g. other vehicles) which make it inherently more dangerous than diving. On the other hand, as Miles (1962) points out, all mishaps occurring under water which could cause unconsciousness but would be non-fatal in air where the victim could lie around waiting for help, are potentially fatal in water owing to the risk of drowning. It is this fact that gives diving its high risk rating, and accounts for the inclusion in all diving regulations of a clause about diving in pairs or with adequate communications and links to the surface.

The psychological problems involved in setting up and carrying out a research programme involving diving should not be underestimated. The first of these is the natural apprehension and caution of the institute director or department head who is ultimately responsible for the personal safety of all his staff, coupled with a reluctance on his part to interfere in the details of their work. Secondly, there is the freedom of approach to specific problems that is expected of, and by, the individual scientist; which is in this case associated in part with the proverbial independence of spirit of the professional diver. The principle to be followed by the director of an institute embarking on underwater research is to choose the man who will lead the work and then to give him a free hand in writing and interpreting the rules, bearing in mind that if the rules are too strict individuals will tend to break them. It is essential that the training programme be so geared to the type of diving that is being performed in the research institute concerned that the chief diver can effectively eliminate all those who might not be able to cope with some aspects of the subsequent diving work: for example, turbid or dark water, confined spaces or the vicinity of fishing nets. There must be selection as well as training of individuals.

Breathing Apparatus

The free diver, unlike the submariner, is not protected from the hydrostatic pressure of the sea and he must, therefore, breathe air (or some more exotic gas mixture), supplied at the ambient pressure of the environment by an automatic valve. The majority of underwater scientists use breathing apparatus based on the original Cousteau-Gagnan aqualung developed during the early 1940s. A supply of air, compressed to approximately one ton per square inch, is carried in steel or alloy cylinders worn on the diver's back and supplied via a demand valve and rubber hoses to the diver's mouthpiece. Exhaled air is released to the sea in such open-circuit diving apparatus. Self-contained apparatus of this kind is manufactured in a variety of forms in many countries throughout the world and these permit the wearer complete freedom of action down to a depth of rather more than 70 m. The capacity of the air tanks is usually about 15 litres at 150 Kg./cm.^{2} which gives an endurance of about 45 minutes at a depth of 20 m., or about 10 minutes at 70 m., but these figures vary considerably with individual performance and with the nature of the physical effort (and hence the breathing rate). Greater endurance is often achieved, though at the expense of mobility, by piping air to the diver from a low pressure compressor or a bank of large cylinders on the surface.

Military divers generally use a more complicated apparatus in which the exhaled air is rebreathed after being passed through a chemical absorbent for carbon dioxide. It is usual with such *closed circuit* apparatus to breathe a gas consisting of nitrogen and oxygen in one of a variety of standard mixtures selected on the basis of the maximum depth to be encountered on any given descent. Although the virtues of

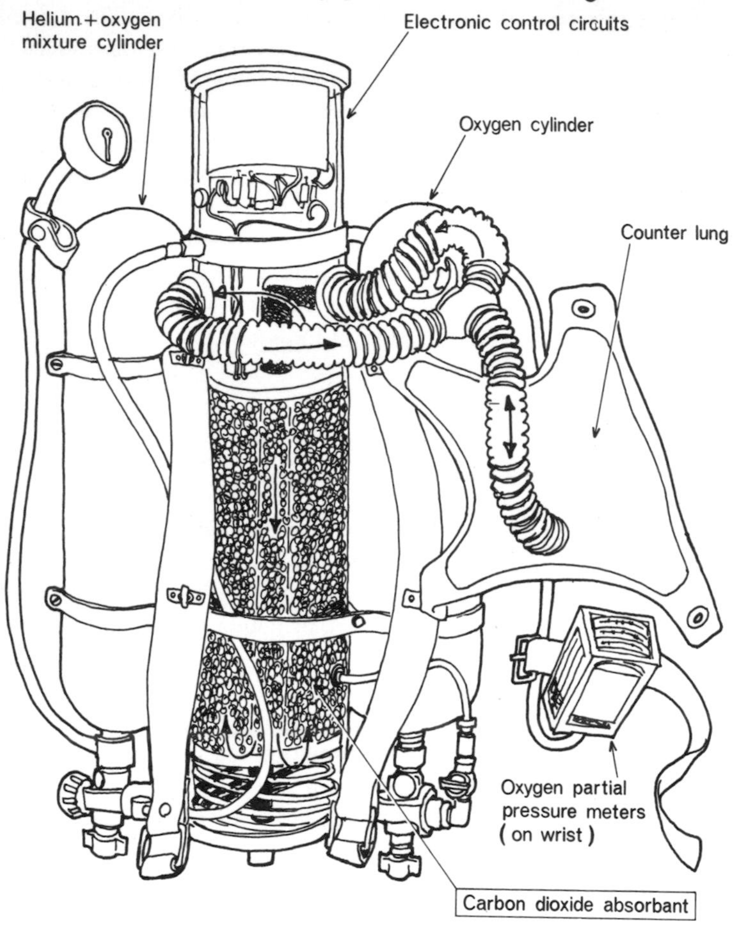

Fig. 1

The Starck-Kanwisher *electrolung*, manufactured by Oceanic Equipment Corporation of Miami, is the first closed-circuit breathing apparatus designed specifically for use by the underwater scientist. The oxygen partial pressure is measured and controlled electronically; helium is used as the inert gas and carbon dioxide is absorbed chemically. The helium and oxygen cylinders of the electrolung each hold about one cubic metre of gas (about $2 worth of helium and 40c worth of oxygen). The average dive to 90 m. uses less than one-third of the helium supply and the oxygen is sufficient for four to six hours, depending upon the diver's physiology and activity levels. (For further information see the report in *Oceans*, **1** (2), 45–8, 1969.)

closed circuit apparatus, which include longer endurance and a reduction in noise, exhaled gas bubbles, and decompression times, make it particularly suited to scientific work, it has been so far used mainly by military and commercial divers, and only rarely by scientists. The reasons are probably the additional complication involved in replacing the chemical absorbant before each dive, and the problem of refilling. An air compressor obviously can no longer be used, but a bank of cylinders of mixtures must be available instead. The same closed circuit apparatus may also be used for helium–oxygen gas mixtures, which make a significant improvement in mental clarity at depth (see Chapter 3) though at the expense of increased decompression problems (e.g. Flemming 1966). At the time of writing, a novel, electronically-controlled closed circuit apparatus designed specifically for scientific diving is being tested by Starck, Kanwisher, and Link in the Caribbean. This device, which is silent, has several hours' endurance and will work to 300 m., promises to revolutionize scientific diving in the not too distant future.

Controlling Heat Loss

Except in deep or tropical waters, a diver's endurance is usually limited by cold rather than by an insufficiency of air. The purpose of his rubber suit is to reduce heat loss to the sea by interposing an insulating layer of air between his skin and the colder sea-water. However, no really satisfactory diving suit is available. Insulation provided by the popular foam neoprene materials decreases significantly with depth as the air bubbles become compressed by the hydrostatic pressure. Betts (1965) has reported that the heat lost from a diver wearing a 4 mm. thick neoprene suit may increase by as much as a factor of four as he descends from the surface to a depth of 30 m. in isothermal water. As the diver returns to the surface the heat loss drops again and the resulting feeling of warmth is frequently interpreted as being due to an apparent rise in water temperature (the so-called 'false thermocline'). It is possible to overcome the depth effect by using a neoprene material filled with hollow glass spheres, whose volume does not decrease at depth, but the residual heat loss is still sufficient to reduce efficiency and endurance.

The ideal solution is to replace the heat lost to the sea, and several workers have recently perfected practical systems based on either incorporating an electric heater into a waterproof (dry) suit (e.g. Barthelemy, Berry, and Michaud 1968) or by piping hot water from a boiler on the surface to a circulation built into the diving suit (e.g.

Krasberg 1968). In the electrical system, the power consumption of 260–400 watts is supplied from batteries worn around the diver's waist. Given adequate heating, divers have worked for many hours in water temperatures as low as 5°C without discomfort or fatigue. Another solution is the constant volume dry suit, of which an example used successfully by biologists in Northern waters is the Swedish *Poseidon Unisuit*.

Underwater Houses

One of the simplest ways for man to go under water is to surround himself with air at ambient pressure supplied by a pump on the surface. In its earliest form, supposed to have been used by Alexander the Great,

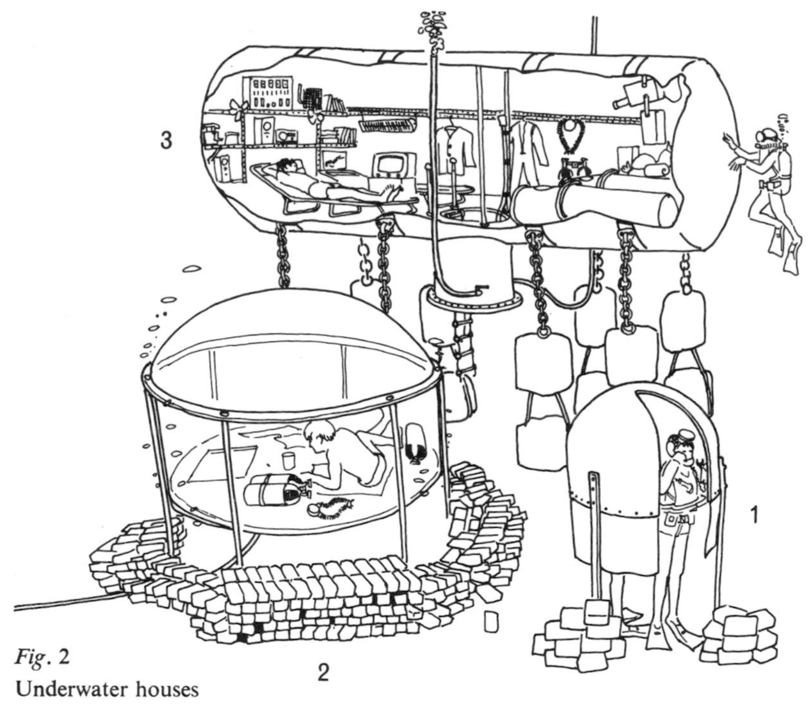

Fig. 2
Underwater houses

1. A simple shelter used as a sea-floor telephone booth and equipment store.
2. A perspex shelter suitable for dry decompression, note-taking, on-site discussions and refreshment. The shelter (and its seven tons of iron ballast) can be erected by four divers in a few hours. (Jones 1968)
3. Cousteau's *Precontinent 1*, the first of a series of sea-floor houses used for underwater living experiments involving saturation diving for several weeks at depths in excess of 100 metres.

this is a diving bell (Fig. 2) which is open to the water underneath. Davis (1951) described many of the early diving bells in his classic text on commercial diving. Recent developments of the basic idea have been used commercially for saturation diving involving periods of several days at depths of more than 60 m. (Krasberg 1968) and for naval deep diving trials. A feature that is common to all these diving bells is the facility for raising and lowering them from a ship to the seabed: underwater houses on the other hand are essentially diving bells that remain in one position on the sea floor. This description emphasizes the dependence of underwater houses upon the surface for their air supply. It has not yet proved feasible to make a totally independent underwater dwelling, though this ideal may become an expensive reality in the next decade (by virtue of the unlimited power afforded by relatively small self-contained nuclear reactors).

The most exciting use for an underwater house lies in saturation diving. Several experiments during the past seven years have demonstrated the feasibility of sustaining divers on the sea bed at ambient pressure for a week or more while they make regular sorties in conventional diving apparatus. The advantage gained by this procedure is the elimination of lengthy decompression at the end of each working dive from the surface in return for a single decompression when the job is completed. Commercial diving companies now use saturation diving techniques as the most economic solution to many long tasks in deep water, especially in offshore oil operations. At the time of writing, no underwater house has been established for use by scientists in the way that research ships or the French 'Bouée Laboratoire' are. However, this requirement will no doubt be met in the not too distant future; meanwhile several of the temporary houses used in underwater living experiments have included scientists amongst their residents (for example, the 1970 Tektite II experiment in the Virgin Islands).

On a less ambitious level, there are many situations in which a simple underwater house (or, more strictly, shelter) may usefully be used for periods so short that inert gas saturation does not occur. Such underwater shelters have two principal uses. Firstly, they may be used as the marine equivalent of the ornithologist's hide—a base to leave equipment, make notes in the dry, telephone the surface, tag live fish and make observations without the need to surface. Apart from the obvious advantage of performing these tasks on site, but in the dry, there is an additional saving in decompression time if all unnecessary ascents to the surface are avoided. However, care should be taken when decompressing after extended occupation of an underwater shelter since such dives almost invariably extend beyond the limiting line of the Royal Navy decompression tables. Jones (1968) has shown that 24 hours in an air-filled shelter at 10 m. depth can lead to bends if neither decompression nor oxygen flushing (Cousteau 1964) is performed.

The second benefit to be gained from the use of a simple underwater shelter is in decompression after a single deep dive. Even after a dive to 60 m., the majority of the time spent decompressing lies within 10 m. of the surface and the tables may conveniently be adjusted to permit a single final stop at 10 m. in place of the series of stops at somewhat shallower levels. This single stop is performed in the comfort of a fixed shelter, where the diver can dry himself and warm up with a hot drink (from a vacuum flask). He then proceeds to make an immediate report on the results of his dive, so overcoming the forgetfulness which follows a lengthy, boring decompression in cold, rough waters.

Finally, the psychological values of a shelter should not be underestimated; the lessening of anxiety, especially at depth and in cold, murky waters may lead to a significant improvement in diver performance (see Chapter 3). In the extreme case, it has been noted (Norman 1967) that an underwater shelter can become a funk hole for an anxious diver, who may become unwilling to leave its relative safety.

Many of the complications inherent in a seabed house designed for extended occupation may be neglected in the simple shelter discussed above. The open-circuit system described by Jones (1968) and illustrated in Fig. 2 is quite adequate yet remains simple to erect and maintain. The primary operating criterion is to ensure that the carbon dioxide concentration does not exceed a partial pressure of 0·05 atmospheres; in Jones's 'Bubble' CO_2 concentration was measured every hour by means of expendable samplers (rather like police alcohol 'breathalysers') manufactured by Drager-Normalair. A mean air inflow of about five litres per minute ensured a safe atmosphere, but in practice the rather noisy air supply could be shut off for brief telephone conversations without dangerous CO_2 build-up.

Transport

Swimming long distances under water is extremely tiring, not to say inefficient, so the natural inclination of a diver wanting to move from one site to another is for him to rise to the surface cover boat which can carry him (on board or hanging alongside) to the new site. This procedure is sound since vertical movement in the sea is made easy by virtue of buoyancy control, while horizontal movement is tiring and may introduce errors of navigation. Nevertheless there are occasions on which a diver needs to cover long distances (greater than 100 m., say) without losing visual contact with the sea floor. Once he has become accustomed to using a mechanical aid to assist his longer journeys, the serious diver may well find that they are worth while on routine working dives covering relatively modest distances.

Towed vehicles

The most widely used device for exploring large areas is the aquaplane (Fig. 3); the simplest version is a board, which, when tilted downwards or sideways, provides a dynamic thrust to counter the corresponding pull on the towing cable. The addition of a broom-handle seat and proper balancing of the towing points permit one-handed control of the flight path. With this aquaplane, which can be made in a few hours from materials available in the field, a diver may be towed at speeds of two or three knots by a rubber boat, the maximum speed being limited by the hydrodynamic forces that tend to tear off the diver's mask.

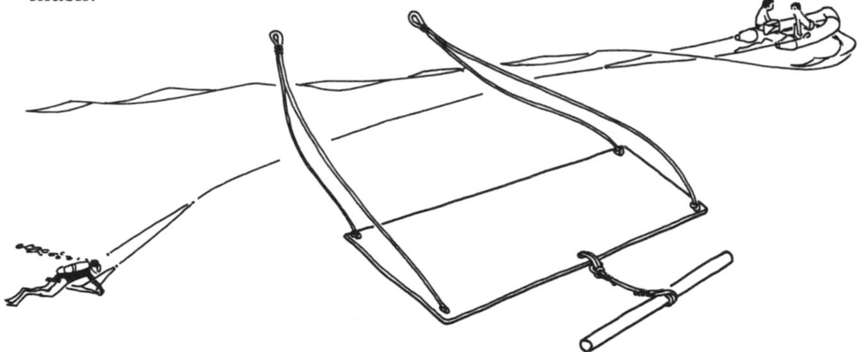

Fig. 3
A simple board suitable for towing at 1–2 knots.

Many designers have proceeded to incorporate improved controls and a face shield into this elementary design in order to improve manœuvrability and pilot comfort (and hence his performance). These improvements do also permit a somewhat higher towing speed, though this can seldom be exploited in water of less than perfect visibility without a risk that the diver will fail to recognize and respond to objects on the seabed. A notable early example of these rather more sophisticated towed vehicles is the French Navy machine used in searching for mines (Fig. 4). More recently the Aberdeen Marine Laboratory has experimented with Cdr. Brookes's *Mobel* for inspecting trawl nets; the excellent manœuvrability of this vehicle permits the operator to position himself above or even inside the mouth of the trawl. Young (1967) reports that a complex Russian towed submersible (containing an atmosphere at normal pressure), the *Atlant 1 bathyplane*, is now in service for trawl net observations. Naturally these heavy and complicated machines need a large towing vessel, usually the same trawler as is pulling the net under inspection.

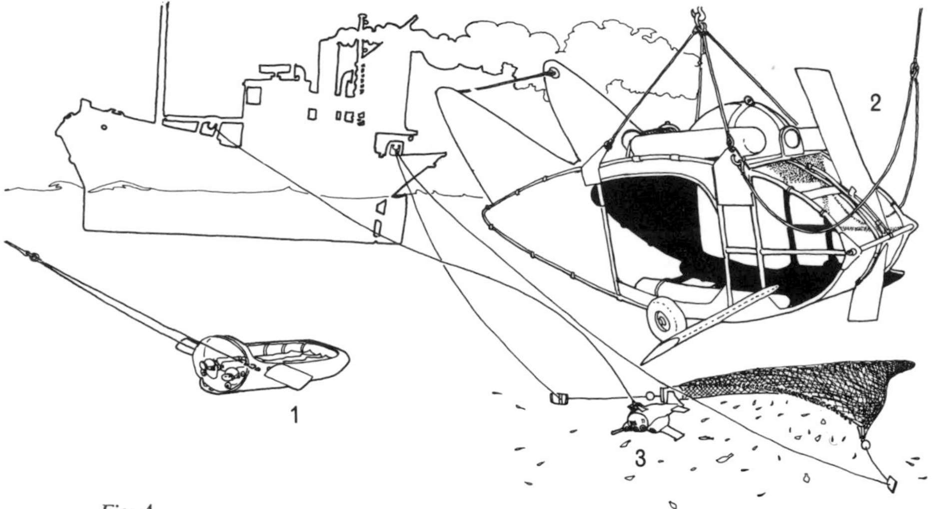

Fig. 4

1. A streamlined, free-flooding, towed vehicle used by the French Navy when searching for mines.
2. One of the more elaborate enclosed vehicles that provide the fisheries scientist with first-hand evidence of the behaviour of trawl-nets.
3. The free-flooding Brooks *Mobell*, used in trawl studies at the marine Laboratory, Aberdeen.

Electric tugs

During the Second World War, Italian divers perfected a two-man chariot for use in attacking Allied shipping in the Mediterranean Sea and subsequently these machines were used by the Royal Navy (Fig. 5). After the war, Rebikoff and Cherney (1955) developed a succession of increasingly heavy underwater cameras and strobe units, which ultimately needed their own electric propulsion unit. This subsequently

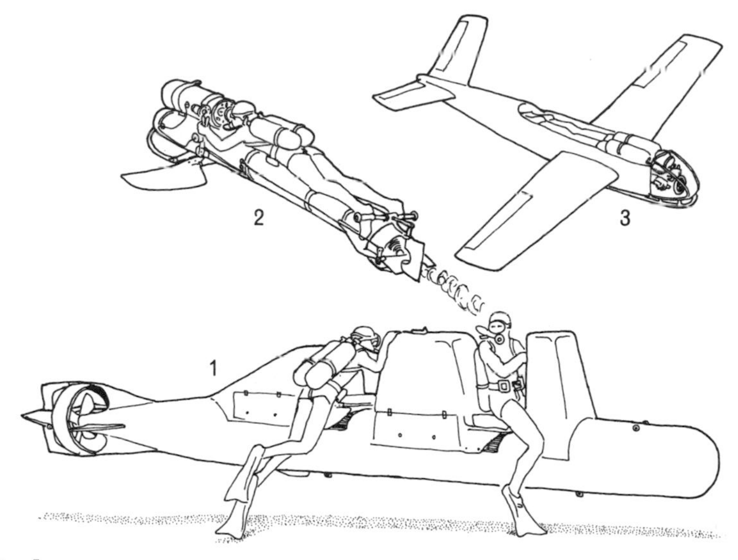

Fig. 5

Self-propelled vehicles for divers.

1. A military chariot of the type used by Italian and British naval divers during the Second World War.
2. Rebikoff's *Pegasus*—a highly manœuvrable electric torpedo fitted with comprehensive instrumentation for navigating in murky water.
3. The Irving underwater glider, designed to gain forward momentum from buoyancy changes controlled internally as in a submarine.

evolved into the remarkably manœuvrable *Pegasus*—later versions of which have been used effectively in tasks as diverse as photogrammetric surveys and pipeline inspection. Military vehicles derived from the *Pegasus* idea are currently being developed in several countries. Regrettably the costly batteries used in such sophisticated machines make them too expensive for most research purposes and a simpler device is needed. Cousteau (1964) used a powerful tug capable of towing up to four divers at their normal swimming speed and his later models have incorporated a seat for the diver. Like all machines in this class, the vehicle is manœuvred by directing it as a whole, rather than by elevators and rudders: the gain in strength and simplicity more than outweighs the slight extra effort needed by the diver. Given adequate power supplies, the gain in diver performance achieved by using these electric tugs for all movement under water fully justifies the extra logistic support involved.

The Irving glider

In a theoretical paper in 1964, Irving calculated the likely performance of a winged vehicle that would harness the buoyancy control possible with compressed air cylinders to achieve quasi-horizontal flight. The calculations showed that flights of over one mile at, say, two knots are possible on a cylinder containing 80 cubic feet of free air. The speed would depend on the angle of glide, but Irving suggests that the best figure would be in the region of 1 in 10. Originally proposed for sport diving, this underwater glider might prove invaluable to the scientist who often has compressed air, but no electricity and who requires silent operation.

The Communication Problem

The inability of divers to converse freely between themselves and with their companions on the surface has been a source of frustration since free diving began and is a particularly serious limitation to underwater research. Despite many attempts to solve the problem there is still no wholly satisfactory commercial apparatus that will permit clear speech communication at will. In part the difficulty has been a failure to recognize the nature of the problem and it is only in the past year or so that we have had an adequate description of one of the principal difficulties, namely the need for the masked diver to adjust the processes by which he normally forms his words. In practice the diver fails to produce the sound that he seeks and consequently the clarity of his speech is limited before transmission, rather than distorted by the transmission process itself (which is now fairly well understood). The introduction of noise from the breathing apparatus further reduces intelligibility.

Speech formation by divers

The processes of normal speech formation have been studied with considerable success during the past few decades, notably at laboratories sponsored by the manufacturers and users of telephone and radio equipment. The analysis of speech wave forms have been related by X-ray studies to the controlled structure of the vocal tract which has then been analysed in terms of an equivalent electrical circuit. The difference between a diver's speech and that of someone speaking in the open air is due principally to the former's enclosed mask or mouthpiece. The strong reflection of sound at the air:water boundary makes the resulting distortion far worse than that for an airman wearing an oxygen mask, whose speech has been subjected to extensive study. One method of analysis is to consider how the mask may be represented in the equivalent electrical circuit mentioned above. In order to produce the desired sound, the speaker tries to reshape his vocal tract to cope with the effect of the mask. Ideally, the adjustment is complete and the desired sound is indeed produced, but more usually the resulting speech only approximates to that sought. The degree of departure from normal sound depends upon the properties of the mask, the natural voice of the speaker (women, with higher pitched voices, tend to have less difficulty than men), and the individual's experience with the mask. In the case of airmen wearing oxygen masks, the speech is rendered adequately intelligible by virtue of careful mask design and by appropriate training and practice, but the same success has not yet been achieved with divers. As a rule, speech is clearer when a diver wears a large mask (see Fig. 6).

Fig. 6
Various masks that permit a diver to speak under water. He encounters least difficulty in speaking and the resulting speech is clearest when the diver wears a mask enclosing a maximum volume of air. Speech distortion is the major difficulty hindering communication between divers.

The effect of wearing a mask is present at all depths. At great depths, where the diver breathes a mixture of oxygen and helium rather than the usual nitrogen–oxygen mixture, there is a new effect due to the increased velocity of sound in helium. The phenomenon of 'helium speech' is now well documented (e.g. Rowarth 1969) and several workers have claimed to have made translators which produce relatively normal speech from the distorted 'Donald Duck' voice of a helium filled vocal tract. However, early optimism that the problem would be solved by a simple displacement of the harmonics of helium speech were confounded when it became clear that intelligibility deteriorated markedly with increasing depth. A system that yields an adequate translation at three atmospheres pressure may be quite ineffective at ten atmospheres, where the use of heliox mixtures becomes obligatory for free divers. Recent helium speech decoders use a non-linear frequency shift for the speech harmonics with considerable success.

Hearing under water

One of the most attractive methods for communicating with divers is to transmit amplified speech directly through the sea. This system, like the loud-hailers used in air, has the great virtue that no special receiver is needed; the diver hears the sound directly. Consequently, several commercial devices (notably the Bendix 'Watercom') exploit this system and a variety of experimental devices have been tested with varying degrees of success.

However, there is still some confusion over the precise mechanism of hearing under water. It has generally been assumed that the acoustic impedance mismatch at the interface between the sea-water and the enclosed air would decrease the effectiveness of the diver's ears, while the closer acoustic matching with the skull would result in the majority of sound reception occurring through bone conduction rather than by the normal aural process. This change would lead to a loss of binaural directional acuity, an effect which would be enhanced by the fivefold increase in speech wavelengths in the sea compared with the air (due to a corresponding increase in the speed of sound in sea-water over that in air). Ray (1968) has pointed out that the loss of binaural acuity will lead to a decrease in the diver's ability to understand speech in an environment such as a rocky sea floor, where there is noise and multiple path distortion. Attempts have been made to overcome this problem by covering the diver's skull with soundproof material in order to restore the relative effectiveness of the ears, but it seems unlikely that such a shield could be made sufficiently compact for routine use. Alternatively it is possible to increase the effective separation and relative sensitivity of the ears by fitting them with a pair of hydrophones set several feet apart, but this is scarcely practical for operational diving.

Despite these objections the best underwater loud-hailers are useful tools needing no special receiver and future development of this system may further exploit the ability of the brain to discriminate between the wanted signal and a noisy background subject to multiple path distortion, which an electrical receiver cannot.

The importance of training

Learning has an important part to play in underwater communication. Professional divers using communication apparatus regularly learn to form their words carefully and they learn which phrases are readily understood and which cause difficulty. Equally, regular use will increase the listener's ability to understand a diver's distorted 'dialect'. The problem facing the underwater scientist is that he seldom attains this degree of proficiency because he dives infrequently. So, even though effective devices are available at a reasonable cost well within most research budgets, they are rarely used. Yet the potential of these devices is enormous; almost every underwater experiment would benefit from a modest level of speech communication between the divers and their companions on the surface. One solution may be to exploit the learning potential by pre-dive training; in this case by a form of language laboratory, programmed to prepare an underwater team for the dialect of diving.

The transmission of sound through the sea

Many early communication devices functioned excellently until they were tested in the open sea, where their range proved far less than had been predicted on the basis of laboratory test. The discrepancy was due to transmission losses in the sea, which may be explained in terms of well understood physical principles. Since the physics of sound transmission through the sea is discussed in great detail in the standard textbooks (e.g. Horton 1957, Albers 1965, Tucker and Gazey 1967) it will be sufficient here to mention briefly the more important processes. The main problem in rocky coastal waters, near the sea floor or against a cliff face where most divers work, is the high level of noise due to animals, waves, and boats. Speech communication between divers in such circumstances is as difficult as the whispered conversation between ornithologists during the dawn chorus near a busy airfield. In quieter waters the problem of signal loss due to refraction by temperature gradients and reflection at the surface becomes important. Scattering and absorption may become significant in rough weather when the uppermost layers of the sea contain bubbles from breaking waves. These bubbles absorb sound at high frequencies just as fog droplets absorb the high pitched sounds emitted by bats.

Causes of transmission losses for sound in the sea	
1. Absorption	In pure sea-water, absorption increases exponentially with frequency, setting a practical limit to low power transmission at about 0·5 MHz.
2. Scattering	Most important in turbid water or in the presence of large concentrations of fish. Serious losses occur at the *deep scatter layer* consisting of swarms of fish which rise to near the surface at night and sink below the depth limit of normal diving during the hours of daylight.
3. Refraction	(a) In isothermal water, the speed of sound decreases with pressure and hence with increasing depth, at a rate of approximately 1·8 m/s. per 100 metres. (b) In the thermocline, the speed of sound increases as the water becomes colder, bending the soundwaves downwards in the summer thermocline, located at about the depths commonly attained by divers (see Chapter 9). A temperature gradient of about 0·07°C/metre counters the pressure effect; higher gradients cause increasingly sharp downward refraction. Communication along nearly horizontal paths is most seriously affected, while vertical paths (e.g. between the surface and a diver directly below) are not.
4. Multiple path distortion	(a) Multiple reflections off the sea surface give the Lloyd's mirror effect, which leads to fading when the receiver is near the surface. (b) Multiple reflections from a rocky sea floor or in caves leading to signal distortion and fading. The loss of signal clarity may be overcome to some extent by exploiting binaural acuity (see 'hearing' above).

Modulated carrier wave systems. An equally popular method for transmitting speech between divers is by means of a modulated acoustic carrier wave. The usual carrier frequencies lie in the range 50–80 KHz, which have wavelengths in the sea of approximately 2·7–1·5 cm. The acoustic absorption of sea-water at these frequencies is still relatively weak and under normal noise conditions (see Horton 1957) ranges of up to a mile or more can be achieved with an acoustic power output of under one watt. Several devices (for example the U.S. Navy 'Aquavox') use a directional transducer, which can be made omni-directional at the expense of a reduced range by reflecting the signal off the sea bed or the surface. Several early systems developed in Britain (e.g. Woods 1962) employed a spherical transducer to give an omni-directional signal, but these have since given way to tubular transducers, which offer an adequate cylindrical symmetry without the poor sensitivity to returning plane signals suffered by the spheres (Tucker and Gazey 1967). Following Horton's advice most early devices employed amplitude modulation, but this has since been replaced almost exclusively by frequency modulation. The single side-band f.m. carrier-wave system has reached a commercially viable state in the French 'Erus' manufactured by La Spirotechnique. This device, selling at about £1,000 for two diver units and one surface unit, has a useful range of several hundred metres under most conditions. While the Erus's level of intelligibility is adequate for the professional diver, it may usefully be improved for the less experienced scientist by replacing the rather small mouthpiece by a full-face mask.

Some other methods. The oldest device for communicating with divers, the telephone, is ideal for the hookah or surface demand diver, who is necessarily linked to the surface by his air pipe, but it usually places an unacceptable restriction on the free diver. A possible compromise, suitable for use in especially difficult conditions, would be to use the ultra lightweight twin wire produced by Sippican for their expendable bathythermograph. This expendable telephone wire would offer reliable communication without placing any restriction on the diver's movements, though at a cost of about $10 per dive. A cheaper solution is for the diver to tow a floating radio transmitter by means of a neutrally-buoyant telephone cable. Several commercial plug-in telephones are also available for conversation between divers.

Divers have also successfully used speech frequency electromagnetic induction to converse over very short ranges. The principal difficulty encountered in designing such underwater radios lies in achieving an efficient antenna. De Sanctis has used a wire aerial suspended above the diver's head by a float, while others have used electrodes on the head and feet. The range of these devices has been a few metres—far shorter than the theoretical limit set by the skin depth for speech frequency radio waves in sea-water. These devices have never proved popular and they cannot compete with the acoustic devices described above.

Finally, when a fairly permanent experimental site is planned, it is worth incorporating a small air-filled shelter of the type described on page 8. The shelter should be large enough to accommodate the heads and shoulders of two divers and should be fitted with a telephone to the surface. This remains the simplest, the cheapest, and the most effective method for on-site discussions between divers and with the surface—a poor, if typical, reflection on contemporary diving practice.

Recording Methods

The simplest method for recording one's observations under the sea is to write or draw with a graphite or wax pencil on a white, double-sided board made of some plastic laminate such as Formica. These records will be sufficiently permanent to withstand normal handling during a dive, but they should be transcribed into a notebook directly after surfacing. Any delay before transcription increases the risk of misinterpreting the often rather erratic notes made under water, where mental efficiency decreases rapidly with increasing depth, cold, and fatigue. The principal symptom of this drop in mental performance is a striking loss of memory suffered by even the most experienced divers. During routine dives it is possible to overcome amnesia by preparing beforehand a list of each task to be undertaken and tables for the clear entry of all measurements required during the dive; these lists and tables are inscribed on the Formica pads. In some cases, however, it is desirable to retain the original records (particularly important in the case of archaeological drawings, for instance): drawings are then made with wax crayons, such as Chinagraph, on a waterproof paper, such as Permatrace or Draftex, attached to the Formica board by screws or rubber bands.

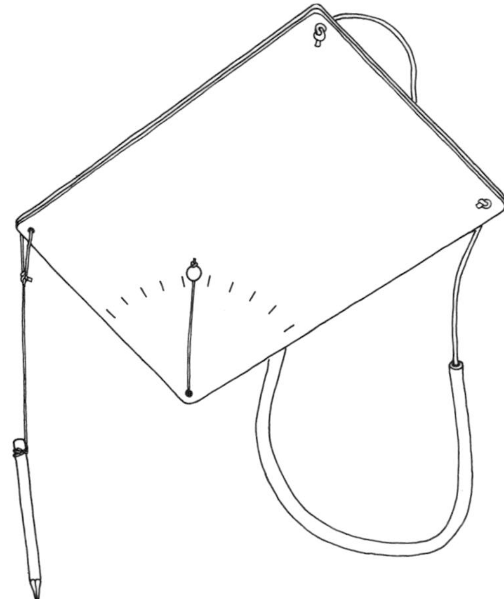

Fig. 7

A Formica notepad used by divers. A scale, a table for experimental data, or (as illustrated) angles of inclination may be permanently inscribed on the board; notes are made with a graphite or wax pencil.

Several attempts have been made to record the diver's spoken commentary on magnetic tape. Modern cartridge tape-recorders are ideal for mounting in a pressure case and are readily fitted with a pressurized bone-conduction microphone of the type used in divers' telephones. However, these apparently very desirable aids have not come into general use. The reason for their limited success with scientists lies in the problem of speech formation discussed on page 12. The cumbersome face-masks needed to achieve an acceptable clarity of speech are unpopular with the scientist who spends only a few weeks diving each year, while the distorted speech from commonly-used masks is difficult to transcribe. The transcription problem is further aggravated by the noise generated in most open-circuit demand valves. Thus the diver's tape-recorder is typical of many of the gadgets described in this chapter; it has obvious potential, but unless the inherent limitations are understood before incorporating it into an experimental programme, the machine may prove a constant source of frustration rather than the expected aid.

Fig. 8
A miniature tape-recorder is used to record a detailed commentary of the diver's observations.

While the direct recording of speech under water has its problems the recording of suitable coded data on magnetic tape is far simpler and may leave the diver free to concentrate on more complex tasks. Suitable instruments (for example, the Plessey digital data logger) record the inputs from six or more sensors, such as water temperature, current speed and direction, or light intensity at several points scattered about the diver's working area. Alternatively the instrument may be attached to the diver to record the physical variables encountered during his dive in a manner analogous to an aircraft flight recorder.

Photography

Undoubtedly, however, the most powerful recording technique available to the diving scientist is photography and in most underwater research it is best to design the experimental programme so that it makes the most effective use of still and motion cameras. These should almost invariably be the first choice, with sketching and other methods introduced to support the photographic record rather than vice versa. Indeed it is usually worth while to use exotic photographic techniques to overcome special difficulties and only as a last resort should photography be abandoned completely. The virtues of photography in scientific research have been discussed by many authors (e.g. Engel 1968); under water it provides a detailed record of the many secondary features that surround the main subject of the diver's attention. This secondary evidence, missed at the time by the bemused diver, often proves to be crucial in the subsequent analysis—no other system of recording has this essential feature. The problems of scientific photography under water are essentially twofold. Firstly, the cameraman must obtain the best results that the conditions permit (and *best* in this context means the most useful for the experiment in hand and not necessarily the most artistic). Secondly, the scientist must make an accurate interpretation of the information on the resulting photographs or motion film. The many books on underwater photography have been concerned with the techniques of taking artistically satisfying pictures to the neglect of the skills needed for their correct interpretation; we shall start to correct this imbalance in the section that follows, but specific examples of the incorporation of photography into underwater experiments will occur throughout this book.

The special problems encountered in taking photographs under water will be apparent from Chapter 4. They include poor contrast, low light intensities and the strong blue or green cast of the ambient light. Much can be done to improve the light intensity and correct the colour balance by the use of bulb or electronic flash and suitable correction filters. Very often the photographer wishes to record the colours which he observed by natural light; this is not so easily done as it might seem for the eye adapts to allow for the green or blue cast of the environment (see p. 134), whereas the film emulsion does not. The best solution would be to develop special colour films for underwater use, but the potential market would not allow this to be economic and the photographer must rely on correction filters instead.

The photographic equipment available to the underwater scientist has been recently summarized by Richter (1968) and Mertens (1970). Cameras (and light-meters) are usually housed in specially built pressure casings but the 35 mm. Calypsophot (now taken over virtually

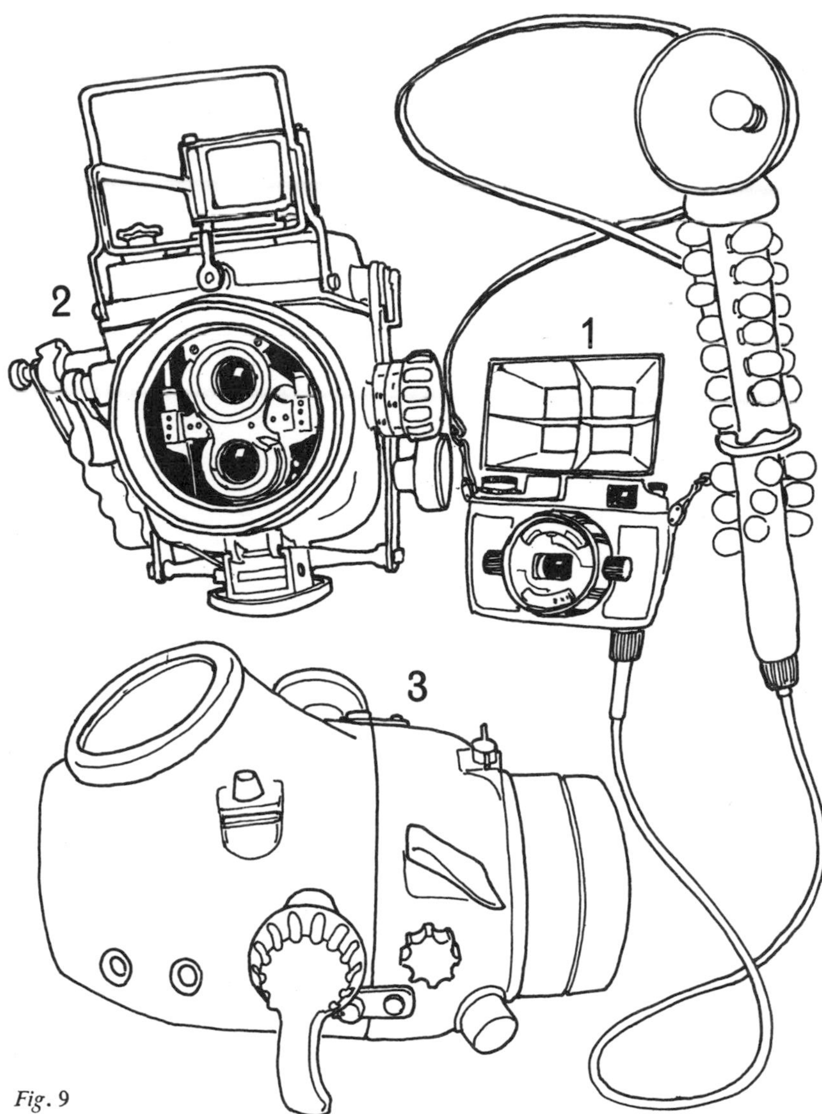

Fig. 9
Cameras used by underwater scientists.

1. The Nikonos is compact and offers 36 exposures on 35 mm. film. It has a 28mm. fully corrected lens.
2. The Rolleimarin has reflex viewing, interchangeable filters and close-up lens and a reputation for reliability. But only 12 exposures on each film.
3. The Hasselblad gives single lens reflex viewing and 70 exposures on perforated 70 mm. film.

unchanged by the Japanese firm Nikon and marketed under the name 'Nikonos') is itself waterproof and needs no housing. It is thus no larger than an ordinary 35 mm. camera and many divers take one with them on every dive 'just in case something should turn up'. The Calypsophot and Nikonos have no reflex viewing or range-finding device and this makes close-up work difficult and rather chancy. However, the recent addition of the fully corrected Nikkor lens has added the optical precision needed to guarantee the Nikonos a place in the basic equipment of every diving scientist (Wakimoto 1967).

When a larger format and greater precision of focusing is needed the Hans Hass Rolleimarin has a well-earned reputation for mechanical and optical reliability, but the number of 6 cm. × 6 cm. exposures is limited to twelve and this is sometimes too few for scientific work. The more recent Swedish Hasselblad 500c is perhaps more versatile than the Rolleimarin and takes 70 6 cm. × 6 cm. exposures on perforated 70 mm. film although at the expense of a heavy and unwieldy package.

Every writer on underwater photography emphasizes the need to keep the camera–subject distance as short as possible in order to reduce the contrast loss due to turbidity, and explains the value of a short focal-length lens, whose wide reception angle ensures a broad field of view even at short ranges. While this solution is quite satisfactory for pictorial photography, it seriously increases the distortions and aberrations inherent in taking pictures through an air–water interface. These lead to loss of definition at the edges of the field of view (due to chromatic aberration and astigmatism) and curvature of the whole field (due to pin-cushion distortion). These faults must be overcome before the camera can become the precision tool demanded by the underwater scientist.

The best-known correction lens system due to Ivanoff (1951a, b), has been used for accurate photogrammetry by Bass (1966) with excellent results. Similar lenses are made by most oceanographic institutes for use in remote control cameras lowered by cable to the ocean floor (see, for example, Hopkins and Edgerton 1961).

Lighting

In shallow water there is often enough ambient light from the sun to permit reasonable exposure factors and an intelligent use of natural lighting (e.g. keeping the sun behind the camera and pointing the camera slightly downwards) can yield results of excellent contrast despite the inevitable flatness of natural light under water. Artificial light, usually an electronic flash, becomes necessary at night, in deep water or when the subject is in shade. A recent development of considerable value to the underwater photographer is the automatic flash, which adjusts its light output automatically to ensure correct exposure over a wide range of flash–subject distances and subject reflectivities (see Woods 1969, for more details).

Scales

It is important in all scientific photography that a graduated scale should appear in each picture. Scales may vary from a simple ruler to the more elaborate six-armed cross illustrated in Chapter 9, which indicates the orientation of the camera to the vertical, or the box scale used for photogrammetry (see pp. 27–8). In some situations it may also be desirable to include a standard colour card to provide a control, for example, when recording fish colours.

Conclusion

Despite all these aids the over-riding factor in achieving high picture quality remains the skill of the photographer. Underwater cameras are not easy to use and quite a lot of skill is needed to judge the subject distance where there is no reflex viewing, or to follow an erratically moving fish on the focusing screen if there is. Nevertheless the successful analysis of photographs on which the success of an investigation so often depends is not possible if the photographs are poor. Science requires the highest picture quality, it is no excuse for poor technique.

Surveying*

Initial location

The initial surveying problem that confronts the diver is the need to identify the precise location of his working site so that others may find it after reading his report. This task is not unique to the diver; the need for accurate navigation at sea is common to all oceanographic investigations.

* This section is based on a draft by Mr. Ian Morrison, late of the Geography Department, Edinburgh University, Scotland.

At sea. Until the recent development of radio location systems (such as Decca, Loran, and Omega) and navigational satellites, the mariner depended upon radar and visual sightings of coastal features (with magnetic compass, rangefinder, or sextant) or astronomical sightings. Nowadays it is reasonable to assume that a properly equipped oceanographic vessel can fix its position at any time and at any place to an accuracy of better than 500 metres.

Inshore. The great majority of diving is carried out close inshore where surface markers fixed by divers over strategic points in the working site may be surveyed from the shore using well-established land techniques including the theodolite, plane table, and alidade, or from the sea using a magnetic compass or, preferably, a horizontal sextant. With the exception of the compass, these methods allow one to establish the locations of a number of major features in the working area to an accuracy of one metre or better. Particular care is needed to ensure that the surface floats used during this initial survey lie directly over the sinkers anchoring them to the selected underwater features; the best plan is to wait for a really calm day at slack tide. The sea-bed markers remaining after the floats have been cut away should be clearly labelled and coated with fluorescent paint to enhance their visibility.

Detailed surveying on the sea floor

Having established a basic grid on the sea floor that has been fixed relative to permanent features on the shore, the diver now proceeds to record the position of individual features within his working area relative to the grid. This part of the survey may be done in various ways, broadly divided into photographic and non-photographic.

Non-photographic methods. Many of the traditional methods of land surveying may be extended to use in the sea, though experience has shown that divers tend to measure lengths rather than angles, whereas the opposite is usually true in air. The principal reasons for this change in emphasis are, firstly, the poor visibility at most sites and, secondly, the maintenance problem incurred when using delicate optical instruments in sea-water. Nevertheless, Warton (1970) has successfully used a standard surveyor's theodolite in the clear Mediterranean sea, dismantling it and soaking the component parts in kerosene overnight after each dive. And an optical rangefinder has been used by Williams and Ainsley (1966), also in the Mediterranean.

On land, the classic method of tape survey involves setting out a right angle for every point of detail measured. Under water, this procedure tends to be either time-consuming or unreliable unless the distances involved are very short indeed. Trilateration is an alternative method of taping which does not require right angles to be set out. Instead the positions of objects are fixed in terms of systems of triangles by measuring the three sides of each triangle. In the usual (land) form of trilateration, several divers must keep track of the work of the others and this can provide a very real problem in the absence of adequate diver–diver communication. Morrison has therefore developed a trilateration technique for single-handed operation (Fig. 10) to avoid the difficulties arising from poor communication and co-ordination. This method also allows the diver to plot the survey directly on to a predrawn graticule inscribed on a Formica board. Besides being quicker than writing down the tape readings and making notes about what these refer to, this gives the diver an immediate visual check for errors and allows him to draw details accurately and comprehensively on the spot

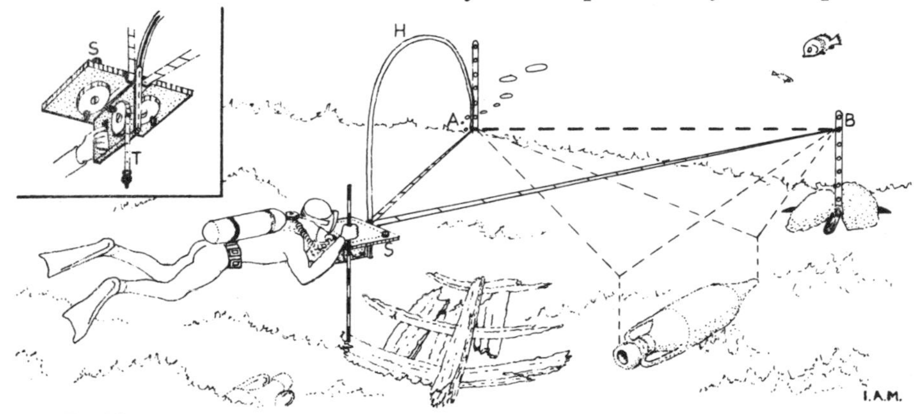

Fig. 10

Morrison's one-man underwater trilateration apparatus. The free ends of the tapes are attached to two of the datum points A, B whose positions have previously been fixed relative to the shore. Spirit level S is attached to the board. The board is raised until it is level with A, when bubbles are no longer emitted from the air hose H.

instead of trying to reconstruct them retrospectively from his notes (but see below the advantages of photogrammetry for filling in the details). The system can be operated in visibility so poor as to preclude completely any co-ordination of several divers. In muddy waters the diver can interpose a clear polythene bag full of clean water between his face-mask and the apparatus to read the tapes and plot the results. Depth gauge and spirit level are used to ensure that measurements are made in a horizontal plane and the height of the apparatus over the selected object is measured with a graduated rod or plumbline. The method works well (by torchlight) in caves, where a three-tape version may be more convenient than adhering strictly to the horizontal plane with the two-tape model.

Occasionally it is worth while constructing a permanent grid made from tapes or rods laid out in a rectangular pattern over the working site. This elaborate and time-consuming procedure has so far only been justified in the case of archaeological surveys (see, for example, Bass 1966) and it is discussed in greater detail in Chapter 7.

Submersibles are often navigated by means of sonar and a series of transponders moored within the working area. A diver version of this acoustic equivalent of the tape-measure trilateration method was tested by Woods in 1962 and several commercial models are now available, though at a cost that deters most underwater scientists. The accuracy of these instruments is set by variability of the speed of sound through sea-water (see Chapter 9), but with care errors should not exceed about one metre over a 100-metre-wide site.

Photographic methods. Photogrammetry minimizes the time spent under water and allows the surveyor to measure additional features that were neglected by the diver. These advantages are so great that photogrammetry is the first choice in almost all scientific surveys; it is discarded in favour of the non-photographic methods described above only when the relevant scale of the subject matter significantly exceeds the range of visibility. And, even then, it is often possible to produce a photo-mosaic based on a number of overlapping photographs, whose position and scale are related to a network of datum points previously established by non-photographic methods. In such cases the survey has three stages: first a small number of datum points within the site are fixed relative to the land using buoys, next a series of secondary markers on the sea bed are fixed relative to the initial datum points by trilateration (the spacing of these secondary markers should be about one-half of the visible range) and, finally, a series of overlapping photographs is taken of the whole area. This sequence has been well tested by archaeologists whose requirement for accuracies approaching one centimetre over a working area of up to 100 metres has provided the major stimulus for the development of underwater surveying methods (see, for example, Throckmorton *et al.* 1969).

The practical details of underwater photogrammetry are derived from well-established land methods, and the reader can easily devise the best system for his work by combining the requirements of underwater photography with the advice gleaned from standard textbooks on photogrammetry (for example, the publications of the American Society of Photogrammetry, see references; Engel, 1968, contains a useful introduction to the subject by Atkinson and Newton). The technique of photogrammetry has been refined to give extreme accuracy provided precision cameras and analysers are used, but the underwater cameras described on page 21 are capable of yielding quite adequate results for most scientific work under the sea. Whenever possible the diver is recommended to use a pair of identical cameras mounted together and accurately aligned on a rigid frame; the resulting stereo pair of photographs is then rapidly analysed with the help of a stereo plotter. This method has the advantage that it reduces to a minimum the time spent both under water and in analysis, though at a cost that may deter the occasional user of photogrammetry.

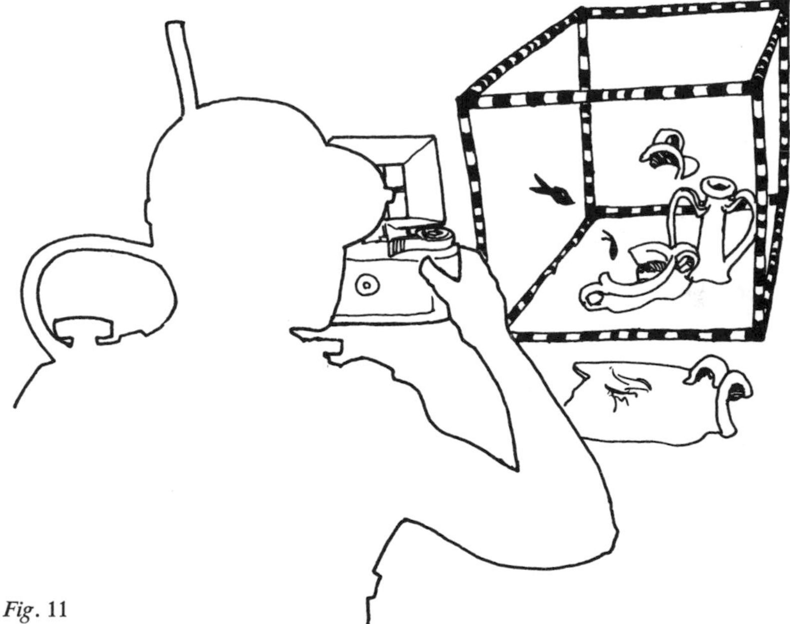

Fig. 11

A box frame included in a single photograph permits the surveyor to reconstruct accurately the positions, shapes and sizes of objects in the picture. Usually two pictures are taken from different viewpoints, but the precise position of the cameras does not have to be known. An alternative frame is illustrated in Chapter 9, Fig. 102b.

It is also possible to carry out accurate photogrammetry without a stereo camera, but only at the cost of lengthy analysis and the need for greater care under water. The technique is to ensure that a three-dimensional scale (Fig. 11) is included in pictures taken by a single camera. Morrison and Williams have successfully modified this old-established method for use by small groups of divers equipped only with a single Nikonos (or similar) camera; the procedure is described in Williams's excellent book (1969). Despite the disadvantage of having to derive the co-ördinates of each object in the pictures by means of time-consuming graphical construction, this method places accurate photogrammetry within the reach of all underwater scientists and especially those who unexpectedly discover on expedition that they must make an accurate map of their site with the limited apparatus available (Morrison 1970).

As in all underwater work, the need for careful preparation cannot be over-emphasized. In the case of photogrammetry this means that the subject should be well lit (by artificial light when necessary) and all the major objects in the field of view of each picture should be labelled with small fluorescent markers. The accuracy of the result can depend critically upon the care with which the scale is orientated in the picture and, finally, it is strongly recommended that all photographs should be in colour, since even minor colour contrasts help subsequent interpretation.

References

Albers, V. M. (1965), *Underwater Acoustics Handbook*. 2nd edition. Pennsylvania: Pennsylvania State University Press.

American Society of Photogrammetry (1966), *Manual of Photogrammetry*. 3rd edition. Falls Church, Va., U.S.A.

American Society of Photogrammetry (1959), *Photointerpretation Manual*. Washington, D.C.

Atkinson, K. B. and I. Newton (1968), 'Photogrammetry'. In *Photography for the Scientist*. Edited by C. E. Engel.

Barthelemey, L., F. Berry, and A. Michaud (1968), 'Cold and diver efficiency.' *Underwater Assn. Rep.*, **2**, 25–33.

Bass, G. F. (1966), *Archaeology Under Water*. London: Thames and Hudson.

Betts, J. (1965), 'Insulating properties of Neoprene.' *Science and Technical Group Bulletin*. British Sub Aqua Club, London.

Cousteau, J. Y. (1964), *Le Monde sans Soleil*. Paris: Hachette.

Davis, R. H. (1951), *Deep Diving and Submarine Operations*. London: Siebe.

Engel, C. E. (Ed.) (1968), *Photography for the Scientist*. London and New York: Academic Press.

Flemming, N. C. (1966), 'Operational diving with oxy-helium self-contained diving apparatus.' (In *Malta '65*.) *Underwater Assn. Rep.*, **1**, 3–12.

Hopkins, R. E. and H. E. Edgerton (1961), 'Lenses for underwater photography.' *Deep Sea Research*, **8**, 312–16.

Horton, J. W. (1957), *Fundamentals of Sonar*. Anapolis: U.S. Navy Institute.

Irving, F. G. (1964), 'An underwater glider.' *Triton*, October 1964.

Ivanoff, A. (1951a), 'On underwater photography.' *J. Optical Soc. America*, **41**, 645.

Ivanoff, A. (1951b), 'Fenêtre pour la photographie sous-marine.' *C.r. hebd. Seanc. Acad. Sci. Paris*, **232**, 1193.

Jones, D. F. (1968), 'A transparent underwater house.' *Underwater Assn. Rep.*, **2**, 46.

Krasberg, A. R. (1968), 'The evolution of functional saturation diving.' *Underwater Assn. Rep.*, **2**, 39–45.

Mertens, L. E. (1970), *In-Water Photography*. N.Y.: Wiley Interscience.

Miles, S. (1962), *Underwater Medicine*. London: Staples Press.

Morrison, I. A. (1970), 'An inexperience photogrammetric approach to the reduction of survey diving time.' *Underwater Assn. Rep.*, **4**, 22–8.

Newton, I. (1730), *Opticks: or a Treatise of the Reflections, Refractions, Inflections and Colours of Light*. (4th Edition). London.

Norman, P. (1967), 'The mermaid who didn't stay down.' *Sunday Times Magazine*, 25 June.

Ray, B. (1968), 'Voice communication between divers.' *Underwater Assn. Rep.*, **2**, 47–52.

Rebikoff, D. and P. Cherney (1955), *Guide to Underwater Photography*. Philadelphia: Chilton.

Richter, H. U. (1968), 'Underwater photography.' In *Photography for the Scientist*. Edited by C. E. Engel. 385–418.

Rowarth, D. (1969), 'Helium speech.' *Hydrospace*, **2**, 26–8.

Throckmorton, P., E. T. Hall, H. Frost, C. Martin, M. G. Walton, and S. Wignall (1969), *Surveying in Underwater Archaeology*. London: Quaritch.

Tucker, D. G. and B. K. Gazey (1967), *Applied Underwater Acoustics*. London: Pergamon.

Wakimoto, Z. (1967), 'On designing underwater camera lenses.' *Photogrammetric Engineering*, **33**, 925–36.

Warton, R. (1971), 'Underwater surveying by theodolite.' *Underwater Assn. Rep.*, **5** (in press).

Williams, J. C. C. (1969), *Simple Photogrammetry*. London and New York: Academic Press. (Including 20 pp. on underwater photogrammetry.)

Williams, J. and H. Ainsley (1966), Unpublished report. University of London.

Woods, J. D. (1962), 'A new device for communication between divers.' In *Underwater Challenge*. Edited by B. Eaton, London: British Sub Aqua Club.

Woods, J. D. (1969), 'A note on the use of automatic flash underwater.' *Underwater Assn. Rep.*, **3**, 41–2.

Young, E. P. (1967), 'Two new Russian submarines.' *Hydrospace*, **1**, 30–1.

Additional sources of information

This first chapter has been concerned with introducing some of the problems of underwater science and with establishing the philosophy that leads to successful *in situ* experiments. It was not our aim to produce a comprehensive catalogue of techniques and apparatus, although we have indicated those we have found useful. The following textbooks and periodicals will provide more detailed and up-to-date information for the underwater scientist.

BOOKS

British Sub Aqua Club Diving Manual. B.S.A.C., London.

U.S. Navy Diving Manual. NavShips 250–880. U.S. Govt. Printing Office. 1952. (Available in U.K. from Hydrographic Office and Admiralty Chart agents.)

Royal Naval Diving Manual. BR 155C. Min. of Defence, London. 1956.

Handbook of Ocean and Underwater Engineering. New York: McGraw-Hill. 1969.

Deep Sea Photography. Edited by J. B. Hersey. Baltimore: Johns Hopkins University Press. 1968.

PERIODICALS

Annual Report of the Underwater Association. Iliffe Science and Technology Publications Ltd. (Guildford). The only journal devoted solely to papers on underwater science.

Marine Sciences Instrumentation (Annual). New York: Plenum Press.

Underwater Science and Technology (Quarterly journal and monthly bulletin). Iliffe, Guildford, England.

Undersea Technology (Monthly). Compass Publications, Arlington, Virginia, U.S.A.

Ocean Industry (Monthly). Gulf Publishing Co., Houston, Texas.

Hydrospace (Quarterly). Spearhead Publications, Bromley, Kent, England.

Aquatic Biology Abstracts,

Societies

The following U.K. societies are of particular interest to the underwater scientist:

Underwater Association 3 St. Michael's Alley, London, E.C.3 (for the professional diving scientist).

British Sub Aqua Club 160 Great Portland Street, London, W1N 5TB (for diving training).

British Society of Underwater Photographers c/o British Sub Aqua Club.

Committee for Nautical Archaeology c/o Institute of Archaeology, 31–4 Gordon Square, London, W.C.1.

Society for Underwater Technology 1 Birdcage Walk, London, S.W.1.

Libraries

The British Sub Aqua Club is building a library concerned with all aspects of diving.

The National Oceanographic library is housed in the National Institute of Oceanography, Wormley, Godalming, Surrey.

The Marine Biological Association Library, The Citadel, Plymouth.

2 Diver Performance

A. D. Baddeley

With the rapid growth of underwater science and technology, increasing demands are made upon the diver, and as a result attention is inevitably shifting from the problem of sheer survival at depth to the question of assessing the diver's ability to perform useful work and deciding how this can be optimized.

In view of the wide range of tasks a diver may be asked to perform, and the many environmental stresses which may affect his efficiency, this is clearly a very wide area of study. With the exception of pressure-chamber studies of inert gas narcosis, little work has been done on diver performance, and any attempt to produce a systematic survey of the field would rapidly become a catalogue of unanswered questions. The present chapter therefore concentrates on two contrasting topics which have been studied relatively thoroughly, namely the effects of inert gas narcosis and the visual performance of divers.

Inert Gas Narcosis

When breathed at pressure, air has an intoxicating effect somewhat similar to that of alcohol. This effect is commonly (though not invariably) attributed to the effects of nitrogen, and hence is frequently termed nitrogen narcosis. Since it limits the diver's efficiency at depths as shallow as 30 m., and may cause stupor and loss of consciousness at depths exceeding 130 m., it is clearly a factor of major importance in deep diving. As such, it has probably in the past received more attention than all other problems of diver performance combined.

Studies of narcosis and human performance fall into two clear categories. The vast majority have used performance tests as an aid to understanding the underlying physiology, and have almost invariably been carried out in dry pressure chambers. In recent years, however, there has been a growing number of studies concerned with the more applied problem of what level of efficiency can be expected of a working diver at various depths and with various breathing mixtures. While preliminary studies of this problem can usefully be done under the safer and more controlled conditions of a pressure chamber, it has become clear that the results of such studies may not be directly applicable to practical diving, and that the only really satisfactory solution is to extend experimentation to the open sea (Baddeley 1966). However, despite the dichotomy between these two types of study, physiological and ergonomic, there are a number of problems and techniques common to both. These will be considered first.

1. *Experimental design*

This is one of the thorniest problems of narcosis studies. The basic difficulty is this; it is always necessary to consider at least two conditions (on surface and at depth), and it is usually desirable to consider more, e.g. several depths or breathing mixtures. Since subjects differ considerably both in their abilities and in their susceptibility to narcosis, reliable effects can only be obtained by either testing large enough numbers of subjects in each condition to even out such individual differences, or else by requiring each subject to undergo all conditions. Since large numbers of divers are rarely available at the right time and place, the second approach is almost invariably used. This immediately introduces the problem of transfer or carry-over effects between conditions.

(a) ***Practice and transfer effects.*** Performance on most tasks will improve with practice. If such improvement is rapid, the effect may be reduced by practising subjects till their performance has reached a stable level before beginning the experiment proper. While this may reduce practice effects, however, it can never be relied on to eliminate them. It may for example be possible to learn ways of coping with a particular task while suffering from narcosis, and since these may be quite irrelevant to performance at normal pressure, no amount of training before the experiment will eliminate practice effects on being required to perform the task under the new conditions. Finally, since performance on even a relatively simple task may continue to improve after thousands of trials extending over years of practice (Crossman 1959) it is clear that while prior training on the experimental task is often advisable, it can only be expected to reduce the effects of practice during the experiment, not to abolish them.

How then should such practice effects be handled? The most conservative approach is to arrange that any practice effects will work against the effect the experimenter is studying.

The drawbacks of such a design are illustrated in Figure 12 taken from an as yet unpublished study by Baddeley and Catton. It represents performance on a reasoning task (Baddeley 1968c) for two groups of subjects; both began with a practice run, but Group A was then tested at progressively greater pressures, while Group B began at 30 m. and was tested at progressively lower pressures. In the case of Group A (pressure increasing) the drop in efficiency at depth shown on this task in other studies where practice is controlled (Baddeley, de Figueredo, Hawkswell-Curtis, and Williams 1968) was completely swamped by the practice effect. In Condition B (pressure decreasing) the practice effect causes an exaggerated effect of pressure. Condition A is similar to the design used by Adolfson (1967) in an otherwise excellent study of performance at pressures ranging from 1 to 13 atmospheres (0 to 130 m.), since his subjects apparently all began with low pressures and worked systematically up to 13 Kg./cm.² (atmospheres absolute) followed by a final test at 1 Kg./cm.² This is not on the whole a good strategy since the effects of practice, fatigue, and depth are inextricably confounded, making interpretation very difficult. The result in Adolfson's case seems to have been a loss in test sensitivity since he finds atypically small performance decrements even at relatively high pressures. It is interesting to note that the only task to show effects at depths less than 60 m. in Adolfson's study was mental arithmetic, the one test on which he attempted to take account of practice effects.

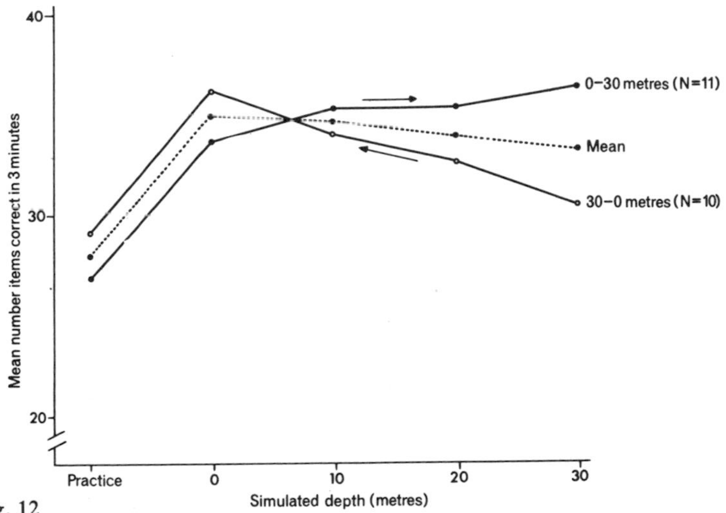

Fig. 12

Performance on a reasoning test in compressed air. One group was tested at progressively increasing pressures, whilst the other began at high pressure and was tested in the reverse order. Data from Baddeley and Catton (in preparation).

A more satisfactory way of handling practice effects is to balance the order in which the various conditions are performed so that no condition is unfairly favoured. Perhaps the neatest design of this sort is to have an equal number of subjects doing the various conditions in each of the possible permutations of order. However, since N conditions will give $N!$ permutations, this procedure becomes impracticable when the number of conditions exceeds 3 (6 permutations) or 4 (24 permutations), since one is unlikely to have sufficient subjects to cover all possible orders of presenting the conditions. In such a situation it is better either to randomize the order of presenting the various conditions, with a different random order for each subject, or to use a Latin Square design in which the order of presentation of the various conditions is counterbalanced (Fisher 1960).

Suppose, then, that we selected a simple design with just two conditions, e.g. surface and 30 m. We allow for practice effects by having half the subjects do the shallow condition first and the deep second, while the other half begin with the deep condition and then do the shallow. Can we now assume that practice effects will not be a problem? Alas, no! Such a conclusion will only be justified if the practice effects are symmetrical, that is if doing the task at the surface will affect performance at pressure just as much as practice at pressure will influence subsequent performance on the surface. While such transfer or carry-over effects may often be symmetrical, this is by no means always the case. This problem is discussed in some detail by Poulton and Freeman (1966) and is illustrated by a pressure chamber study by Poulton, Catton, and Carpenter (1964). They required their subjects to sort playing cards into the four suits and timed each sorting response. In their initial experiment, their subjects were tested twice, once at normal pressure, and once at 3·5 $Kg./cm.^2$, half beginning at the surface and half at pressure. Their results were scored in terms of slow responses and showed that subjects tested first at pressure made many slow responses, and continued to make slow responses when subsequently tested at normal pressure. Other subjects who sorted cards for the first time at normal pressure made relatively few long responses, and made very few more when subsequently tested at pressure. In other words pressure apparently affected the learning of the task, which in turn carried over to a subsequent test under different conditions. Overall analysis of their data indicated an interaction between pressure and order of test, but no significant effect of pressure alone. Separate analysis of the first trial, the only one which was not contaminated by carry-over effects, however, indicated that there *was* a significant effect of pressure. What apparently happened was that the learning of the task was very sensitive to pressure, while subsequent performance was not. Poulton and Freeman (1966) cite a large number of studies in which such asymmetrical transfer effects occurred, and note that they usually reduce the sensitivity of the experiment by masking a genuine effect.

Can such unwanted transfer effects be avoided? Unfortunately, only by testing subjects in only one condition each, and hence requiring prohibitively large numbers. The best strategy is probably to use a Latin Square design (Fisher 1960) ensuring that all conditions occur equally often in each position in the testing order. Table 1 gives one

TABLE 1. *Latin square design for five conditions (A–E)*

	Order of Test				
Subject No.	1st	2nd	3rd	4th	5th
1 and 6	A	B	C	D	E
2 and 7	B	D	A	E	C
3 and 8	C	A	E	B	D
4 and 9	D	E	B	C	A
5 and 10	E	C	D	A	B

Note that: (1) Each condition occurs once in each column. (2) To complete the design some multiple of 5 subjects must be tested; in this example 10 subjects are used.

such Latin Square for five conditions, others are given in Fisher and Yates (1963, p. 86). An asymmetrical transfer effect will be reflected in the difference between the relative levels of performance on the five conditions overall, and their relative levels of difficulty considering only column 1 of the Latin Square (i.e. for each subject using only the first condition he performed). If the picture presented by these uncontaminated first tests is different from the overall picture, then asymmetrical transfer effects are probably responsible, and will show up as an interaction between conditions and testing order if the data are subjected to an analysis of variance (Fisher 1960). If this happens, a separate analysis based only on the uncontaminated first items is advisable. However, the occurrence of asymmetrical transfer effects is by no means inevitable, and provided the experimenter is aware that they *may* occur and is prepared to discard the contaminated part of his data if they do, there is no reason why Latin Square designs with repeated testing of the same subject should not be used.

Although a Latin Square design is probably the most economical in requiring relatively few subjects, it runs into two practical problems. First, such a design is very rigid, since the number of subjects tested has to be some multiple of N (where N is the number of conditions) to complete the Latin Square. This lack of flexibility can be very frustrating in diving experiments where the probability of one or more subjects dropping out owing to factors such as ear trouble is high, and the pool of available subjects often small. The second problem arises if one hopes to test at several depths on the same day. Consider for example Adolfson's study in which subjects were tested at pressures of 1, 4, 7, 10, and 13 Ats.a. A Latin Square design might require one subject to be tested in the order 10, 4, 13, 1, and 7, a sequence which would present enormous decompression problems. Even if one could work out appropriate decompression schedules, the time involved would be prohibitive. If the five tests can each be given on a separate day this problem does not arise, but if the subjects are only available for one session, and one *must* test at several depths, there being too few subjects for a separate groups design, then the best compromise is probably to start half the subjects at the shallowest depth and test them at successively greater depths while starting the other half at the deepest condition and thus reversing the order of testing. Such a design gambles on practice effects not being too large or complex, but at least the data give some indication of whether this assumption is justified: in the case of the reasoning task data shown in Fig. 9 this was not a successful design; however, a simpler digit copying task used in the same study gave curves for the two test orders which were sufficiently similar to suggest that carry-over effects were not too great. In other words the gamble worked for one task but not the other.

(b) ***Placebo effects.*** In testing the effects of a drug, it is usual to give one group a placebo, a completely inert substance otherwise indistinguishable from the drug. Sometimes subjects will show an effect from the placebo, presumably due entirely to their belief that it *was* a drug. If one is interested in the effect of the drug itself, then performance in the placebo condition should be used as the baseline for comparison.

A placebo effect of this type was observed in the previously mentioned study by Baddeley and Catton. Two groups of subjects performed a digit copying task (Legge 1965) twice, first as a practice and then as a test. For one group the test was performed in a pressure chamber at normal atmospheric pressure. For the second group the pressure was raised enough to make ear clearing necessary, and was then surreptitiously lowered by bleeding out the air till pressure was again at virtually 1 at. The variable measured was writing size, which tends to increase under the influence of a drug such as nitrous oxide (Legge 1965), or at depth (Baddeley *et al.* 1968). There was a statistically significant increase in writing size for the placebo group which thought itself at pressure.

An alternative approach to the placebo problem is to use oxy-helium as a placebo condition when studying nitrogen narcosis (Bennett, Poulton, Carpenter, and Catton 1967). This is perhaps being somewhat conservative, since there is some evidence that oxy-helium may itself impair performance (Bennett 1966; Baddeley and Flemming 1967), but nevertheless it is a useful additional control if available.

A placebo aims to prevent the subject knowing what condition he is doing, so as to minimize the effects of suggestion. When possible, it is also desirable for similar reasons that the experimenter should be unaware of the condition while he is testing the subject. This can be achieved using the double-blind procedure, whereby the conditions (e.g. breathing mixtures) are controlled by a second experimenter while the tester and subject are both unaware of the condition involved.

In open sea studies it becomes much more difficult to disguise the depth or, in the case of an experienced diver, to keep him unaware of gross differences in his breathing mixture; the difference between air and oxy-helium for example is very obvious at 60 m. (Baddeley and Flemming 1967). Indeed it is arguable that a placebo condition is inappropriate in open sea tests which generally aim to simulate the practical diving situation, where the diver would be fully aware of both his depth and his breathing mixture.

(c) ***Type of subject.*** Ideally one should use experienced divers familiar with working at the depths concerned. In practice this is frequently not possible, and it is therefore comforting to note that three studies in which the experience of the subjects varied considerably showed no correlation between experience and performance at depth (Baddeley 1966; Baddeley *et al.* 1968; Poulton, Catton, and Carpenter 1964). However, none of these studies used full-time professional deep divers, so the question of experience cannot be completely eliminated.

2. *Test design*

A performance test is basically a measuring instrument, and as such it should have two basic qualities; it should be sensitive enough to pick up relatively small effects, thus allowing accurate comparisons, and it should be a valid indicator of some reasonably important capacity.

(a) ***Sensitivity.*** In order to be sensitive, a test must be reliable, that is, it should give the same score when applied to the same person on two comparable occasions. Factors which affect reliability include the following. (i) Excessive complexity: a test which allows a wide range of possible strategies, like a game of chess, is unlikely to give rise to consistent day-to-day performance. (ii) Randomness: a test in which there is an element of luck is likely to be unreliable. This includes certain motor skill tests such as the commonly used task in which the subject attempts to pick up ball-bearings with a pair of tweezers and drop them down a glass tube. A subject can increase the speed of this task by being less careful about placing the ball in the tube, but with the relatively short test runs normally used (Bennett 1966), this produces rather erratic performance, with the score fluctuating markedly from trial to trial. (iii) Practice effects: a test which shows a marked improvement with practice is less likely to be sensitive since differences between conditions are likely to be masked by practice effects and differential learning rates among the subjects. (iv) Number of components: if we assume a test to comprise a number of sub-tests, each of which gives an imperfect measure of the underlying performance variable, then the greater the number of sub-tests, the more reliable the estimate of the underlying variable is likely to be. (v) The subject's attitude or strategy should not be too important (unless of course one is trying to measure this), since such factors are likely to vary from trial to trial, subject to subject, and day to day.

While reliability is essential to a good test, the reliability of diver performance tests is rarely quoted, and, one suspects, seldom measured. This can be done by simply correlating subjects' performance on successive trials (see Baddeley *et al.* 1968). This can best be done on subjects at normal pressure prior to the main experiment; if this had been standard practice, as it should, one suspects that a large number of tests that are still commonly used would have disappeared long ago.

Needless to say reliability is not in itself enough to ensure a sensitive test. The test must be an accurate indicator of some factor or capacity of importance, which is likely to be influenced by narcosis.

(b) ***Validity.*** A test is valid in as much as it measures the underlying capacity which it claims to measure. Ideally the validity of a test is assessed by correlating performance on the test with some independent measure of the capacity in question. In the case of an applied problem where the type of job involved is known, performance on the test should correlate highly with performance of the job. Where such an independent assessment is not available, however, one may have to fall back on 'face validity', general agreement that a test involves reasoning or manual dexterity or whatever is the capacity involved.

(c) ***Convenience.*** Because of the decompression problem, time at depth is at a premium. It is therefore advisable to use tests that are as short as is possible without sacrificing too much reliability. Such tests are being developed (e.g. Baddeley 1968b), but there is still a great need for short, sensitive tests. Furthermore, if the test is to be suitable for open sea use, it must be robust and simple enough to operate without placing excessive demands on the experimenter who may himself be suffering from narcosis.

In conclusion, it is clear that the study of inert gas narcosis and performance involves some very tricky problems of experimental design and test development. Unless we are aware of the pitfalls and are prepared to continue to improve our techniques, progress will be very slow.

Pressure-Chamber Studies

This section gives a brief survey of the results of pressure-chamber studies of diver performance. For a more detailed account the reader is referred to Bennett (1966), Adolfson (1967), and Jennings (1968).

1. *Oxygen poisoning and performance*

When breathed at depths exceeding 12 m., pure oxygen is likely to produce convulsions and unconsciousness (Donald 1947). The question of whether high partial pressure of oxygen would also impair performance was studied by Frankenhauser, Graf-Lonnevig, and Hesser (1960). They compared the performance on reaction time and a mirror-drawing task, of subjects breathing pure oxygen at 3 Kg./cm.^{2} with their performance breathing air at normal pressure. They found no drop in efficiency, and in the case of one subject who convulsed while doing the mirror-drawing task, his performance was quite unaffected up to the time he began to convulse.

2. *Nitrogen narcosis*

The vast majority of performance studies have used oxy-nitrogen mixtures, usually air. Problems studied include the following.

(a) ***Minimum depth for narcosis.*** Estimates of this range from 10 m. (Poulton, Catton, and Carpenter 1964) to 80 m. (Case and Haldane 1941). However, most workers would probably accept 30 m. as a useful cut-off point. Many studies have shown an effect at 30 m. (e.g. Behnke, Thomson, and Motley 1934; Kiessling and Maag 1962; Baddeley 1966; Bennett, Poulton, Carpenter, and Catton 1967), and failures to find an effect at depths exceeding 30 m. are probably due mainly to the use of insensitive tests and inadequate experimental designs. Since the effects of narcosis show fairly clearly at 30 m., it seems very probable that a mild degree of narcosis will begin at rather shallower depths. Indeed, two studies have observed a reliable performance decrement at 10 m. though in one of these (Baddeley and Catton, in preparation) this was largely a placebo effect, while the other (Poulton *et al.* 1964), has proved difficult to replicate (Bennett, Poulton, Carpenter, and Catton 1967). Nevertheless, it seems likely that with the increasing sophistication of experimental techniques, it will become possible to show performance decrement at depths shallower than 30 m., though whether such decrements will prove large enough to be of practical significance remains to be seen.

(b) ***Time at depth.*** The rate of onset of nitrogen narcosis is still in doubt. A study by Bennett, Dossett, and Ray (1964) showed that the onset is very swift; subjects compressed to a pressure equivalent to 160 m. showed slower reaction times within the 40 sec. test run, though a similar study at 130 m. did not show a reliable decrement. Using physiological measures Bennett and Glass (1957) and Bennett and Cross (1959) found the time of onset of narcosis to be inversely proportional to the square of the pressure (i.e. $P\sqrt{T} = K$, where $P =$ Pressure, $T =$ Time and K is a constant representing the individual subject's susceptibility to narcosis). An attempt to generalize Bennett's results to diver performance was made by Kiessling and Maag (1962), who showed a reliable decrement at 30 m., but no change in level of decrement over successive 12 min. periods. It is possible, however, that the narcosis had already reached a maximum during the first period, so the question of rate of onset remains an open one. A related problem is that of whether the effects of narcosis change over longer periods; if a diver lives under pressure for several days, will he adapt, or will there be a cumulative effect causing him to get steadily worse? The development of saturation diving makes this a question of considerable practical importance, though as yet there appears to be no experimental evidence for oxy-nitrogen mixtures, though there is some data on oxy-helium (Bennett 1966) suggesting the subject adapts over time.

(c) ***The nature of the task.*** There are almost certainly differences in the susceptibility of different tasks to the effects of narcosis, with the more complex forms of behaviour tending to be affected first. Thus Kiessling and Maag (1962) found the percentage impairment in performance on a reasoning task to be greater than that on reaction time or manual dexterity. The concept of complexity is too vague to be satisfactory, and it is to be hoped that it will prove possible in the next few years to produce a 'profile' showing the effects of narcosis on a series of standardized tasks covering the range of human performance.

Adolfson (1967) has shown that the amount of physical work the subject is doing has a marked effect on the amount of decrement at pressure. He required his subjects to perform arithmetic and manual dexterity tasks while pedalling a bicycle ergometer (a stationary bicycle used by physiologists when they want their subject to work at a standard rate). Exercise increased the impairment in manual dexterity at pressure but did not affect arithmetic performance. Adolfson attributes the effect of exercise to a build-up of carbon dioxide, though in order to fit his data he has to assume that this has a differential effect on his two tasks. A simpler explanation might be to assume that the manual dexterity task is hindered more by pedalling than is mental arithmetic; by being made more difficult the manual task becomes more sensitive and hence shows a bigger impairment at depth.

3. *Oxy-helium, hydrogen, and neon*

There is abundant evidence that divers are much less impaired at depth when either helium or hydrogen is substituted for nitrogen (End 1938; Case and Haldane 1941; Zetterstrom 1948). Little work has been done using oxy-hydrogen mixtures, however, since these gases are explosive when combined in certain proportions, and current deep diving relies almost entirely on oxy-helium. There is some disagreement about whether helium impairs performance at depth. Experiments carried out during saturation dives have in general shown no impairment (Bowen, Anderson, and Promisel 1966; Hamilton, MacInnis, Noble, and

Schreiner 1966; Cousteau 1966; Krasberg 1967). Although none of these reports presents convincing performance data from an adequately designed experiment, it nevertheless seems probable that any decrement in the performance of an adapted subject is relatively small at depths shallower than 100 m. A drop in efficiency at depth for divers breathing oxy-helium has, however, been shown by Bennett (1966)—150 m., Adolfson (1967)—100 m., and by Baddeley and Flemming (1967)—65 m. These studies differ from those obtaining negative results in two important respects: they were all pressure-chamber studies using reasonably sensitive performance tests, and they all compressed their subjects rapidly and began testing soon after reaching pressure. Bennett's results suggest that there is a genuine effect of helium which causes trembling and dizziness, but which gradually wears off after a few hours at pressure. Unfortunately, Bennett's data can be interpreted as a simple improvement with practice rather than an adaptation to pressure. In an attempt to avoid this problem, we have run a control group of comparable subjects on the tasks used by Bennett. Performance did improve with practice, but probably not to the extent that occurred in Bennett's experiment, thus giving some support to the claim that there is a genuine adaptation to the adverse effects of helium at pressure. Whether helium will impair the performance of an adapted subject is still in doubt, but with the rapid development of deep diving techniques, we shall certainly know within the next few years. If oxy-helium does prove narcotic, there will no doubt be a rapid development of interest in neon, which preliminary tests have shown to be a promising alternative (Bennett 1966), and possibly also a reconsideration of hydrogen.

Open Sea Studies

If one wishes to study nitrogen narcosis under controlled experimental conditions, then a pressure-chamber study is the most satisfactory approach. The question then arises as to whether the results of such studies should be generalized to performance in the open sea, where the diver may be subjected to a host of additional stresses such as weightlessness, cumbersome equipment, narrowed vision, cold, and anxiety.

This problem was studied by Baddeley (1966) who compared the performance of divers at three depths, surface, 3 m., and 33 m., both in a dry pressure chamber and in the open sea. The screwplate test of manual dexterity was used; this comprises a 30 × 15 cm. brass plate containing 32 holes. The subject transferred 16 2 B.A. cheese-head brass nuts and bolts from one end of the plate to the other, and the time taken, number of screws lost and number left loose were all recorded. In the open sea condition, the subject was seated on a folding canvas chair and stabilized by weight belts across his lap. His performance was timed by an underwater experimenter using the second hand of a diving watch. This is a task where timing errors could easily be made, especially at depth where the experimenter would also be suffering from narcosis. A pre-arranged code of pulls on the experimenter's lifeline was therefore used to signal the beginning and end of the run to a second experimenter on the surface, who also timed the run using a stopwatch. For subsequent studies, the experimenter's task was made much easier by using a stopwatch in an underwater case, which relieves the experimenter of the difficult job of continuously monitoring both the subject and his diving watch. The open sea testing was carried out from the deck of a Royal Engineers Z craft in Famagusta Bay, Cyprus, in calm August weather and good visibility conditions. Figure 13 shows the mean time taken to complete the test in the various conditions. In the dry pressure chamber, the effect of depth was statistically significant but small (6%) and as such resembled the results of Kiessling and Maag (1962). In contrast to this, the open sea result showed a marked drop in speed (28%) at 3 m., presumably due to the various stresses associated with working under water, while at 33 m. a drop in speed of 49 per cent occurred. The important feature of this result is the fact that the effects of working under water and of narcosis are not simply added. If this were the case the two lines connecting the 3 m. and 33 m. points in Fig. 13 would be parallel. In fact the slope of the line in the open sea condition is much steeper, implying that the effect of narcosis is exaggerated by the stresses associated with working in the open sea. Because of this it would not have been possible to predict performance at depth in the open sea from the pressure-chamber results. If we wish to draw conclusions about practical diving, it appears that we must depend on open sea experiments. It is interesting to note that a similar discrepancy between expectations based on dry pressure-chamber studies and the results of underwater testing were found for both oxygen poisoning (Donald 1947), and decompression sickness (Taylor 1965).

Subsequent experiments showed a comparable interaction at 65 m. for both air and oxy-helium diving (Baddeley and Flemming 1967), a result which agreed quite closely with data collected during the U.S. Navy Sealab II project (Bowen, Anderson, and Promisel 1966). A

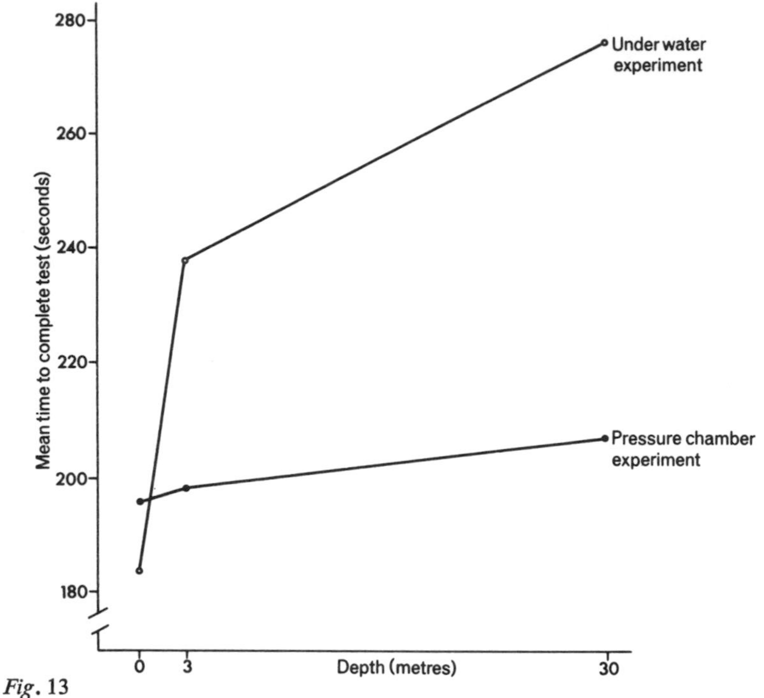

Fig. 13
Time to complete the screwplate test of manual dexterity under dry pressure chamber and open sea conditions.

comparable discrepancy between training and subsequent performance was observed in astronauts when a manœuvre which took 30 seconds during training took 30 minutes when carried out during a space walk (Anon. 1966).

It seems then that the narcotic effect of air or oxy-helium interacts with the stresses associated with diving to produce an exaggerated performance decrement. The problem was to decide which stress or combination of stresses was responsible for the interaction. Weightlessness? narrowed vision? encumbering equipment? anxiety? cold? The task seemed hopelessly complex until a fortunate accident occurred. An open sea experiment was carried out at 33 m. using the screwplate test and showed no interaction (Baddeley, de Figueredo, Hawkswell-Curtis, and Williams 1968). Presumably a crucial variable had been changed. The only obvious difference was that the test involved shore diving under ideal conditions instead of the rather stressful open sea boat diving involved in the previous studies. In other words the level of anxiety of the subject seemed to be the crucial variable (for a more detailed discussion see Baddeley 1967). This hypothesis fits in well with both anecdotal evidence, and the results of Bennett (1966) who found that tranquillizing drugs tended to reduce, and arousing drugs to increase narcosis effects. It is also consistent with Adolfson's (1967) failure to find an interaction when divers were tested under water in a wet pressure chamber, since this was presumably safer and less anxiety-provoking than a deep open sea dive. A really adequate test of this hypothesis would require independent measures of anxiety level such as heart rate and galvanic skin response. Techniques for recording such physiological variables are not yet sufficiently developed to be applied in open sea situation, but considerable work is going on in this area, and such recordings will probably become standard practice within the next few years.

Visual Performance

The human eye becomes extremely long-sighted in water. The refractive power of the cornea is lost, because its refractive index is similar to that of water, and the lens alone cannot accommodate sufficiently to focus an image on the retina. Galton (1876, p. 87) invented underwater spectacles to compensate for this; and modern contact lenses are also available (Faust and Beckman 1966). Many different types of masks, goggles, helmets and optical devices have been invented (Tailliez *et al.* 1949). Most divers wear a face-mask with a single flat plate of glass, which allows the eye to operate in air while looking into water. Almost all systems will produce some kind of distortion compared to air vision, but only the flat face-mask will be considered here.

Because water is an optically denser medium than air, light rays from an object in water are refracted away from the normal when entering the air inside a diver's face-mask. The lens accepts a thin pencil of rays from any point on the object, which appear to diverge from a point approximately on the normal. A virtual image is thus formed at approximately three-quarters of the physical distance (Fig. 14). Snell's Law states that $\sin i/\sin r =$ a constant (1·33 in the case of air and water, where i is the angle to the normal in air and r the angle in water). For small angles, i/r is approximately equal to $\sin i/\sin r$ (Fig. 15). Objects of small angular size viewed normal to the face-plate therefore subtend an angle approximately 4/3 larger at the eye, and are located optically at about three-quarters of their distance. The nearer the eye is to the

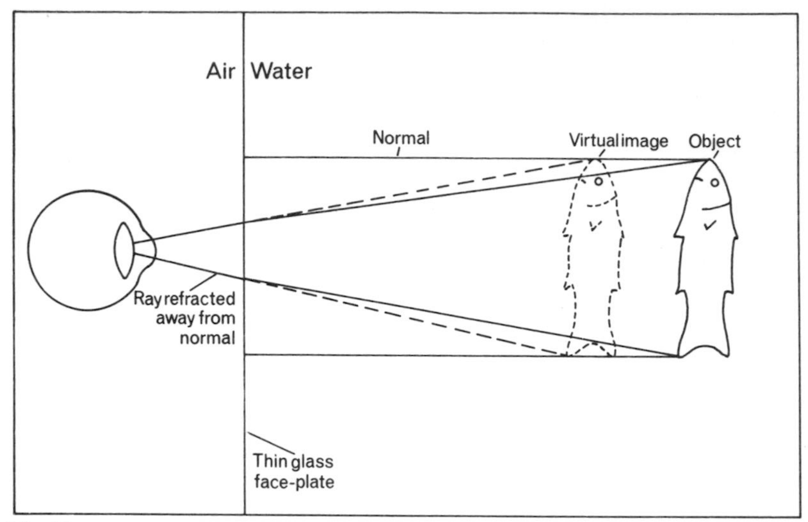

Fig. 14

Rays passing from water into air are refracted away from the normal, since the refractive index of water is 1·33 times that of air. The lens accepts a thin pencil of rays (represented by only one ray) from any point on the object. Optically the rays appear to diverge from a point approximately on the normal, at about three-quarters of the physical distance from the air/water interface. Objects thus subtend an angle at the eye approximately four-thirds larger than in air.
(Redrawn from Ross 1967, with permission of the Editor of the *British Journal of Psychology*.)

face-plate, the greater the angular magnification and reduction of distance. Rays of light at the periphery of the field of view must strike the face-plate more obliquely in order to be accepted by the lens: they make a larger angle to the normal, and are therefore refracted disproportionately more than those at the centre. This produces 'pincushion' distortion, a square appearing to be bowed inwards (Fig. 16). This is a two-dimensional representation of the distortion. In the third dimension, the centre appears farther away than the extremes. The extent of the curvature is shown in more detail in Fig. 17. For objects of large angular size the distortion is quite noticeable. Moreover, the optical location of peripheral points is imprecise, causing a blurred image. A fuller discussion of these optical problems is given by Southall (1933). The effects of refraction are not the only important changes in visual information under water. Of equal importance are the reduction in light intensity, brightness contrast and colour contrast, due to the absorption and scattering of light (see Chapter 4).

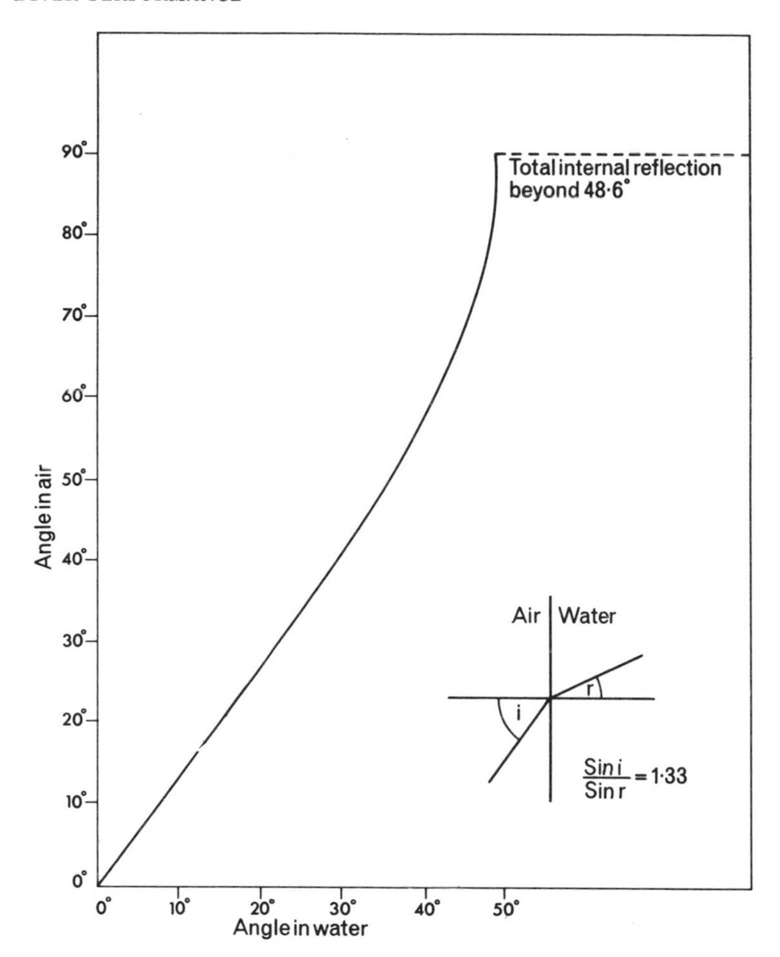

Fig. 15

The angle between the ray of light and the normal in the optically less dense medium (air) is conventionally termed the angle of incidence (i), and the corresponding angle in the more dense medium (water) the angle of refraction (r). Snell's Law states that $\frac{\sin i}{\sin r}$ equals a constant, the relative refractive index of the two media. In the case of air and water this is 1·33. The graph shows that $\frac{i}{r}$ is approximately constant for small angles, but not for large angles. When $i = 90°$ ($r = 48{\cdot}6°$) sin $i = 0$, and total internal reflection of the ray in water occurs. (Diagram H. E. Ross.)

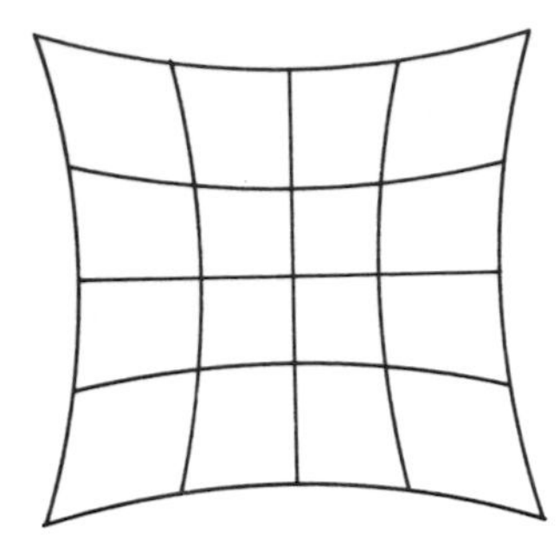

Fig. 16

A square viewed through a face-mask appears bent inwards (pin-cushion distortion), and the centre appears farther away than the edges. These effects are due to the increased refraction of oblique rays at the periphery of the face-mask.

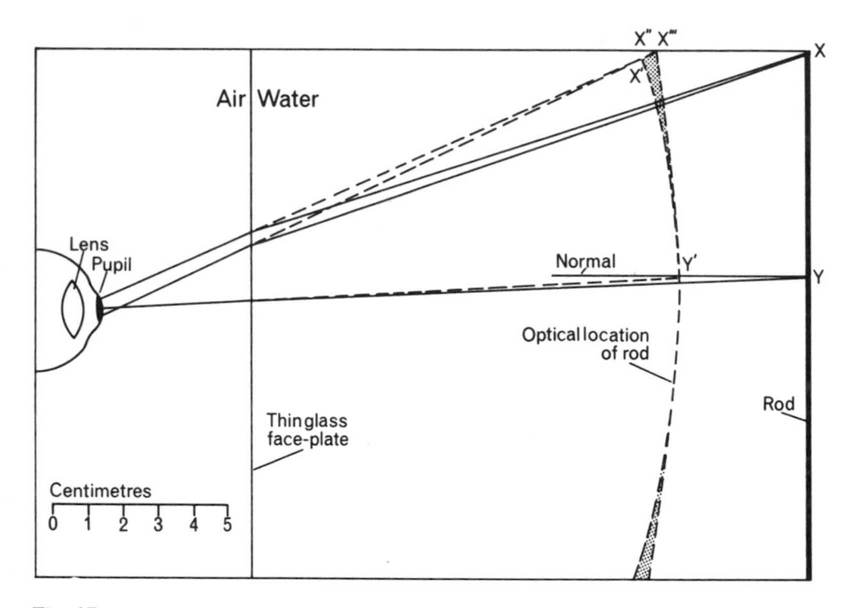

Fig. 17

Points seen through the periphery of the face-plate are optically nearer than those in the centre of the field of view. This diagram shows the optical distortion which occurs for a straight line, with the eye at the typical distance behind the face-plate. Points at the periphery also appear blurred, because the pencil of rays accepted by the lens does not appear to diverge from a single point. Thus the optical location of X lies within the points X', X'' and X'''. Both the curvature and the blurring increase with the aperture of the pupil, and with a reduction in distance between the eye and the face-plate.

(Reproduced from Ross (1969), with permission of the Editor of *Ergonomics*.)

1. *The visual field*

Because of light refraction at the water–air interface (see page 47) the diver's visual field is much more restricted under water. The standard diving mask exaggerates this effect by further restricting peripheral vision, with the extent of the tunnelling depending on the type of mask. The absence of peripheral vision is inconvenient since it limits the observational field, interferes with visual diver-to-diver contact, and may also cause postural instability (Dickinson and Leonard 1967). The wide range of masks available reflects the failure to come up with a satisfactory solution to this problem, and this in turn has led to a number of studies aiming at the comparison of the visual field provided by the various types of mask using the technique known as perimetry.

The first of these studies was carried out by Barnard (1961) who compared the visual field for diving goggles with that of both a standard face mask and a mask designed to give wide-angle vision. He found the goggles produced the smallest visual field and the wide-vision mask the largest, though only at the expense of distortion over much of the visual field due to the use of a curved face plate. This latter point was confirmed in a study by Weltman, Christianson, and Egstrom (1965) which compared the visual fields allowed by five types of face mask. Their report is particularly interesting since it discusses in some detail the problems of carrying out experiments under water.

The apparatus used by Weltman *et al.* is shown in Fig. 18. It comprises a supporting frame on which was mounted a shoulder yoke which could be rotated so as to hold the diver in any orientation. The diver was held by a quick release device which allowed him to free himself in case of emergency. The frame was fitted with a 1-m. diameter arc, which could be rotated around the diver's head to form an enveloping sphere whose centre lay midway between the diver's eyes. The diver's head position was fixed by clamping his mouthpiece.

Mounted on the arc was a small battery-operated target light which the experimenter moved from the centre of the diver's visual field outwards. The subject attempted to follow the light with his eye (the two eyes were tested separately), and as it disappeared signalled by tapping the supporting frame with a spanner, whereupon the experimenter noted the light's position and recorded the result on a plastic writing board with a lead pencil. Observations were made in steps of 30° so as to cover the whole visual field. The experiment was performed at a depth of 2½ m. in a swimming pool (see Fig. 19).

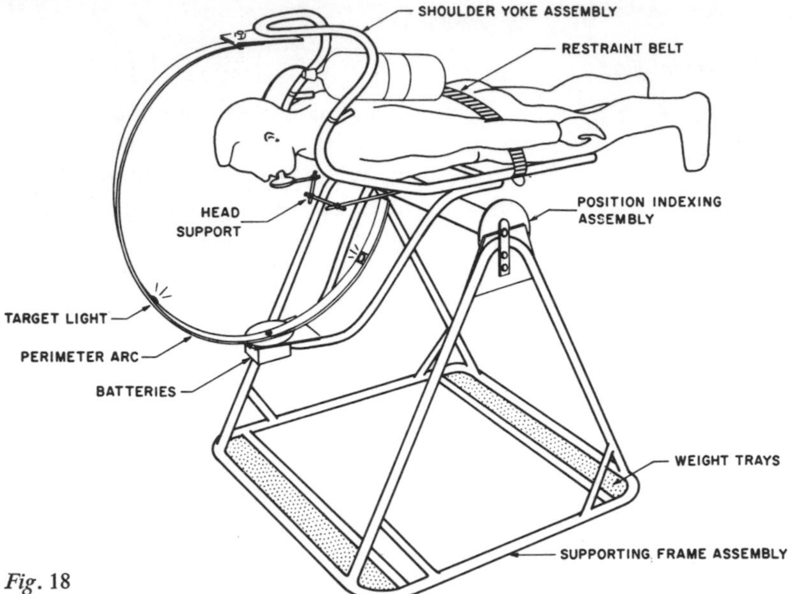

Fig. 18
The diver restraint device used by Weltman *et al.* (1965) fitted for underwater perimetry.

Three standard masks were tested initially, an oval mask, a kidney-shaped mask, and a kidney-shaped mask with a deep receptacle for the nose to minimize the distance between the face plate and the eyes. Two further masks, a full-face mask and a wide-vision mask with wrap-around face plate, were subsequently examined using a simplified procedure and two experienced subjects.

The visual fields obtained from the three standard masks are shown in Fig. 20. The greatest overall visual field is given by the recessed kidney-shaped mask, though unfortunately this mask has the smallest binocular field, which might be expected to have an adverse effect on space perception (see Chapter 3). Visual fields for the two additional masks are given in Fig. 21, where results for the two subjects are plotted separately. Both masks appear to increase the visual field, although since results from the two eyes were not recorded separately it is not clear how much of the field is binocular. The wrap-around mask, however, proved unsatisfactory because of the distortion and double vision it produced, and, on balance, the authors suggest the full-face mask would probably be best suited to the working diver, although the interdependence of mask and air supply is a drawback if the diver wishes to use both aqualung and snorkel on the same dive.

Fig. 19
Measuring a diver's visual field using underwater perimetry (Weltman *et al*. 1965).

Weltman *et al.* conclude with some observations on the problems of experimenting under water. One major difficulty was that of communication. The most satisfactory way of communicating to the subject was by underwater loud-speaker. However, this equipment was unreliable and the experimenters frequently had to rely on memorized instructions, hand signals, and the tedious writing of messages on the plastic data sheets. The second point they make is that careful planning is essential, for even in a swimming pool both subject and experimenters soon become cold and tired. It is important to keep the experimental area orderly, to use oversize apparatus for easy handling, and to mark it with large contrasty letters to facilitate identification and data collection.

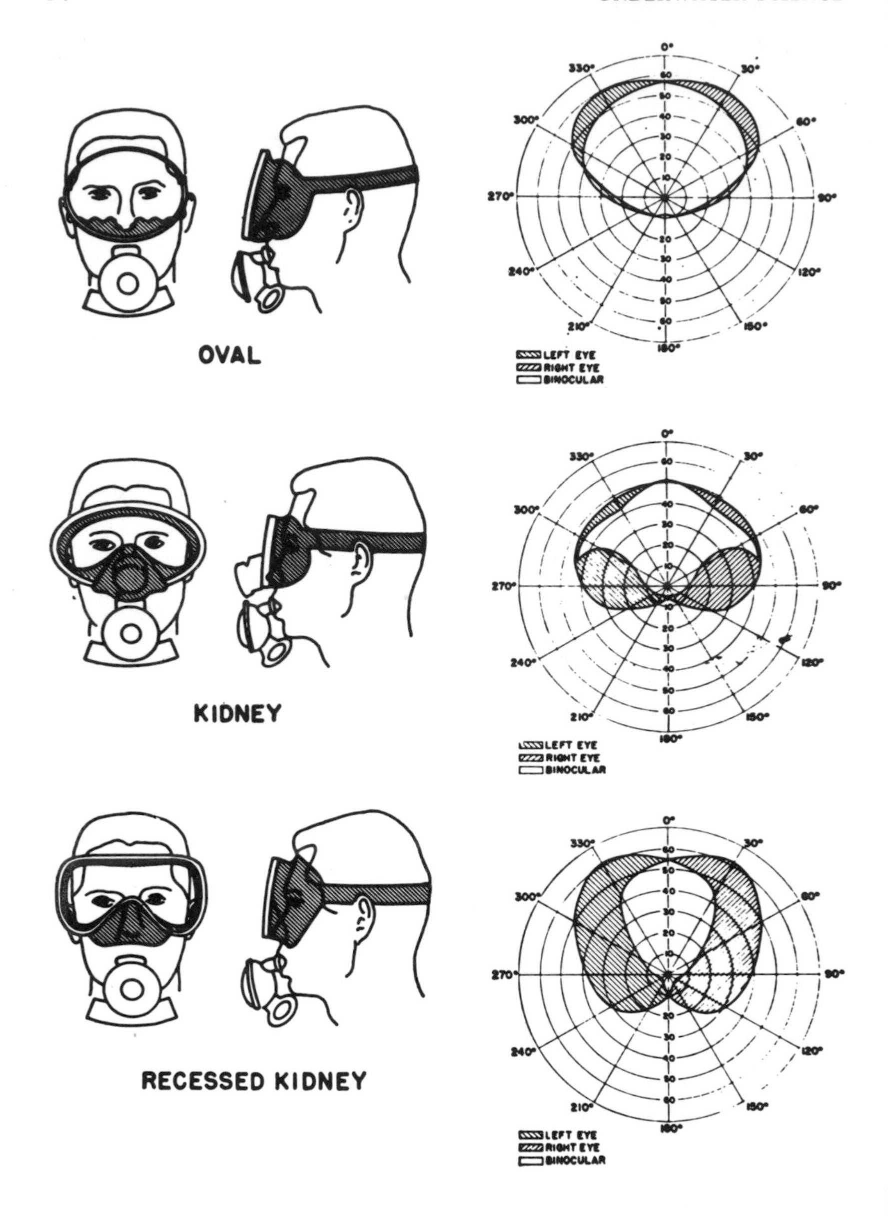

Fig. 20 Median visual fields for three standard face-masks.

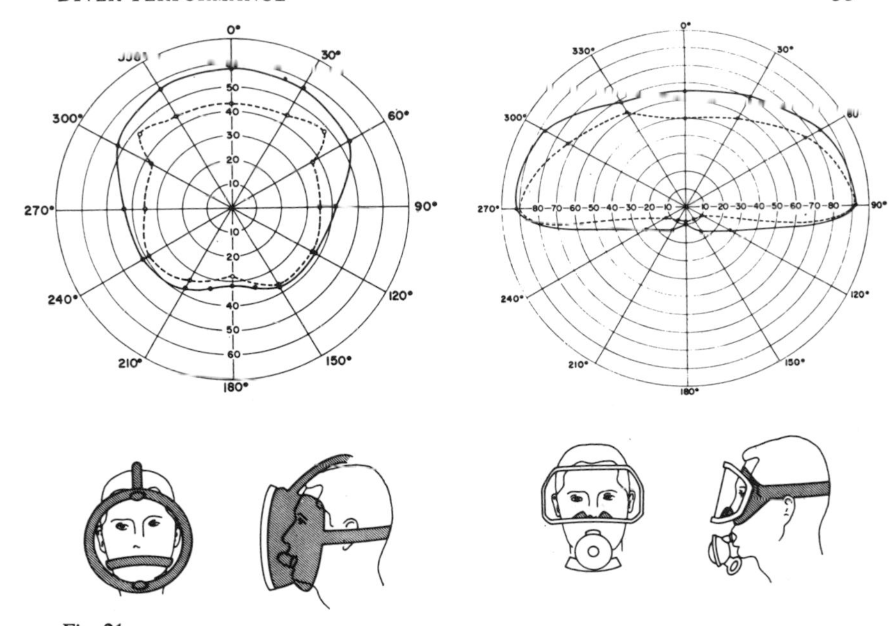

Fig. 21

Median visual fields for full face-mask and wrap-around-mask. Data from the two subjects tested plotted separately (Weltman *et al*. 1965).

It is clear then that none of the five masks tested by Weltman *et al*. is really satisfactory. An alternative solution to the problem is to replace the mask with contact lenses. Although the design of suitable lenses presents a number of technical problems, experimental lenses of various types have been produced, and in at least one case (Faust and Beckman 1966) have been evaluated fairly thoroughly. This study included a comparison using perimetry between the contact lens system and an oval face mask. Their technique differed slightly from that used by Weltman *et al*. and seems to give rather smaller visual field values. When compared to performance in air, their results show a mean reduction in visual field of 10–14 per cent when wearing the contact lenses under water, compared to a reduction of 44–46 per cent when wearing the oval face mask. They also found a consistent improvement in visual acuity when subjects were tested under water wearing contact lenses. Unfortunately, contact lenses do not protect the eye in the way that a mask does, and when worn in sea-water all subjects reported a stinging sensation in the eyelids ranging from mild to unbearable. Until this problem is solved the value of contact lenses is severely limited.

A device called the Rebikoff 'Visiorama' has recently been advertised which uses lenses to counteract the narrowed visual field. This sounds a very promising development, but as yet we know of no objective performance data for divers using this system.

The studies so far described have been concerned with the *optical* limitations of a diver's visual field. A recent study by Weltman, Egstrom, and Christianson (1967) shows that there may be a further *attentional* narrowing of the diver's field of awareness under certain conditions. In this study the diver was required to perform two jobs at the same time, one of which occupied the centre of his visual field while the other was peripherally located. The authors were interested in whether divers would show evidence of 'perceptual narrowing', a tendency to be unaware of peripheral events which has been claimed to occur in subjects under stress (Watchell 1967). Such an effect should produce relatively greater impairment on peripheral than on central tasks. Two central tasks were used, adding rows of figures and monitoring a voltmeter. The pointer of the voltmeter was deflected 47° at regular 1 sec. intervals; the subject's job was to detect occasional larger (75°) deflections, and to press a switch within 10 seconds when this occurred. The peripheral task involved detecting an irregularly occurring dim light mounted on the subject's face-plate. The subjects (novice divers) were tested on land, in an experimental diving tank and in the open sea at a depth of about 8 m. Their results showed that subjects fell into two quite distinct groups, 'Steady' subjects whose performance on the peripheral task was constant across conditions, and 'Unsteady' subjects who showed a marked deterioration on the peripheral but not the central tasks in the tank condition, and an even greater deterioration in the open sea. The subjects who showed this perceptual narrowing effect seemed to be the most anxious, though unfortunately since the authors had not expected two distinct sub-groups to emerge they had not taken any objective measures of anxiety. It would be desirable to repeat this study using some objective measures such as heart rate or galvanic skin response to try to predict in advance which divers will show perceptual narrowing.

2. *Visual acuity*

Because of the refractive index of water, the human eye is incapable of focusing adequately under water. This problem is solved by using the face mask to trap a layer of air in front of the eye, which has the incidental effect of making objects appear to be only three-quarters of their actual distance away (see Fig. 14). Since the visual angle subtended by a given object is increased, a diver should have better acuity for detail under water. A number of studies have tested this.

The first published data on underwater visual acuity comes from the study by Faust and Beckman (1966) which evaluated underwater contact lenses. Acuity was measured using a photographically reduced copy of the Snellen letter chart frequently used by opticians, mounted on the window of a diving tank and viewed from a distance of 3 m. The subjects were tested wearing both a standard oval face mask and the contact lens system, and performance was compared with acuity on land (presumably measured using a normal chart and the standard distance of 6 m.). Acuity was best wearing the contact lenses, but subjects also had somewhat better acuity when wearing a face mask under water than they had on land, though performance was not as good as would be expected on the basis of the one-third increase in the visual angle subtended by the letters.

A similar result was obtained by Kent and Weissman (1966) who tested twenty divers both under water in a submarine escape training tank and on land. The targets in this case were Landolt Cs, circles with a break located in one of four positions (North, South, East, or West), viewed from a distance of 5 m. In addition to testing their subjects in clear filtered water, Kent and Weissman took the additional precaution of increasing the level of illumination of the target so as to compensate for the light absorbed during transmission through the water. Like Faust and Beckman, they found that acuity tended to improve when tested under water, with the improvement not being as great as would be expected in view of the reduced *optical* distance of the targets under water. They attribute this discrepancy to the apparently greater tendency for the subject's face mask to steam up under water.

In a more recent study, Christianson (1968) tried to minimize misting by having the subject's air intake blow across the inside of the face plate. He tested his subjects in a filtered fresh water tank, specially designed for studying diver performance. Like Kent and Weissman he used black Landolt Cs on a white background, with illumination adjusted so as to give constant luminance in all conditions, as viewed by the subject. Target distance was arranged so that the *optical* distance of the stimulus was constant, thus the target was viewed from 3 m. in air and approximately 4 m. in water. By underwater standards, the instrumentation was very sophisticated, with the target being rotated automatically to one of the four gap positions behind a shutter which then opened to expose the stimulus for a predetermined length of time. The subject was tested in a standing position and signified his response by turning a knob to one of four positions. Both the subject's response and the time it took him to decide were automatically recorded and displayed on the experimenter's console in an adjacent room. Six divers were tested intensively, with each subject making 100 judgements about each of 15 targets, with gaps ranging in size from 25 seconds to 2 minutes of arc. All subjects were tested in four conditions: (i) diver and target both in air; (ii) diver in water, target in air; (iii) diver in air, target in water; and (iv) both diver and target in water. Christianson found no difference in acuity scores across the four conditions, and since targets under water were one-third farther away than targets in air (so as to equate their *optical* distance), this implies that acuity would have been one-third better under water for targets at the same physical distance. Unlike Faust and Beckman, and Kent and Weissman, Christianson therefore found an improvement in visual acuity of the magnitude expected on the basis of the one-third increase in visual angle subtended by the target. However, in terms of decision time, the underwater conditions were reliably slower, with the mean time per decision increasing by 72 per cent between conditions 1 and 4.

The three diving tank studies described all agree in suggesting that visual acuity is better under water than in air, though not to the extent that might be expected on the basis of the magnification of the target. However, testing was carried out in filtered fresh water, and in the Kent and Weissman and the Christianson studies, with special illumination of the target to offset loss of light during transmission through the water. The next two experiments (Baddeley 1968a) were concerned with the question of what level of acuity might be expected under the less artificial conditions of open sea diving, given conditions of good visibility.

The first study tested divers on land and under water at depths of 10 and 30 m. Two tests were used, the Snellen letter chart and a task involving acuity for black lines against a white background. The Snellen charts were standard except that they were made of white plastic rather than cardboard, and since several subjects were able to read all the letters at the standard distance of 8 m. on land, the viewing distance was increased to 10 m. for all conditions. The second task involved a black painted aluminium sheet containing sixteen 7-cm. diameter holes and backed by a white painted aluminium sheet. Across each hole, a black thread or wire ranging in thickness from 0·025 to 2·5 mm. was stretched in one of four orientations (vertical, horizontal, and the two diagonals). The subject wore his mask in all conditions, worked from a sitting or kneeling position, wrote down on a roughened Formica board as many letters as he could read, and noted the orientation of each line. Subjects were instructed to guess rather than omit an item; this is an important precaution to take since otherwise any change in performance may reflect either a real change in acuity, or simply a change in the level of caution with which the subject is making his judgements (Swets, Tanner, and Birdsall 1961).

In the underwater condition the test stimuli were suspended from a wire frame supported by two air-filled buckets and were viewed against a water background (see Fig. 54). Visibility (see p. 111) was approximately 30 m. Subjects were instructed on land and then performed the tests at their own rate. This was made possible by the fact that once the vision tests were set up the experimenter did not need to adjust or change anything during the run, a feature which makes testing very much faster and smoother. The maxim that the less the experimenter

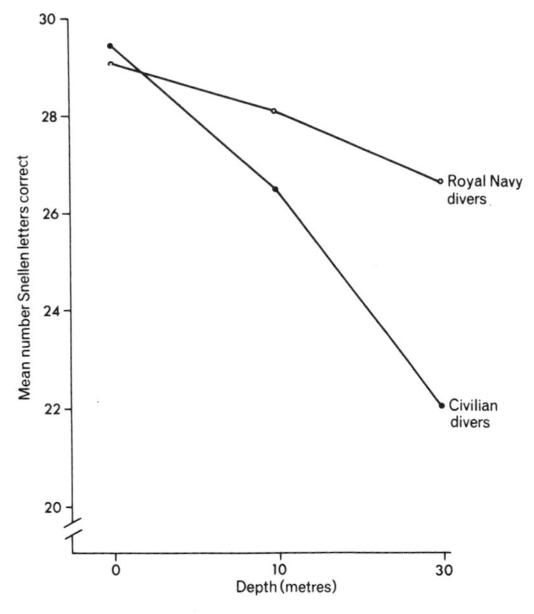

Fig. 22
Visual acuity (Snellen letter chart) as a function of depth for divers breathing air (●——●) and a 40:60 oxy-nitrogen mixture (○——○).

needs to do during the experiment the better, is doubly true under water. The three conditions (land, 10 m., and 30 m.) give six possible permutations of testing order and one diver breathing a 40 per cent oxygen–60 per cent nitrogen mixture and one diver breathing air was tested at each. This balanced order of testing ensures that differences between conditions are not due to simple practice effects.

Results from the Snellen letter chart are shown in Fig. 22, where acuity is measured in terms of letters correctly reproduced since this proved much more sensitive than the more conventional measure of completely correct lines. Both groups of subjects showed a statistically reliable drop in acuity under water, and those breathing air showed a further drop at 30 m. Line acuity (Fig. 23) showed a similar

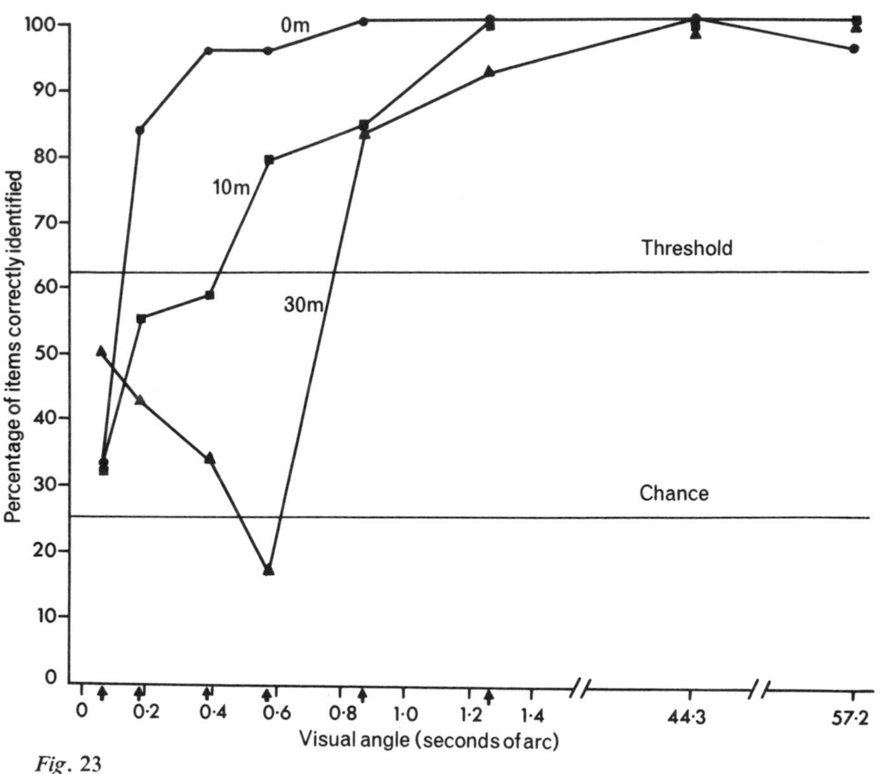

Fig. 23

Per cent detection of lines of different thickness as a function of depth. Data from six divers breathing air. Arrows represent thickness of target wires.

pattern, with the mean threshold (defined as the point midway between perfect performance and the chance guessing level) rising from a minimum visible angle of 0·2 sec. on land to 0·4 sec. at 10 m., and in the case of the air breathing group still further to 0·8 sec. at 30 m. This drop in acuity is even more marked if one takes into account the magnification effect.

This study leaves two unanswered questions; why is acuity so bad? and why is there a further impairment at depth for divers breathing air? The existence of slight impairment in visual performance at a depth of 30 m. was subsequently confirmed in a dry pressure chamber with subjects breathing air (Baddeley 1968a). The absence of an effect when subjects breathe an oxygen-rich mixture has interesting implications for mechanism involved and is worth further investigation.

The major problem, however, remains: why acuity should be worse under water, despite the magnification effect. The most likely explanation is that the loss of definition due to the absorption and scattering of light from the test stimuli by suspended particles outweighed the advantage of magnification. If so, the adverse effect should be reduced as the distance between the stimulus and the diver's eye is reduced, since this should reduce the amount of light lost during transmission.

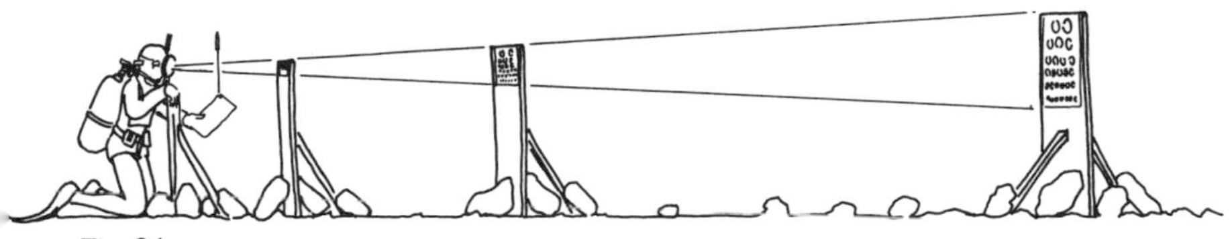

Fig. 24
Sighting range used to test visual acuity under water for targets at different distances. Targets at each distance subtend the same angle at the diver's eye.

A further experiment examined this hypothesis by testing acuity at three distances.

Fourteen subjects were tested on land and under water with targets set at 12 m., 4 m., and 1 m. The stimuli were sets of Landolt Cs arranged as on an optician's chart, with large Cs at the top and small Cs at the bottom. The charts were reproduced photographically in three sizes such that the large stimuli set at 12 m. subtended the same angle as the medium stimuli at 4 m. and the small stimuli at 1 m. (see Fig. 24).

Testing was carried out in the Blue Lagoon and off Marfa Point, Malta, at a depth of 5 m. in conditions of good visibility (about 30 m.).

The results are shown in Fig. 25 in terms of number of items correctly reported. Whereas on land, acuity is constant across the three target distances, underwater acuity drops substantially between the 4 and 12 m. conditions. However, performance at 1 m. is no better than at 4 m., both being slightly, though not significantly, worse than performance on land. The poor performance under water at 12 m. suggests that the absorption and scattering of light *is* a major limitation on underwater acuity. However, acuity of the 1 m. target is still far below the level expected in view of the increase in the visual angle subtended by the stimuli under water, suggesting some further limitation on acuity under water.

A notable feature of these results was the difference between subjects with poor acuity (presumably due to short-sightedness), who in the near condition tended to show better acuity under water, and good acuity subjects whose acuity did not improve under water. Re-examination of the data of Faust and Beckman (1966), and Kent and Weissman (1966) showed a similar tendency for the improvement in performance under

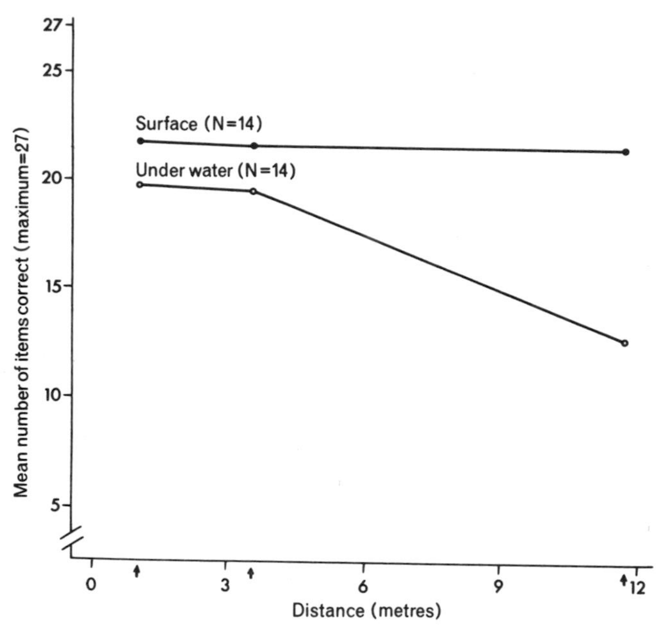

Fig. 25
Visual acuity (Landolt Cs) as a function of target distance on land and under water.

water to be limited to poor acuity subjects, suggesting that a focusing problem may be partially responsible for the failure to find the expected improvement in acuity under water (see Baddeley 1968b).

Conclusion: The Ergonomist and the Diver

What then is the role of the ergonomist in diving? Since there is as yet little information on diver performance, the ergonomist's main contribution to diving lies in his skill and experience in carrying out experiments involving people. The most obvious use for this skill lies in comparing different types of equipment. The work by Weltman *et al.* (1965) comparing various types of face mask is a good example of this role, which basically involves providing a consumer's guide to available equipment. The drawback of this approach is that it *is* limited to available equipment. The ergonomist can of course make suggestions, but unless he is able and prepared to design and develop new equipment himself, his recommendations may have little effect. Since there are so few trained ergonomists with diving experience it would in any case seem most sensible to use them as ergonomists, and not as amateur engineers. There are two possible answers to this problem, the first is to train the engineer as an ergonomist: courses with this aim have been created in recent years, but nevertheless this is a relatively long-term solution. A more immediate solution lies in co-operation between the engineer designing diving equipment and the ergonomist.

Problems in the relationship between the ergonomist and the engineer designing underwater equipment arise largely from non-communication. Since the two disciplines share a common aim, it seems likely that their development will lead to greater co-operation. In the case of inert gas narcosis, the problems of the relationship between the diver and the ergonomist are rather more complex. This shows up most clearly if we consider the pioneering work in deep diving typified by the Sealab and Conshelf projects. The aim of such projects is to *demonstrate* that man can live and work at great depths for long periods. Such pioneering work demands considerable financial backing, and as such is dependent on 'selling' the project, which in turn depends on showing as dramatically as possible what *can* be done. The ergonomist on the other hand is interested in questions rather than demonstrations, and in order to answer his questions in a way which convinces fellow scientists he is likely to use an experimental design which may make inconvenient demands on the day-to-day running of the project, to use tests which may not have any obvious relevance to the normal work of a diver, and to be interested in differences which may seem small and rather trivial to the layman. While the long-term aims of these two groups are common, since both wish to maximize the efficiency of the diver at depth, their short-term aims may be diametrically opposed. The project organizer is concerned to demonstrate the feasibility of useful work at depth, and as such naturally hopes for no narcosis effects. From the ergonomist's point of view, however, negative results are less interesting, since they *may* (and no doubt often do) simply reflect the insensitivity of his tests. In principle this conflict is a healthy one, since, given reasonable communication, the two groups will tend to keep each other up to the mark. The danger, however, is one of a complete breakdown of communication leading to hostility and mutual rejection.

How can this be avoided? As ergonomists we can do two things. First recognize that the need to 'sell' a major pioneering project such as Sealab or Conshelf may in the first instance make demonstrations and films more important than experiments. A step in this direction is the proposal by the ergonomists planning to work on *Sealab III* to observe actual salvage operations rather than use performance tests as attempted in *Sealab II*. That is not of course to say that performance tests should not be used, since it seems unlikely that simple observation of work in progress can ever give us the controlled and precise data needed for accurate comparisons, it simply accepts that a pioneering demonstration project performed in front of the film and TV cameras is not the best place to collect objective scientific data. The second contribution the ergonomist can make to this problem is to take a long, hard look at his basic criteria. One suspects that often the only criterion applied to a particular problem is that the results should be suitable for publication in an academic journal, which often means simply that statistically significant differences should be observed. Ergonomics is basically an applied science, however, and as such should be concerned with usable answers to practical problems. In other words we must select tests which bear as close a relationship as possible to the practical situation, and we should attempt to give quantitative answers. If we are, for example, trying to decide whether it is worth providing a diver with oxy-helium rather than air, it is not sufficient simply to know that he is reliably better on one than the other; the question at issue is whether the gain is enough to justify the additional cost and inconvenience involved, and since the answer will probably depend on the type of job he is doing, the data should be based on an appropriate task. To develop a comprehensive battery of valid, reliable performance tests, together with norms which would give an indication of the importance of any decrement shown would clearly be an enormous task. One answer to this problem, however, is to use selection tests of the kind used for predicting a person's capacity for doing various types of jobs. Such tests are already available in large numbers; they normally have high reliability and validity, and are accompanied by population norms so that any drop in efficiency can be related to the range of performance encountered in the population at large. It may of course prove impracticable to adapt selection tests to the rigorous conditions of open sea experimentation, but in view of the enormous potential saving in time and effort, it is surely worth trying.

Acknowledgements

I am grateful to Dr. G. Weltman and to *Human Factors* for permission to reproduce illustrations from *Visual Fields of the Scuba Diver* (Weltman, Christianson, and Egstrom 1965).

References

Adolfson, J. (1967), *Human Performance and Behaviour in Hyperbaric Environments*. Stockholm: Almqvist and Wiksell.

Anon. (1966), 'Experience with Gemini solves EVA problems.' *Aerospace Med.*, **37,** 1284–6.

Baddeley, A. D. (1966), 'Influence of depth on the manual dexterity of free divers: a comparison between open sea and pressure chamber testing.' *J. Appl. Psychol.*, **50,** 81–5.

Baddeley, A. D. (1967), 'Diver performance and the interaction of stresses.' *Underwater Assn. Rep.*, **2,** 35–8.

Baddeley, A. D. (1968a), 'Visual acuity of divers in the open sea.'

Baddeley, A. D. (1968b), 'Visual acuity underwater: a review.' *Underwater Assn. Rep.*, **3,** 45–50.

Baddeley, A. C. (1968c), 'A 3-min. reasoning test based on grammatical transformation.' *Psychon. Sci.*, **10,** 341–2.

Baddeley, A. D. and M. J. Catton (in preparation).

Baddeley, A. D., J. W. de Figueredo, J. W. Hawkswell-Curtis, and A. N. Williams (1968), 'Nitrogen narcosis and performance under water.' *Ergonomics*, **11,** 157–64.

Baddeley, A. D. and N. C. Flemming (1967), 'The efficiency of divers breathing oxy-helium.' *Ergonomics*, **10,** 311–19.

Barnard, E. E. P. (1961), 'Visual problems under water.' *Proc. Roy. Soc. Med.*, **54,** 9–10.

Behnke, A. R., R. M. Thomson, and E. P. Motley (1934), 'The psychologic effects from breathing air at four atmospheres pressure.' *Am. J. Physiol.*, **112,** 554–8.

Bennett, P. B. (1966), *The Aetiology of Compressed Air Intoxication and Inert Gas Narcosis*. Oxford: Pergamon Press.

Bennett, P. B. and A. V. C. Cross (1960), 'Alterations in the fusion frequency of flicker correlated with electroencephalogram changes at increased partial pressures of nitrogen.' *J. Physiol.*, **151**, 28–9.

Bennett, P. B., A. N. Dossett, and P. Ray (1964), 'Nitrogen narcosis in subjects compressed very rapidly with air to 400 and 500 ft.' *R.N. Personnel Research Committee*, U.P.S. 239.

Bennett, P. B. and A. Glass (1957), 'High partial pressures of nitrogen and abolition of blocking of the occipital alpha rhythm.' *J. Physiol.*, **138**, 18–19.

Bennett, P. B., E. C. Poulton, A. Carpenter, and M. J. Catton (1967), 'Efficiency at sorting cards in air and a 20 per cent oxygen-helium mixture at depths down to 100 ft. and in enriched air.' *Ergonomics*, **10**, 53–62.

Bowen, H. M., B. Anderson, and D. Promisel (1966), 'Studies of divers' performance during the Sealab II project.' *Human Factors*, **8**, 183–99.

Case, E. M. and J. B. S. Haldane (1941), 'Human physiology under high pressure.' *J. Hyg.*, **41**, 225–49.

Christianson, R. A. (1968), 'A study of visual acuity underwater using an automatic Landolt ring presentation technique.' *Ocean Systems Rept.*, X8–128/020.

Cousteau, J. Y. (1966), Paper presented at the B.S.-A.C. Brighton Conference.

Crossman, E. R. F. W. (1959), 'A theory of the acquisition of speed skill.' *Ergonomics*, **2**, 153–66.

Dickinson, J. and J. A. Leonard (1967), 'The role of peripheral vision in static balancing.' *Ergonomics*, **10**, 421–9.

Donald, K. W. (1947), 'Oxygen poisoning in man.' *Brit. Med. J.*, **1**, 172.

End, E. (1938), 'The use of new equipment and helium gas in a world record dive.' *J. Ind. Hyg. Tox.*, **20**, 511–20.

Faust, K. J. and E. L. Beckman (1966), 'Evaluation of a swimmer's contact air-water lens system.' *Milit. Med.*, **131**, 779–88.

Fisher, R. A. (1960), *Design of Experiments*. Edinburgh: Oliver and Boyd.

Fisher, R. A. and F. Yates (1963), *Statistical Tables for Biological, Agricultural and Medical Research*. Edinburgh: Oliver and Boyd.

Frankenhauser, M., V. Graff-Lonnevig, and C. M. Hesser (1960), 'Psychomotor performance in man as affected by high oxygen pressure (3 atmospheres).' *Acta Physiol. Scand.*, **50**, 1–7.

Galton, F. (1876), *The art of travel: or shifts and contrivances available in wild Countries*. London: Murray.

Hamilton, R. W. Jr., J. B. MacInnis, A. D. Noble, and H. R. Schreiner (1966), 'Saturation diving to 650 feet.' *Ocean Systems Tech. Mem.*, **B 114.**

Jennings, R. D. (1968), 'A behavioural approach to nitrogen narcosis.' *Psych. Bull.*, **69**, 216–24.

Kent, P. R. and S. Weismann (1966), 'Visual resolution under water.' *U.S. Nav. Sub. Med. Center Rept. No. 476*.

Kiessling, R. J. and C. H. Maag (1962), 'Performance impairment as a function of nitrogen narcosis.' *J. Appl. Psychol.*, **46**, 91–5.

Krasberg, A. R. (1967), 'The evolution of functional saturation diving.' *Underwater Assn. Rep.*, **2**, 39–45.

Legge, D. (1965), 'Analysis of visual and proprioceptive components of motor skill by means of a drug.' *Brit. J. Psychol.*, **56**, 243–54.

Poulton, E. C., M. J. Catton, and A. Carpenter (1964), 'Efficiency at sorting cards in compressed air.' *Brit. J. Indust. Med.*, **21,** 242–5.
Poulton, E. C. and P. R. Freeman (1966), 'Unwanted asymmetrical transfer effects with balanced experimental designs.' *Psych. Bull.*, **66,** 1–8.
Ross, H. E. (1967), 'Water, fog and the size–distance invariance hypothesis.' *Brit. J. Psychol.*, **58,** 301–13.
Ross, H. E. (1969), 'Adaptation of divers to curvature distortion under water.' *Ergonomics*. (In Press.)
Southall, J. P. C. (1933), *Mirrors, Prisms, and Lenses*. 3rd edition. Ch. 4. London: Macmillan.
Swets, J. A., W. P. Tanner, and T. G. Birdsall (1961), 'Decision processes in perception.' *Psych. Rev.*, **68,** 301–40.
Tailliez, P. F., F. Dumas, J. Y. Cousteau, J. Alinat, and F. Devilla (1949), 'La vision en plongée'. In *La Plongée en Scaphandre*, pp. 18–22. Paris: Elzevir.
Taylor, H. J. (1965), In *The Physiology of Human Survival*. Edited by O. G. Edholm and A. L. Bacharach. London: Academic Press.
Watchell, P. L. (1967), 'Conceptions of broad and narrow attention.' *Psych. Bull.*, **68,** 417–29.
Weltman, G., R. H. Christianson, and G. H. Egstrom (1965), 'Visual fields in the Scuba diver.' *Human Factors*, **7,** 423–30.
Weltman, G., G. H. Egstrom, and R. H. Christianson (1967), 'Perceptual narrowing in divers: a preliminary study.' *Human Factors*, **8.**
Zetterstrom, A. (1948), 'Deep-sea diving with synthetic gas mixtures.' *Milit. Surg.*, **103,** 104–6.

3 Spatial Perception Under Water

Helen E. Ross

Man is able to appreciate the spatial relationships of objects in the world around him because his brain is capable of analysing and correlating the various sensory messages which arrive from the different sense organs. Normally we perceive the world through a combination of several different senses. For example, we may see a dog, hear it bark, and also touch it. Usually the brain analyses all the sensory messages appropriately, so that they confirm each other. Sometimes in unusual circumstances the brain makes an inappropriate analysis in one or more sense modality, and we suffer a 'perceptual illusion'. A common example is relative movement in a train: when another train at a station starts moving it is difficult to tell whether one's own train or the other is in motion.

The way in which the brain analyses the sensory messages must depend to some extent on past experience. For example, Turnbull (1961) reported that a Pygmy youth who had lived all his life in a dense forest was unable to appreciate the size and distance of cattle on an open plain, and thought they were insects which grew in size as he approached them. However, the brain is very adaptable, and if it finds itself to be consistently wrong it will relearn and make a more appropriate analysis.

Our perceptual experience begins on land. Under water the sensory information is different, so it is not surprising that we are at first misled. An example is the changed apparent distance of objects seen in water. However, as we become more experienced we learn to interpret the available information more appropriately, and our perception of the underwater world becomes tolerably accurate. The changed environment leads to two sorts of errors: *constant errors* and *variable errors.* Constant errors are due to the consistent misinterpretation of a sensory cue, and they disappear fairly readily with experience. Variable errors are inconsistent and unpredictable: they are probably due to a reduction of the usual sorts of sensory information, aggravated by underwater stresses such as cold, anxiety, and nitrogen narcosis (see Chapter 2). Variable errors also decrease with practice, except where the sensory information is irretrievably lost. For example, colour discrimination deteriorates with depth as the illumination becomes increasingly monochromatic (see Chapter 4); and directional sensitivity to the source of a sound deteriorates because the phase and intensity differences at the two ears are seriously reduced. However, information once thought irretrievable is often regained by the ingenuity of man: Bauer and Torick (1966) have invented hydrophones which restore the normal phase and intensity differences to the ears. Other instruments, such as a depth gauge and compass, are in common use amongst divers. The diving scientist will of course use all possible instruments when making measurements under water. Nevertheless, the perception of divers without such aids is itself a subject of scientific inquiry.

Visual Judgements of Size, Shape, and Distance

As described in Chapter 2, the diver normally wears a face-mask because the human eye is long-sighted when immersed in water. The face-mask allows the eye to operate in air, but introduces other disadvantages: peripheral vision is largely excluded by the sides of the mask and optical images of objects are distorted in various ways. Experienced divers overcome these disadvantages to some extent. They pay more attention to peripheral visual stimuli than novices (Weltman and Egstrom 1966); and they show greater adaptation to optical distortion (Ross 1969).

Curvature distortion

The face-mask produces 'pin-cushion' distortion under water, a square appearing bowed inwards (see Fig. 16). In the third dimension the centre of the field of view appears farther way than the outsides (see Fig. 17). This is a transformation rather than a loss of sensory information. It is well known that people who wear distorting lenses adapt to their effects, provided that they can gain sensory information about the nature of the distortion (Rock 1966). As a diver moves his head around, the shapes of objects are transformed consistently; and even if he does not consciously notice the changes he nevertheless adapts (Ross 1969). The extent of the adaptation can be measured by using a flexible rod which the diver can adjust through any degree of curvature (Fig. 26). The diver can set the rod to appear straight, or bent towards or away from him. Ross (1969) measured the curvature perception of divers in air, in water, after half an hour in the water, and in air after emerging. A straight line appeared to curve away in the centre in water, though not to the full optical extent. After half an hour it appeared less curved—about 25 per cent of full adaptation had occurred. On emerging from the water the

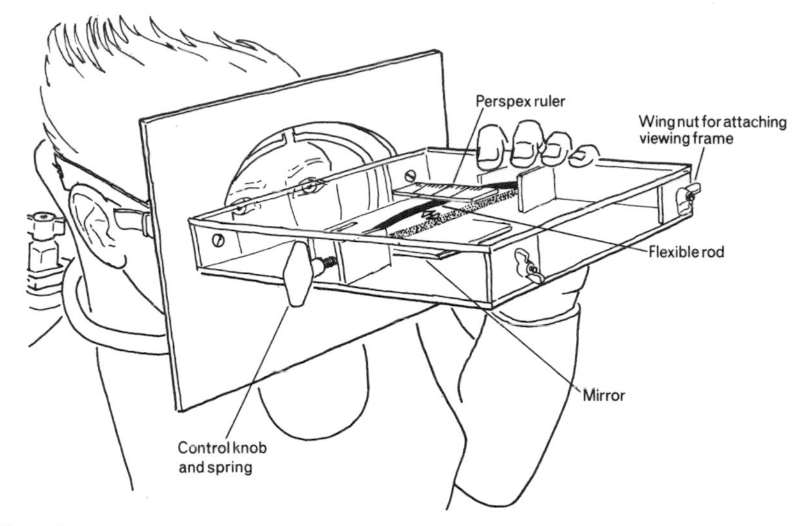

Fig. 26

The diver places his face-mask at the viewing hole and turns the control knob until the rod appears bent towards him. An additional viewing hole can be placed on the other side of the apparatus for viewing the 'bent-away' position. Alternatively a mechanism can be added for changing the direction of bending of the rod. The curvature of the rod is read from the ruled scale when no parallax is visible in the mirror.

divers showed a negative after-effect of the same size: a straight line in air appeared to bend slightly towards them. Significant differences also existed between novice and experienced divers immediately on entering the water. Novices saw the line as curved to its full optical extent, while experienced divers saw it as less curved. The learning which takes place must have two aspects: one of these is concerned with *how* to compensate for the optical distortion, and the other with *when* to compensate (e.g. when wearing a face-mask under water). The latter aspect is analogous to a switch which brings in an appropriate compensating mechanism. This type of adaptation is known as 'situation-contingent' There is some controversy as to its nature (Foley and Abel 1967).

Relation between size and distance

As an object recedes into the distance it subtends a smaller angle at the eye. The brain compensates for this, so that objects do not appear to change much in size as their distance changes. This is known as 'size-constancy'. We need to know the distance before we can know the size of an unfamiliar object. A given retinal angle can represent a small near object or a large far object (Fig. 27). In practice the relation between size and distance judgements rarely approaches the geometrical ideal (Epstein *et al.* 1961), but it is quite clear that changes in apparent distance affect the perception of size.

Refraction at the face-plate upsets the normal relation between image-size, object-size, and distance. Objects are optically nearer in water, but not larger (in Fig. 14 the fish does not change in size). If their apparent size is determined by the same perceptual rules as in air, they could be seen at 3/4 of their distance and the right size, or at their physical distance and 4/3 larger, or with some other proportionate combination of size and distance. On the other hand, since perceptual

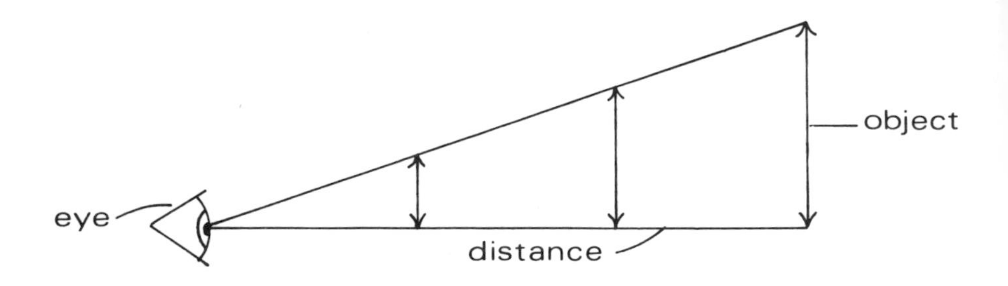

Fig. 27

The angle subtended by an object at the eye cannot alone determine the physical size of the object. The same angle can represent a small object at a near distance or a larger object farther away. The apparent enlargement of objects under water cannot be explained simply on the basis of their increased angular size. Other cues such as apparent distances or 'known size' must be taken into account.

learning is known to occur, they might be seen at their true distance and of their true size, with a new ratio between size and distance. This is a complicated problem to investigate because adaptation might take place to sizes or to distances or to both; moreover, adaptation will probably occur during the course of a diving session, and there are probably differences between novices and experienced divers.* Neither

* The expected types of adaptation have been found in recent experiments on size and distance judgements by Ross, Franklin, Lennie, and Weltman at the University of California at Los Angeles; and in recent experiments on hand–eye co-ordination by Kinney, Luria, and Weitzman at the Naval Submarine Medical Center, New London, Groton, Connecticut.

size nor distance judgements can be understood in isolation, and the problem will be discussed again in the sections below.

Distance judgements

The majority of the distance cues are related to the optical distance of the object, so it might be supposed that novice divers would initially see objects at about three-quarters of their distance in air. However, distance cues concerned with the colour and brightness of objects bear no fixed relation to the optical distance, and may suggest that objects are much farther away than in air. Thus, before perceptual learning occurs, objects may appear too far rather than too near.

Most of the distance cues indicate which objects are farther and which nearer, but do not contain enough information to specify precise distances. We have to combine all the available cues in order to make good distance estimates. The cues related to the optical distance are: muscular cues of accommodation and convergence (effective only over short distances, and conveying little precise information); the visual cues of binocular disparity, linear perspective, and texture gradients, and the overlap of far objects by near objects; and the proprioceptive-visual cue of movement parallax, nearer objects appearing to move more than farther objects as the head or body moves. These cues are described in detail by Graham (1965). Most cues can be misleading in isolation. Under water many cues are absent or reduced, so the remaining cues can give rise to mistakes. For example, Ross (1966) placed objects of different sizes at the same distance, but under water the smaller objects appeared farther away (Fig. 33). Similar results were found by Woodley (1968).

It might be thought that stereoscopic vision should enable divers to perceive relative distances correctly. However, stereoscopic judgements are affected by differences in brightness contrast between the objects, objects of low contrast appearing too far away (Fry *et al.* 1949). Individuals may show various biases under conditions of low visibility (Palmer 1960). It now seems likely that the brain correlates brightness-disparities rather than shape disparities between the images in the two eyes to give stereoscopic perception (Kaufman and Pitblado 1965). It is not surprising, then, that stereoscopic acuity should be poor when contrast is low (Jonkers and Kylstra 1963). In the low-contrast conditions which usually obtain under water we might expect to find both increased variable errors and also some constant errors. In very clear water, however, we might expect the increased angular size of objects to lead to an improvement in performance. Improvements in performance on standard visual acuity tests in clear, warm swimming baths have been found by Faust and Beckman (1966), Kent and Weismann (1966),

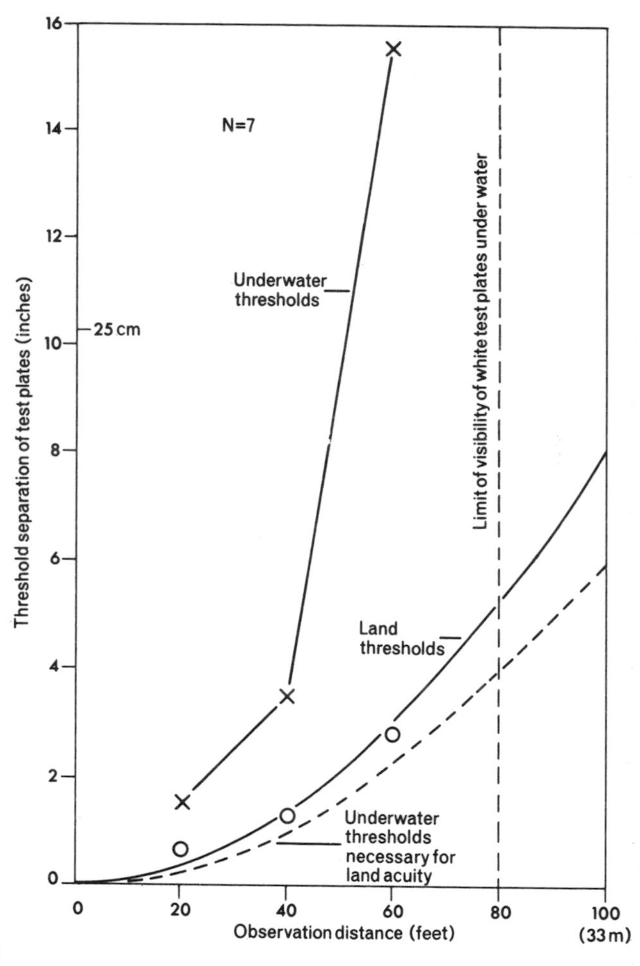

Fig. 28

For constant stereoscopic acuity, the minimum detectable separation in depth between two objects is proportional to the square of the viewing distance. This is shown for land acuity in the middle curve. The lower curve shows the separation necessary for comparable underwater acuity, taking into account the angular magnification. The upper points show the threshold separations actually obtained under water. Stereoscopic acuity is clearly worse under water, and deteriorates with viewing distance.

and Christianson (1968). In the sea there are so many factors militating against good acuity that performance deteriorates (see Chapter 2). Ross (1967a) found that stereoscopic acuity was considerably worse than in air, when divers were tested at a depth of 20 m. in water with a black-body visibility of about 25 m. (Fig. 28). The apparatus for this experiment is shown in Fig. 29. Luria (1968) also found a reduction in stereoacuity in a swimming pool, both constant errors and variable errors increasing with the turbidity of the water.

In addition to causing a loss of sensory information, low brightness contrast is also involved in a transformation of distance cues. Brightness

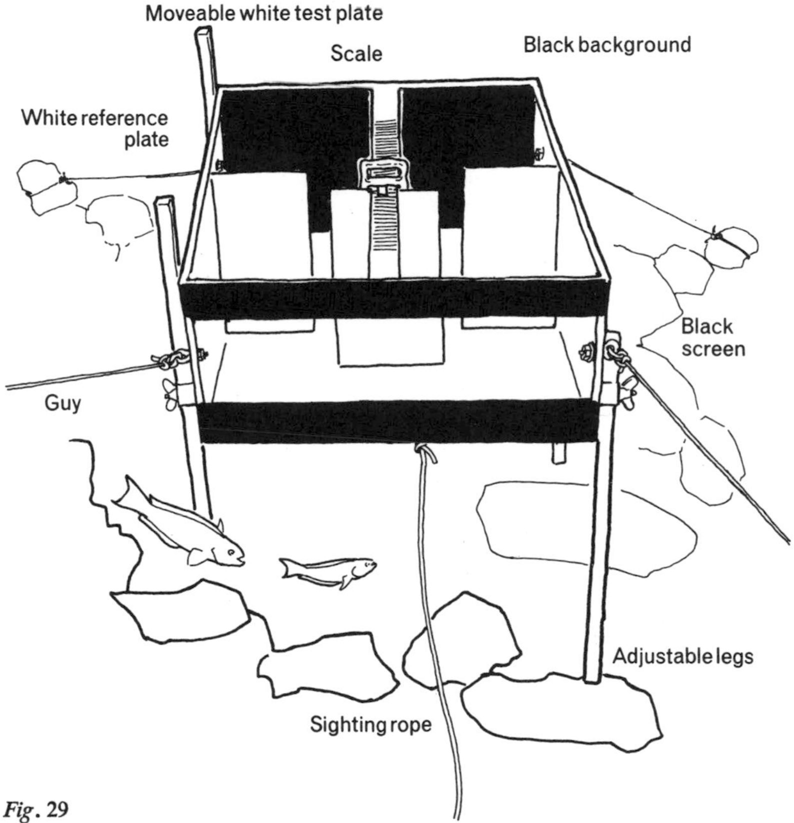

Fig. 29

The subject views the three white plates from a fixed distance along the sighting rope, his head level with the front of the apparatus. The experimenter adjusts the position of the centre plate along the scaled rack, and the subject decides whether it is behind or in front of the outer plates.

contrast can be a cue to distance: given knowledge of the original brightness of an object and the attenuation coefficient of the atmosphere, it is possible to calculate the distance of the object from its brightness contrast (see Chapter 4). The atmosphere is never perfectly clear, and as objects recede into the distance their colour and brightness approximate to that of the sky background, until they can no longer be seen. This reduction of colour and brightness contrast with distance is sometimes called 'aerial perspective' (Berkeley 1709; Helmholtz 1856). When atmospheric conditions change, so does the rate of change of contrast with distance. We are initially poor at adapting to such transformations, and see distant hills as near in very clear weather and as far in misty weather. Visibility under water is usually comparable to a thick fog on land (see Chapter 4), and leads novices to overestimate distances. Ross (1968) found that when targets of different brightness were suspended at the same distance in a glass tank full of murky water, targets of lower brightness contrast were judged farther away. There was a linear relation between judged distance and the logarithm of the brightness contrast, regardless of whether the targets were brighter or darker than the water background (Fig. 30). This is the expected type of mathematical function, since the logarithm of the brightness contrast changes linearly with distance for all atmospheric conditions. It remains to be investigated how quickly people adapt to waters of different visibility.

A similar effect was found by Woodley (1968) in an open-water experiment. Grey squares were judged farther away than white and black chequered squares, at least at viewing distances beyond 15 ft. when the contrast of the grey squares was very low.

Aerial perspective must be a compelling cue, and adaptation fairly slow, since most investigators have found overestimation rather than underestimation of underwater distances. Luria *et al.* (1965) had their subjects look through a porthole into a large tank of water, and estimate the true distance of targets in the water with reference to a standard target in air (placed at 60 cm.). The comparison target was positioned randomly, at 30 cm. intervals, from 130 cm. to 5 m. The subjects made similar judgements in air, out of doors. Non-diving subjects (20) overestimated all underwater distances, the overestimation increasing with distance (Fig. 31). The results for a small group (6) of experienced divers showed no significant differences between air and water judgements. The experienced divers also showed less variability of underwater distance judgements than the non-divers. The results suggest that both variable and constant errors are reduced with experience. Ross (1966)

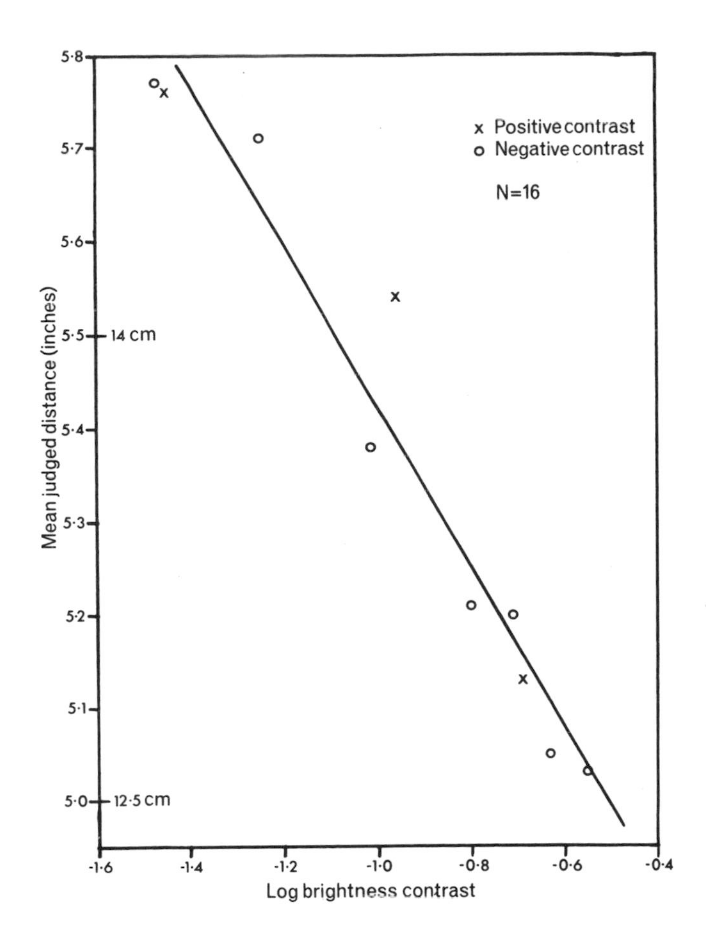

Fig. 30

Objects at the same distance (10 in.) but of different brightness are judged to be at different distances. Those nearest the brightness of their background appear farthest away, while brighter or darker objects appear nearer. There is a linear relation between judged distance and the logarithm of the brightness contrast $\left(\text{Contrast} = \dfrac{\text{Object brightness} - \text{background brightness}}{\text{Background brightness}}\right)$. The brightness measurements in this experiment were made with a spot photometer, looking into a tank of water from the viewing position.

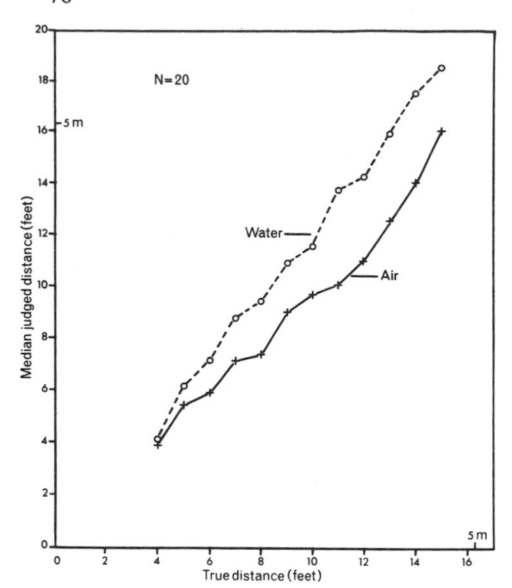

Fig. 31

Results for 20 non-diving subjects show overestimation of distance when looking into a tank of water, in comparison with estimates in air. Subjects judged the distance of the target with reference to a two-foot target shown in air. Six experienced divers showed no difference between air and water judgements, suggesting perceptual learning.
(Redrawn from Luria, Kinney, and Weissmann 1965, by permission of the Commanding Officer, U.S. Naval Submarine Center, Groton, Connecticut.)

Fig. 32

Mean distance estimates for 16 divers of mixed experience show that underwater estimates are greater than three-quarters of the land estimates. Overestimation under water increases with distance, and small targets are judged farther away than large targets.
(Redrawn from Ross 1967b, with permission of the Editor of the *British Journal of Psychology*.)

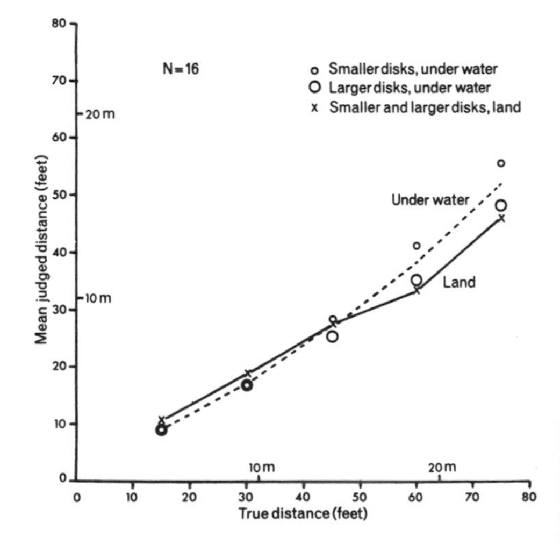

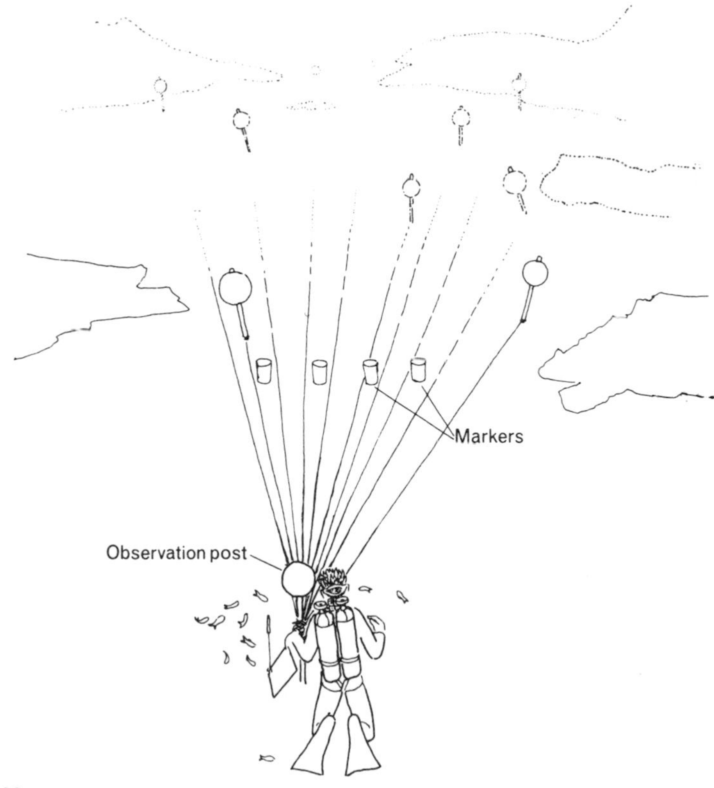

Fig. 33

The subject writes down an estimate of the size and distance of each disc, counting them from left to right. The ropes and cylindrical markers help to identify the positions of the discs—a necessary precaution in low visibility water.

used a range of targets placed at various distances from 5 to 25 m. from an observation post (Fig. 33), and asked divers to estimate their absolute distance. The experiment was performed both in the sea and on land, with 16 subjects of varied experience. Subjects showed slight underestimation of near distances in water, but increasing overestimation of far distances (Fig. 32). Variability of judgements was significantly greater under water than on land. Nichols (1967) used an adjustment method. He moved a ball towards or away from the subject by a pulley system attached to the edge of the swimming baths. This experiment was performed with a luminous ball in semi-darkness. The subjects (20) were required to signal when they believed the ball to be at a distance of 2 or 5 m. The ball was stopped too far for 2 m. and about the same as on land for 5 m., suggesting that only very near distances are underestimated. Novices showed greater constant errors and variable errors than experienced divers in water, but not on land. Ross (1967b) also used an adjustment method. Using a trombone arrangement similar to that of Hemmings and Lythgoe (1964) (see Chapter 4, Fig. 47), 20 subjects were trained to set a target at 130 cm. from their noses in air. The mean setting when attempting to reproduce the same physical distance in water was 113 cm., and for the same apparent distance 118 cm. The expected setting on the basis of the optical effect is 144 cm. The only experimenter to use familiar objects seems to have been Woodley (1968). He found that beer bottles were judged to be at much the same mean distances as on land (though with higher variability). An array of white squares gave higher variability of distance judgements than the beer bottles, and the distances were mainly underestimated by inexperienced divers. The underwater distances in these experiments ranged from 1 to 10 m.

These experiments by various authors suggest that slight underestimation of true distance may occur at near distances, but that errors tend to overestimation with increasing distance.

The finding of overestimation under water has led to further research on factors causing overestimation in air. Luria *et al.* (1965) found that in a 'reduced cue' situation (when subjects looked through a small window in a room draped with black curtains) distances were overestimated. Ross (1967b) investigated the enhanced cue of aerial perspective, and found that subjects overestimated distances in a fog in comparison to clear-weather estimates.

Size judgements

It is often said that objects appear too large under water because they are magnified. It is not self-evident that this is the correct explanation. The main factors affecting apparent size are the apparent distance of the object, and the relative size of the other objects in the visual field. The relation between size and distance (Fig. 27) is discussed above, objects appearing larger as their apparent (but not physical) distance increases.

The relative size of surrounding objects can affect apparent size independently of changes in apparent distance. This is the probable explanation of many of the geometrical illusions. An example is shown in Fig. 34, where the circle on the left appears larger than the circle on

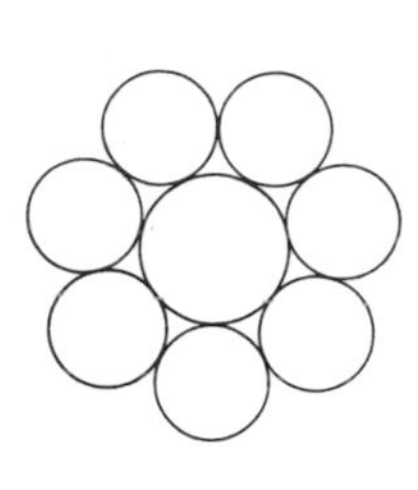

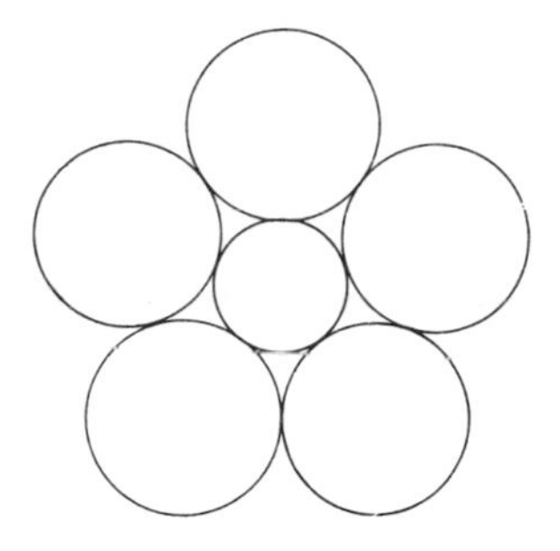

Fig. 34

The inner circles are the same size, but the one on the left appears slightly larger. This is because apparent size is partly determined by the relative size of surrounding objects. If relative size were the main determinant of apparent size, objects should not appear enlarged under water, since all objects are magnified in approximately the same proportion.

the right, because its size is judged relative to large rather than small circles. This effect is sometimes called 'size-contrast'. Since all objects in water are magnified by about the same extent their relative sizes remain unchanged, and there is no reason why any object should appear too large.

However, there remains another possible factor. We may remember the normal angular size of a familiar object at a certain distance, and be aware that its angular size is too large for its optical distance in water. It certainly seems to be the case that objects of known size—such as one's body or a Coca-Cola bottle—appear too large under water, though one is not aware that anything is wrong with the size of plants and animals never seen on land. Even for unfamiliar objects the diver will generally have some information that a change of image-size has occurred. If he touches an object there will be conflict between visual and tactile size-cues. If he moves his head or swims towards an object there will be a new relation between body movements and change of image-size. Under these conditions considerable adaptation should occur within 30 min., objects appearing near their normal size (Rock 1966, Chapter 5).

It is quite possible, then, that magnification *may* affect apparent size under water, but carefully-designed experiments are necessary to prove the point. Rock (1966) discusses the logical requirements of such designs. If magnification is an important factor it will probably be found that adaptation occurs during the course of a dive, and that experienced divers show greater initial adaptation than novices.*

* See footnote on p. 72.

To perform an experiment on apparent size, the size of objects must somehow be varied. The simplest way of varying the size of an object whilst keeping its proportions constant is to use a square partly covered by a movable right-angled section (Fig. 35). The subject can then adjust the size of the square under water until it appears the same size as a test square at a different distance, or the same as the remembered size of an object in air. Alternatively, a set of objects (such as squares, triangles, or discs) of different sizes may be used. If the distance of an object is varied, this alters the image-size but not the the object size. This method of altering size should be avoided, as size estimates are confounded with distance estimates, and are difficult to interpret unless distance estimates are obtained at the same time (Ross 1965a).

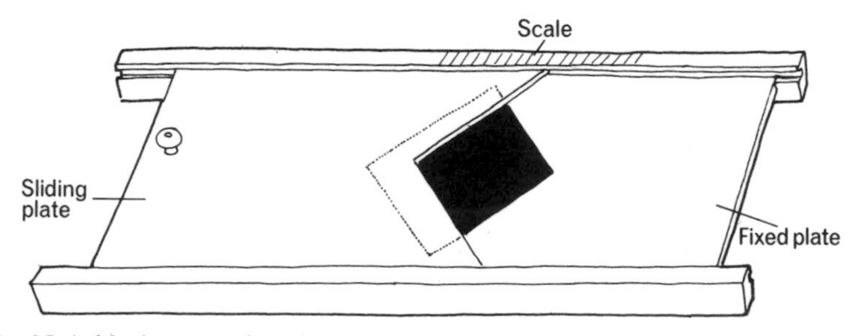

Fig. 35 A black square is painted on a white plate. The visible area of the square can be adjusted by moving a covering plate with a right-angled section.

Luria *et al.* (1965) used a series of comparison squares which the subjects viewed in air and compared with a white 10 cm. square target set at 160 cm. or 4 m. in a tank of water. As a control condition the target was also viewed in air. The mean match made by 25 subjects was: (at 160 cm.) 8·3 cm. in air, and 9·4 cm. in water; (at 4 m.) 8·3 cm. in air, 9·6 cm. in water. The average ratio of water to air judgements was 1·13—slightly less than might be expected on the basis of the optical magnification (about 1·33). There were no significant differences between novices and experienced divers, but the results might be explicable if all subjects adapted to magnification during the course of the experiment. If the subjects' distance estimates (Fig. 31) are considered, the size estimates are surprising. Water estimates were greater than air estimates, and very much greater than the optical distance. If normal size–distance relations hold, water size estimates should be more than 4/3 of those in air. The results suggest that some perceptual learning occurred for the changed relation between size and distance.

Ross (1966) set up a range of different sized discs at different distances (Fig. 33), and asked subjects to estimate the true sizes and distances. Sizes were considerably overestimated in comparison with land judgements, and the degree of overestimation increased with distance. Distance estimates in the same experiment also increased with distance (Fig. 32). Size estimates were directly related to distance estimates, the two increasing together with distance under water and decreasing on land (Fig. 36). Ross (1967b) also found overestimation of both size and distance in a fog on land. These results suggest that overestimation of distance is the main factor responsible for overestimation of size in water, optical magnification being relatively unimportant. As with the results of Luria *et al.*, the overestimation of size was not as great as might be expected from the relative distance estimates. Perceptual learning must have occurred during the course of the experiment or on previous dives. Overestimation of size due to low contrast was also found by Woodley (1968). He suspended a white board and a small preserved shark at a distance of 20 ft. from the divers. The shark was

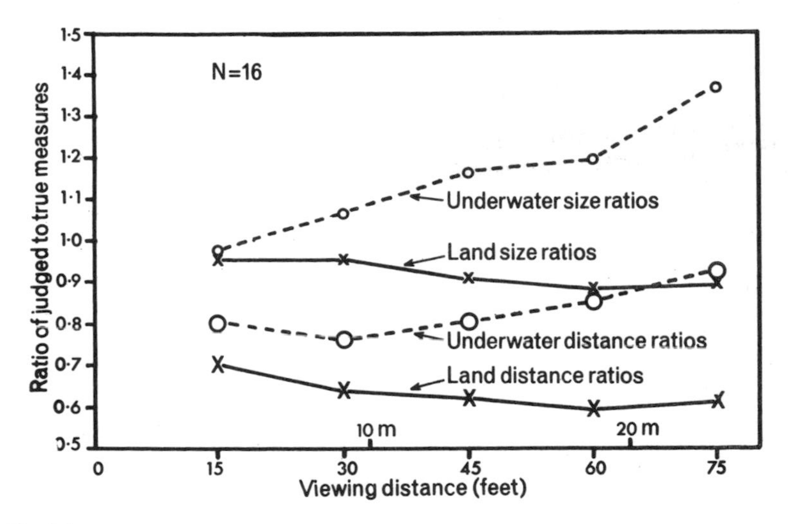

Fig. 36
The ratio of judged to true size shows the same trends as the distance ratios. On land both decrease with distance; under water they increase. The underwater distance ratios are based on the optical distance. If the relation between size and distance judgements were strictly geometrical, the size and distance ratios should be the same. The relation between them changes from air to water, suggesting perceptual learning. (Redrawn from Ross 1967b, with permission of the Editor of the *British Journal of Psychology*.)

grey and inconspicuous (aided by its natural counter-shading). The divers judged the shark to be 8½ per cent longer than the white board, though they were both the same length (2 ft. 5 in.).

Judgements in the vertical plane

Most experiments on space perception are performed with the subject upright and looking straight ahead. If a person looks upwards or downwards on land, or if the normal orientation of his body is changed, his perception of size and distance is altered. This is easy to observe if one looks through one's legs, or down from a height: objects generally appear small and far away. This is a curious observation, because it breaks the geometrical relation shown in Fig. 27. If objects appear too far away they should appear too large, not too small. Alternatively, if some other factor makes them appear small this could account for their greater apparent distance. Another type of explanation is that size–distance relations are learned in the horizontal direction, and this learning does not transfer perfectly to other directions (Van de Geer and Zwaan 1964).

Under water it is almost as common to look up or down as horizontally. Judgements of the distance of the surface or the sea-bed can be important, as a diver may rely upon his visual estimates in planning a dive. If the distances are farther than he thinks, this could be dangerous. There is some evidence that underestimation does occur. This might be due to changed visual cues, as there is little evidence that changed bodily orientation has any effect under water (Ross 1965).

The surface and the sea-bed tend to appear nearer than they actually are, the underestimation increasing with distance. Ross, King, and Snowden (in press) found that the depth of the sea-bed was slightly underestimated for depths down to about 7 m., when the estimates were made from the surface in clear water over a rocky bottom. The true depth was measured by a weighted tape-measure. Divers descending deeper report more serious underestimation, especially in very clear water. If distance is underestimated, then (according to the argument, presented above) size estimates should be approximately correct. Ross, King, and Snowden performed a further experiment to investigate this. The experimenter placed white squares of various sizes one at a time on a bed of dark seaweed. Four subjects simultaneously judged the size and depth of the square from various heights in midwater (true height was calculated from a depth gauge). The heights of the divers above the sea-bed ranged from 3 to 23 m., and their distance estimates showed moderate underestimation at most distances. The median ratio of estimated to true depth was 0·895, considerably greater than the optical ratio of 0·75. The mean size estimates of the squares (whose sides ranged from 30 to 60 cm.) were very accurate. The relation between size and distance ratios was very different from the results of the horizontal experiment shown in Fig. 36. It is difficult to say whether this difference is of any importance, since different subjects were involved and it was impossible (without the use of a balloon or helicopter) to obtain any comparative data in air. However, the experiment does show that magnification does not always lead to the overestimation of size.

The distance of the surface is probably more difficult to gauge than that of the sea-bed. The diver sees a cone of light of constant angular size, but the proportion of the surface that he can detect varies greatly with wave conditions and the angle of the sun. Near the surface, and particularly in still water, the diver may be aware of the unusual shape of the surface: it is not flat but dome-shaped, for the same reason as a straight line appears curved in water (Chapter 2, Figs. 16 and 17). The surface may thus appear to slope down in front, and this may contribute to the sense of disorientation which some divers experience on surfacing (see below).

Some novices have great difficulty in estimating the height of the surface. The author as a novice ascending from 33 m. in clear water felt that the surface was always about 10 m. away, and was alarmed to watch a snorkeller on the surface apparently grow from the size of a sixpence to normal size as she ascended. Ross, King, and Snowden obtained estimates of surface height from five divers of moderate experience in clear water, with the divers at various depths from 3 to 20 m. On the first test the divers showed underestimation which increased with depth—apart from an anomalous reading at 20 m., which was probably due to the sight of the sea-bed in a familiar area. On repeating the test two days later, estimates were very accurate, showing that divers can improve quickly on this type of judgement (Fig. 37).

Spatial Judgements involving Proprioceptive Information

Many types of judgement depend upon bodily (proprioceptive) information from the pressure and joint receptors, the vestibular system, the muscular control system, and other internal receptors. Tactile size judgements, motor skills, geographical orientation and knowledge of the vertical are examples of this type of judgement.

Under water some cues are transformed, some are lost and others are

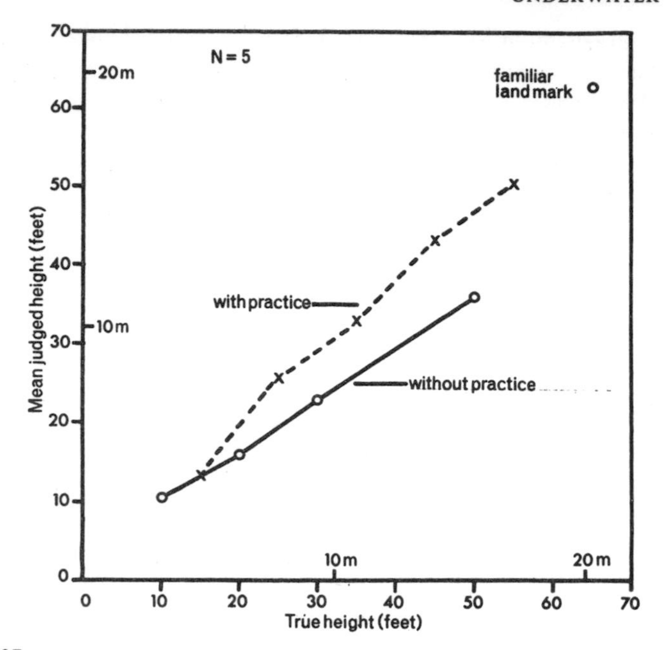

Fig. 37

Results for five divers of moderate experience show an initial tendency to underestimate the true height of the surface in open water. However, with practice, or near familiar landmarks, their judgements may become fairly accurate.

gained. A transformation of weight occurs, objects weighing less than on land because of the upthrust of the water. A diver normally adjusts his weight to achieve approximately neutral buoyancy. The usual pressure cues at the joints and the soles of the feet are therefore much reduced, even when he touches the ground. However, different parts of his body have different densities, and since gravity operates under water he can never become truly weightless. He is not, therefore, subject to such extreme sensory changes as astronauts orbiting in space, or airmen in zero-g flights (Loftus and Hammer 1961; Simons and Gardner 1963). The resistance of the water as he moves, and the behaviour of his aqualung and other apparatus, give him additional cues.

Knowledge of the vertical

Knowledge of the vertical on land is dependent partly on visual cues, partly on postural cues from the pressure and joint receptors, and partly on the gravitational response of the otoliths of the vestibular system.

All cues normally relate to the gravitational vertical, but experiments have been performed in which they conflict. Individuals vary in the extent to which they rely on the different cues. The subject is reviewed by Howard and Templeton (1966, Chapters 7–9).

In open water the visual cues are much reduced: man-made geometry is generally absent, and plants do not follow the vertical as precisely as on land. In empty water and low-visibility water there may be no cues except the brightness-gradient, the surface normally (but not always) appearing brighter than the bottom (Duntley 1960). On a dark, moonless night there will be no light available apart from the artificial light of torches and ships, and sometimes the bioluminescence of animals.

The usual postural cues are much reduced under water, but the drag of the aqualung and other equipment may compensate for this to some extent. However, aqualungs vary in buoyancy, and a diver may be disturbed when wearing a lung to which he is unaccustomed. The diver can also notice changes in air-flow from the demand-valve as he changes in orientation; but again, this varies with the type of regulator. Depending on the location of the exhaust outlet, the diver can also see or feel his exhaust bubbles floating upwards. If in doubt, he can note the direction in which his weight-belt hangs; and if still uncertain he can drop his belt and float to the surface.

The vestibular system continues to respond to gravity under water. By testing blindfold divers or swimmers under water, and reducing postural and breathing cues to a minimum, many authors have attempted to measure the sensitivity of the otoliths in isolation (James 1882, Stigler 1912, Knight 1958, Margaria 1958, Padden 1959, Schock 1960, Brown 1961, Diefenbach 1961, Walsh 1961, Schöne 1964, Nelson 1967). The studies have varied widely in experimental procedure and in the mathematical analysis (if any) of the observed errors to the vertical. It is not surprising that the reported errors range from 4° to 180°. In the majority of studies in which measurements were made the subject was attached to a tilting chair or table, rotated through some angle, and required to set a pointer to the vertical. The problem then arises whether to measure the absolute error from the vertical disregarding the direction, or whether to measure the mean error and standard deviation taking the sign into account. The latter procedure seems preferable, as it distinguishes between bias and sensitivity. However, the absolute error is important in considering whether a diver can swim towards the surface. In most studies the diver and pointer were movable in only one

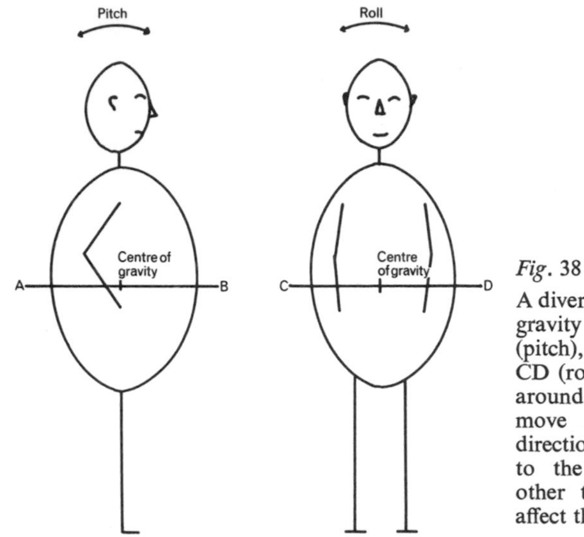

Fig. 38
A diver can rotate about his centre of gravity in the median plane AB (pitch), and also the mid-frontal CD (roll). In addition he can rotate around his mid-body axis (yaw), and move in a straight line in any direction. Pitch and roll give errors to the gravitational vertical, but other types of movement do not affect the misalignment of the trunk.

plane at a time. This is unrealistic, since a diver is normally free to move in three-dimensional space, and can make errors in any plane (Fig. 38). The most sophisticated experiment, both technically and statistically, is that of Nelson (1967). He attached his subjects to a two-gimbal tilt-table, and required them to position themselves in various orientations to the vertical. From the two protractor readings he calculated both constant errors and elliptical probable errors (a measure of sensitivity). He noted a significant pitch-forward bias in the median plane, exceeding 30° in the nominal head-down position. Measures of sensitivity ranged from 15° to 40°, being worse with the head down than the head up.

It can be concluded from this and previous studies that knowledge of the vertical is worse under water than on land (especially for labyrinthine-defective subjects), but improves with practice; it is worse with the head down than up; and there are large differences between subjects and experimental conditions.

Most authors have been concerned with the measurement of the sensitivity of the vestibular system, or the partial simulation of weightlessness for space travel. They have not been concerned with the practical abilities of the diver in open water. This question was investigated by Ross, Crickmar, Sills, and Owen (in press) in clear water off Malta. They photographed a diver and plumb-line from a profile and back view simultaneously, while the diver attempted to align his body and point his finger to the vertical (Fig. 39). The angular deviation from the vertical was measured from the photographs, and the true (or maximum) deviation was calculated from the apparent deviations along the two lines of sight (Fig. 40). The mean deviations for five divers of moderate experience ranged from 8° to 33°, depending upon the

Fig. 39

One diver holds a plumb-line (out of sight of the subject) to indicate the vertical in the photographs. The subject turns a somersault, and then attempts to orientate himself to the vertical, pointing upwards or downwards. When he is in position, two photographs are taken simultaneously from the side and rear. For some trials the subject wears a blackout or semi-transparent screen over the face-mask. The order of the trials is written on a Formica board, carried by the subject.

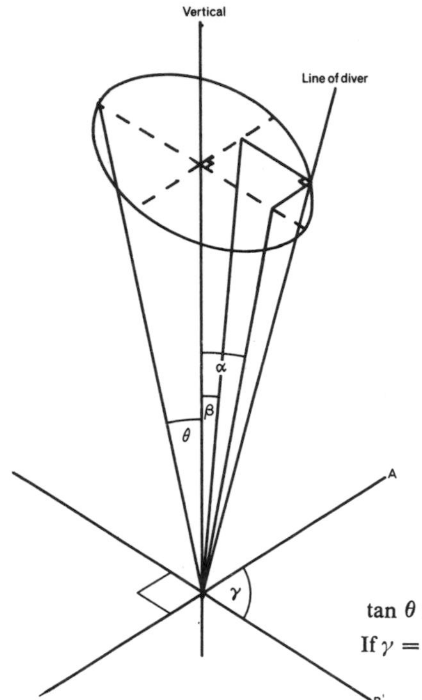

Fig. 40

The angle that the diver makes with the vertical (θ) can be represented by a solid angle whose projection is shown in the diagram. If two photographs of the diver are taken simultaneously from positions A and B, the apparent inclination of the diver from A is α, and from B is β. If the angle γ between the two cameras is known (drawn as 90° in the diagram), then θ can be calculated from the following expression:

$$\tan\theta = \sqrt{\tan^2\alpha + \tan^2\beta + 2\tan\alpha\tan\beta\cos\gamma}$$

If $\gamma = 90°$ (i.e. the cameras are at right angles) then

$$\tan\theta = \sqrt{\tan^2\alpha + \tan^2\beta}$$

experimental condition. Errors were greater with the diver inverted than upright. This may have been partly due to the difficulty in balancing upside down; but similar results were found by other authors with the subject attached to a tilt-table. Errors were less with vision than blindfold, showing that the diver made use of some visual information such as rocks, plants, fish, or the brightness-gradient (the plumb-line and photographers were out of his sight). Divers typically showed a pitch-forward bias when upright, but some subjects changed to a pitch-backward bias when inverted (perhaps for ease of balance). Most subjects showed a left-bias of the trunk, forearm, and finger, probably due to the use of the right hand for pointing (left-handed divers showed a right-bias). Errors of the trunk and finger were about equal in magnitude.

In a second experiment the divers wore a sheet of scratched perspex over their face-masks during the sighted condition. This was intended to imitate the 'white-out' that occurs in low-visibility water, but in practice the perspex acts as an integrator and destroys the brightness-gradient. The diver has to move his head in order to see a change in brightness. He receives little visual information through his face-mask, and what he receives may be misleading. Under this condition visual performance was *worse* than blindfold. This result suggests that poor visual information may degrade superior information from other sensory sources. Alternatively divers may be distracted by the attempt to search for non-existent visual cues. Possibly divers in black or low-visibility water would do better to shut their eyes and concentrate on bodily cues. Clearly further experiments are necessary on this problem.

Errors in the second experiment were less than for comparable conditions in the first experiment, showing improvement with practice. Similarly Nichols (1967) found that experienced divers were better than novices at judging the horizontal in a swimming-bath experiment.

The errors found by Ross *et al.* for blindfold divers appear to be lower than those found by Nelson, indicating that the free diver can make use of non-vestibular bodily cues. Knowledge of the vertical, though poor by land standards, would seem to be sufficiently accurate to allow a diver to find his way to the surface. Errors greater than 90° occurred on only one or two occasions, when the diver was overbalancing upside down. However, the results do not necessarily hold for all possible conditions. The diver, though he turned a somersault between trials, was always in control of his own bodily position. He was not subjected to passive rotations, as may happen in turbulent water. He also opened his eyes between trials to read the next instruction, so was never without vision for more than a minute or two. It is probable that divers have difficulty in monitoring their bodily orientation for long periods without vision, especially when they are moving rather than stationary. Certainly experienced divers have reported disorientation whilst swimming in dark water: 'I became disoriented,' said a colleague, 'thought very hard about it, attempted to swim to the surface and hit my head on the bottom.' A novice also reported attempting to descend whilst underweight, and being surprised to find himself back on the surface a few minutes later. Another problem is adaptation to body-tilt or visual tilt (e.g. a sloping sea-bed), the tilted position coming to represent the normal position. This effect was noted by an experienced diver while swimming down a shot-line: the line seemed horizontal after a while, but suddenly returned to the vertical when he caught sight of the bottom. Similar adaptation effects have been found on land (see Howard and Templeton 1966, Chapter 7).

Weight perception and motor skills

The reduced weight of the body and other objects in water initially disturbs the diver's ability to manipulate objects appropriately. Adaptation seems to occur to some aspects of the transformation, but efficiency is often low due to difficulties in applying torque, and to cold, anxiety, and other factors (see Chapter 2). Experiments on motor skills during zero-g flights show an initial tendency to reach too high when aiming at objects, but the effects disappear with practice, indicating adaptation to weightlessness (Beckh 1954, Gerathewohl *et al.* 1957). Morway *et al.* (1963) found that divers immersed in water for eighteen hours showed an initial tendency to reach too high when aiming vertically, but not horizontally. The error declined with time. The subjects' force estimates were different from those in air, and showed no adaptation.

Objects in water may be weightless, or positively or negatively buoyant. No experiments appear to have been done on judgements of positive buoyancy, though Ross (unpublished) noted that subjects were initially confused when asked to compare the pull of hydrogen-balloons in air. Probably this type of judgement could be learned. Weightless objects can be judged by their inertial mass when moved. Rees and Copeland (1960) asked subjects to compare the mass of objects by pushing them over a frictionless air-cushioned table. They found that mass-discrimination under these conditions was only half as sensitive as weight-discrimination under normal gravity conditions. Kama (1961) also found that the ability to position objects accurately on the frictionless surface was slightly reduced. However, in these experiments the arm was operating under normal gravity, whereas under water the weight of the arm is much reduced. Water also provides some resistance both to arm movements and to object movements, so one cannot conclude much from these experiments about the judgement of weightless objects under water. Changes in arm-weight (whether heavier or lighter) disturb the level of adaptation of the perceptual system, and cause an initial deterioration of weight-discrimination (Gregory and Ross 1967). A diver's discrimination should therefore be poor on first entering the water, but should improve over the first quarter of an hour or so. Immediately on leaving the water his land performance should deteriorate, since his arm feels unusually heavy.

According to Weber's Law, the minimum detectable difference in weight between two objects decreases in proportion to their absolute weight. Since objects weigh less in water, one might expect them to be *more* discriminable than on land. However—as with visual acuity—it is unlikely that the theoretical level can be realized because of all the other changes which occur under water.

Objects weigh less under water, but do they necessarily feel lighter? Experiments on 'weight-constancy' (Katz 1921, Fischel 1926) show that an object feels much the same weight regardless of how it is lifted: we seem to make a judgement about the object itself, and discount irrelavancies in the manner of handling. Preliminary experiments suggest that a similar effect occurs in water: we seem to be able to compensate for the density of the medium, so that objects have almost the same apparent weight as in air.

Geographical orientation

In order to maintain a sense of direction when visual cues are few a diver must be able to estimate how far he has swum and the angles through which he has turned. Unlike a man lost in a fog, he can get lost in three-dimensional space. However, the experiments described here are restricted to horizontal direction-finding. People lost in a mist, or blindfold, tend to walk in circles, generally veering to the right (see Howard and Templeton 1966, Chapter 10). Schaeffer (1928) claimed that subjects showed veering whether they walked, swam, or drove a car. Lund (1930) claimed that asymmetries of leg length were responsible, most people having a slightly longer left leg. Christianson *et al.* (1965) noted inequalities of thrust force between the right and left legs in divers, which should cause veering. Howard and Templeton suggest that slight asymmetries of vestibular 'tonus' may be responsible.

Ross, Dickinson, and Jupp (in press) investigated the problem by requiring their subjects to swim blindfold round two sides of a triangle (almost equilateral), and then find their way back to the starting post (Fig. 41). The subjects also walked blindfold round a similar triangle on land. The routes followed by the subjects are shown diagrammatically in Fig. 42 (each route was sketched on a scaled drawing by an observer). There was no evidence of veering on land over the short distance required (6 m.), and no significant deviation from the required direction of travel. Subjects walked in a straight line for about the right distance. Under water they tended to turn through too small an angle around the outside of the third post (thus swimming wide of the triangle) and some subjects showed marked veering and irregularities. Most subjects swam considerably farther under water, and one subject was unable to find his way round two sides of the triangle.

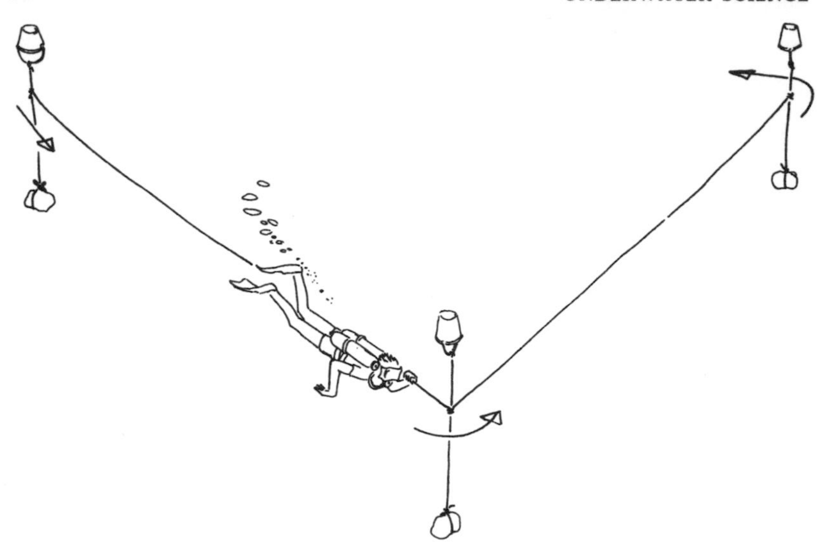

Fig. 41
A rigid support can be formed by roping a buoyancy bucket to a heavy stone—this is an effective substitute for a post under water. Two sides of a triangle are formed by tying a rope to three of the supports. The blindfold subject swims along the rope, then attempts to complete the missing side of the triangle. An observer records his route on a scaled chart.

Failure to return to the starting post is partly due to veering, but could also be due to misperception of angles and distances. Ross *et al.* investigated the angular problem by requiring blindfold divers to swim round segments of a circle, holding a rope attached to a central post. Subjects turned through too *small* an angle. This is partly a distance problem, since the angular error increased with the radius of the circle (160 cm. and 4 m.). On land, however, subjects turned through too *large* an angle. The difference between land and water angle estimation may help to explain the direction of the angular differences in the triangle test. The distance problem was further investigated by requiring divers to swim for certain distances along a 33-m. rope suspended between two poles. The actual distances were read from a tape-measure by an observer. On land subjects walked the required distances fairly accurately. Under water experienced divers were accurate, but novices swam too far, especially at short distances (Fig. 43). Similar results occurred with vision, but since the visibility was only about 5 m. there were few visual cues. This is another example of perceptual learning

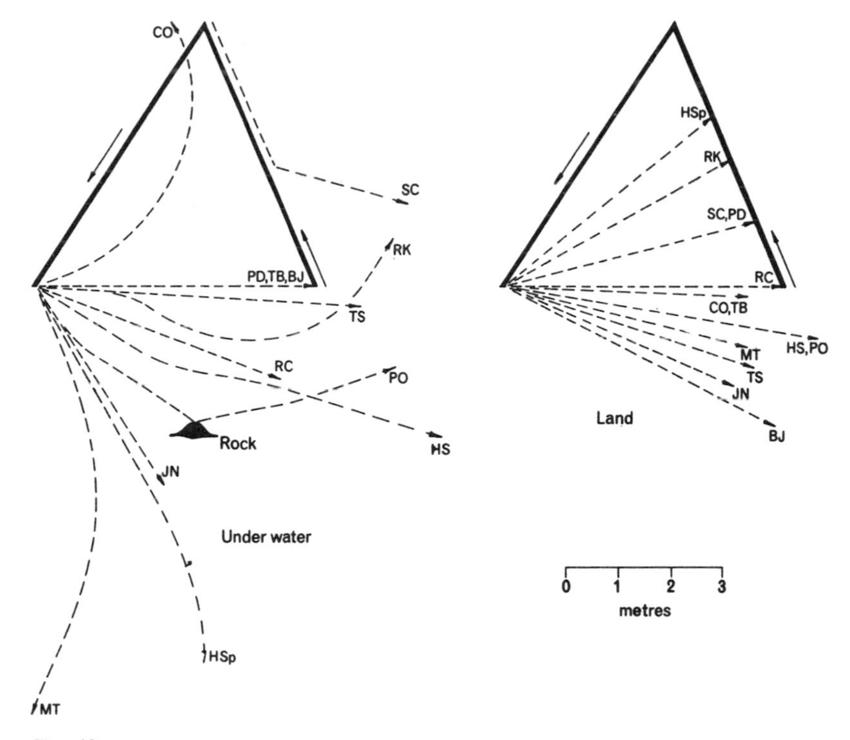

Fig. 42

This diagram shows the routes taken by blindfold subjects in the triangle completion test. On land subjects walked in a straight line for about the right distance, though the direction was often inaccurate. Under water the initial directions were little worse, but divers tended to swim too far and to be unable to follow a straight course. One diver (PO) changed direction after hitting a rock. Another diver (SC) was unable to follow the rope round the first post of the triangle, and turned back on his tracks. (Redrawn from Ross *et al.* 1970, *Human Factors.*)

under water. It confirms the tendency to swim too far in the triangle experiment, which was particularly evident in inexperienced divers.

Bodily movement and visual movement

We cannot know whether an object has moved in relation to its surroundings unless we know whether our eyes, head, and body have moved. Generally the brain corrects quite well for bodily movements, so that objects appear stable unless they are actually moving. There are, however, problems about involuntary and passive movements. If the

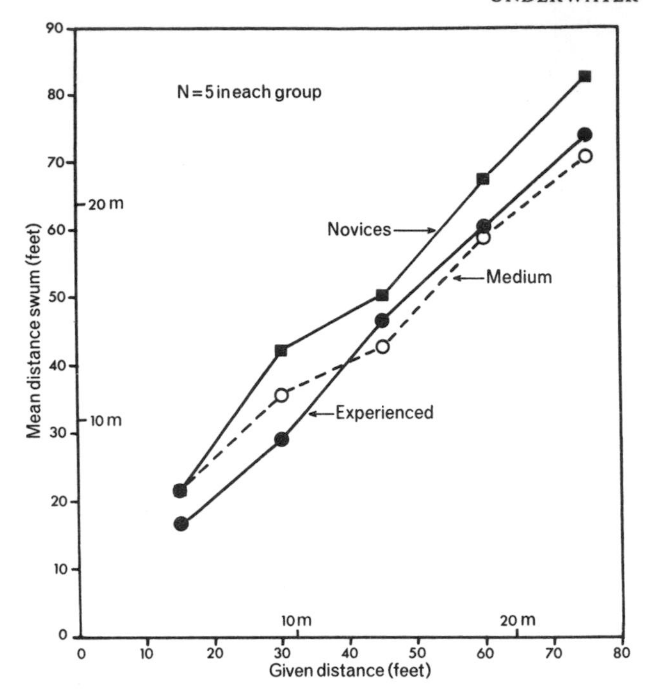

Fig. 43

Experienced divers can estimate the distance they have swum quite accurately, but novices tend to swim too far for a given distance. There were no differences between the groups on land.

eye-ball is pushed ('involuntary' movement) the world appears to swing; and little correction is made for passive movement in a vehicle. A diver normally swims around actively, and is in full control of his movements. Nevertheless, he may drift sideways owing to currents, be tossed around in turbulent water, or sink and rise because of changes in buoyancy. If movement is sudden he will have proprioceptive information that it has occurred. He will probably also be aware of slow sinking and rising because of pressure changes in the ears, and temperature and brightness changes. Can this type of information be used in maintaining the stability of the visual world? In order to investigate the problem it is necessary to have an object whose position relative to the diver is known. This could be an object fixed to the sea-bed, or fixed relative to the diver's body, or fixed relative to his eyes. Perhaps the most useful technique is to obtain an after-image from a flash-gun, and then view it against a blacked-out face-mask. The after-image is fixed on the retina, and it normally appears to move with the intended direction of gaze, but not with passive eye-movements. An alternative technique is to place a glowing object (such as a betalight*) in a blacked-out face-mask: this remains fixed relative to the head, but not relative to the eyes. If the two are viewed simultaneously, the after-image moves relative to the betalight for any type of eye-movement. Since movement is relative it is often difficult to describe unambiguously what is seen. Verbal descriptions need careful analysis. The literature on these effects is often highly confusing (see Howard and Templeton 1966, Chapter 16). Ross and Lennie (1968) used both techniques to investigate movement perception in divers. They noted that both after-images and betalights appeared to move with the diver during intentional movements, but with a slight lag at first. During slow passive sinking and rising (controlled by a Fenzy life-jacket) there was *no* apparent movement of the betalight. When a diver attempts to swim up but actually sinks, the light appears to rise (and vice versa). These results suggest that divers do not correct well for slow passive movement, even though there is sensory feedback concerning their movement. It is possible that experienced divers might eventually learn to do so.

Vertigo

Vertigo is another type of disorientation or dizziness which afflicts some divers occasionally, and seems to be associated with failure to clear one or both ears satisfactorily. It is more frequent on the ascent than the descent, and it also afflicts pilots (Lundgren 1965, Lundgren and Malm 1966, Terry and Dennison 1966). It is probably due to movement of the endolymph in the semicircular canals during attempts at pressure equalization. Changes in temperature (caloric stimulation) may also be a factor, causing convection currents in the endolymph.

If vertigo is due to a specific type of vestibular stimulation, it might be expected to produce apparent movement of an after-image or betalight in a reportable direction. Preliminary experiments during fast ascents suggest that no apparent movement occurs, despite strong bodily sensations of spinning and dizziness (Ross and Lennie 1968). If this result is confirmed it suggests that vertigo is due to 'noise' in the sensory system rather than a specific false signal.

* Obtainable from Saunders-Roe and Nuclear Enterprises Ltd., North Hyde Road, Hayes, Middlesex.

Acknowledgements

The author's work reported in this chapter was supported by grants from the E. M. Pratt Musgrave Fund (Cambridge University), the Browne Fund of the Royal Society, the Royal Geographical Society, the Gilchrist Educational Trust, and the University of Hull; and by gifts from many firms and individuals. I am indebted to the divers who took part in the experiments; to my colleagues for advice; and to Mr. P. Lennie for help in revising the manuscript.

References

(Most U.S. military reports are obtainable from the Clearinghouse for Federal Scientific and Technical Information, U.S. Department of Commerce, Springfield, Virginia, 22151.)

Bauer, B. B. and E. L. Torick (1966), 'Experimental studies in underwater directional communication.' *J. Acoust. Soc. Amer.*, **39**, 25–34.

Beckh, H. J. A. von (1954), 'Experiments with animals and human subjects under sub- and zero-gravity conditions during the dive and parabolic flight.' *J. Aviat. Med.*, **25**, 235–41.

Berkeley, G. (1709), *An essay towards a new theory of vision*, Sec. 67–78. London: Dent, 1910.

Brown, J. L. (1961), 'Orientation to the vertical during water immersion.' *Aerospace Med.*, **32**, 209–17.

Christianson, R. A. (1968), 'A study of visual acuity under water using an automatic Landolt ring presentation technique.' Rep. X8–128/020. Ocean Systems Operations, 350 South Magnolia, Long Beach, California 90802.

Christianson, R. A., G. Weltman, and G. H. Egstrom (1965), 'Thrust forces in underwater swimming.' *Human Factors*, **7**, 561–8.

Diefenbach, W. S. (1961), 'The ability of submerged subjects to sense the gravitational vertical: A pilot study.' CAL Rep. OM–1355–V–1. Cornell Aeronaut. Lab., Buffalo, N.Y. 14221.

Duntley, S. Q. (1960), 'Improved nomographs for calculating visibility by swimmers (natural light).' U.S. Navy, Bureau of Ships, Contract No. bs-72039, Rep. 5–3.

Epstein, W., J. Park, and A. Casey (1961), 'The current status of the size–distance invariance hypotheses.' *Psychol. Bull.*, **58**, 491–514.

Faust, K. J. and E. L. Beckman (1966), 'Evaluation of a swimmer's contact air-water lens system.' *Military Med.*, **131**, 779–88.

Fischel, H. (1926), 'Transformationserscheinungen bei Gewichtshebungen.' *Zeit. f. Psychol.*, **98**, 342–65.

Foley, J. E. and S. M. Abel (1967), 'A study of alternation of normal and distorted vision.' *Canad. J. Psychol.*, **21**, 220–30.

Fry, G. A., C. S. Bridgman, and V. J. Ellerbrock (1949), 'The effect of atmospheric scattering on binocular depth perception.' *Amer. J. Optom.*, **26**, 9–15.

Gerathewohl, S. J., H. Strughold, and H. D. Stallings (1957), 'Sensori-motor performance during weightlessness: eye–hand co-ordination.' *J. Aviat. Med.*, **27**, 7–12.

Graham, C. H. (Ed.) (1965), *Vision and Visual Perception*. New York: Wiley.

Gregory, R. L. and H. E. Ross (1967), 'Arm weight, adaptation, and weight discrimination.' *Percept. Mot. Skills*, **24**, 1127–30.

Helmholtz, H. (1856), *Handbuch der physiologischen Optik*. Vol. **3**, Sec. 30. Translated by J. P. C. Southall. New York: Dover, 1962.

Hemmings, C. C. and J. N. Lythgoe (1964), 'Better visibility for divers in dark waters.' *Triton*, **9**, No. 4, 28–31.

Howard, I. P. and W. B. Templeton (1966), *Human Spatial Orientation*. London: Wiley.

James, W. (1882), 'The sense of dizziness in deaf mutes.' *Amer. J. Otol.*, **4**, 239–54.

Jonkers, G. H. and P. H. Kylstra (1963), 'Brightness contrast and colour contrast in stereoscopic vision acuity.' *Ophthalmologica*, **145**, 139–43.

Kama, W. N. (1961), 'The effect of simulated weightlessness upon positioning responses.' Wright-Patterson Air Force Base, Ohio. AMRL–TD R–61–555.

Katz, D. (1921), *Zur Psychologie des Amputierten und seiner Prothese*. Leipzig.

Kaufman, L. and C. Pitblado (1965), 'Further observations on the nature of effective binocular disparities.' *Amer. J. Psychol.*, **78**, 379–91.

Kent, P. R. and S. Weissman (1966), 'Visual resolution underwater.' U.S. Naval Submarine Medical Center, Groton, Connecticut 06340. Rep. No. 476.

Knight, L. A. (1958), 'An approach to the physiologic simulation of the null-gravity state.' *J. Aviat. Med.*, **29**, 283–6.

Loftus, J. P. and L. R. Hammer (1961), 'Weightlessness and performance: a review of the literature.' Wright-Patterson Air Force Base, Ohio. ASD Tech. Rep. 61–166.

Lund, F. H. (1930), 'Physical asymmetries and disorientation.' *Amer. J. Psychol.*, **42**, 51–62.

Lundgren, C. E. G. (1965), 'Alternobaric vertigo—a diving hazard.' *Brit. Med. J.*, **2**, 511–13.

Lundgren, C. E. G. and L. U. Malm (1966), 'Alternobaric vertigo among pilots.' *Aerospace Med.*, **37**, 178–80.

Luria, S. M. (1968), 'Stereoscopic acuity under water.' U.S. Naval Submarine Medical Center, Groton, Conn. 06340. Rep. No. 510.

Luria, S. M., J. A. S. Kinney, and S. Weissman (1965), 'Estimation of size and distance underwater.' U.S. Naval Submarine Medical Center, Groton, Connecticut 06340. Rep. No. 462.

Margaria, R. (1958), 'Wide range investigations of acceleration in man and animals.' *J. Aviat. Med.*, **29**, 855–71.

Morway, D. A., R. G. Lathrop, R. M. Chambers, and L. Hitchcock (1963), 'The effects of prolonged water immersion on the ability of human subjects to make position and force estimates.' U.S. Naval Air Development Center, Johnsville, Warminster, Philadelphia 18974. Rep. No. NADC–MA–6115.

Nelson, J. G. (1967), 'The effect of water immersion and body position upon perception of the gravitational vertical.' U.S. Naval Air Development Center, Johnsville, Warminster, Philadelphia 18974. Rep. No. AD–658–507 (NADC–MR–6709).

Nichols, A. K. (1967), *A study of some aspects of perception in the underwater situation*. Unpublished dissertation for the Institute of Education, University of Leeds.

Padden, D. A. (1959), 'Ability of deaf swimmers to orient themselves when submerged in water.' *Res. Quart. Amer. Ass. Hlth. Phys. Educ. Recr.*, **30,** 214–26.

Palmer, D. A. (1960), 'Variations in binocular acuity and their influence on contour measurements.' *Photogrammetric Record,* **3,** 357–69.

Rees, D. W. and N. K. Copeland (1960), 'Discrimination of differences in mass of weightless objects.' Wright-Patterson Air Force Base, Ohio.

Rock, L. (1966), *The Nature of Perceptual Adaptation*. New York: Basic Books.

Ross, H. E. (1965), 'The size-constancy of underwater swimmers.' *Q. J. Exp. Psychol.*, **17,** 329–37.

Ross, H. E. (1966), 'Size and distance judgements under water and on land.' (In *Malta '65*), *Underwater Assn. Rep.*, **1,** 19–22.

Ross, H. E. (1967a), 'Stereoscopic acuity under water.' *Underwater Assn. Rep.*, **2,** 61–4.

Ross, H. E. (1967b), 'Water, fog and the size–distance invariance hypothesis.' *Brit. J. Psychol.*, **58,** 301–13.

Ross, H. E. (1968), 'Judging distance under water.' *Triton*, **13,** 64–6.

Ross, H. E. (1969), 'Adaptation of divers to curvature distortion under water.' *Ergonomics*.

Ross, H. E., S. D. Crickmar, N. V. Sills, and P. Owen (in press), 'Orientation to the vertical in free divers.' *Aerospace Med.*

Ross, H. E., D. J. Dickinson, and B. P. Jupp (in press), 'Geographical orientation under water.' *Human Factors*.

Ross, H. E., S. R. King, and H. Snowden (in press), 'Vertical distance judgements under water.' *Human Factors*.

Ross, H. E. and P. Lennie (1968), 'Problems in the perception of movement under water.' *Underwater Assn. Rep.*, **3,** 55–8.

Schaeffer, A. (1928), 'Spiral movement in man.' *J. Morph.*, **45,** 293–398.

Schock, G. J. D. (1960), 'Perception of the horizontal and vertical in simulated subgravity conditions.' *U.S. Armed Forces Med. J.*, **2,** 786–93.

Schöne, H. (1964), 'On the role of gravity in human spatial orientation.' *Aerospace Med.*, **35,** 764–72.

Simons, J. C. and M. S. Gardner (1963), 'Weightless man: a survey of sensations and performance while free-floating.' Wright-Patterson Air Force Base, Ohio. Tech. Rep. AMRL–TDR–62–114.

Stigler, R. (1912), 'Versuche über die beteiligung der schwereempfindung an der orienterung des menschen im raume.' *Arch. Physiol.* (Bonn), **148,** 573–84.

Terry, L. and W. L. Dennison (1966), 'Vertigo among divers.' U.S. Naval Submarine Medical Center, Groton, Connecticut 06340. Rep. No. AD–635 518 (NSMC SR–66–2).

Turnbull, C. M. (1961), 'Some observations regarding the experiences and behaviour of the Bambuti Pygmies.' *Amer. J. Psychol.*, **74,** 304–8.

Van de Geer, J. P. and E. J. Zwaan (1964), 'Size-constancy as dependent upon angle of regard and spatial direction of the stimulus-object.' *Amer. J. Psychol.*, **77,** 563–75.

Walsh, E. G. (1961), 'Role of the vestibular apparatus in the perception of motion on a parallel swing.' *J. Physiol.*, **155,** 506–13.

Weltman, G. and G. H. Egstrom (1966), 'Perceptual narrowing in novice divers.' *Human Factors*, **8,** 499–506.

Woodley, J. D. (1968), 'Judgment of size and distance at Cow Bay.' (Unpublished paper.) Department of Zoology, University of West Indies, Kingston 7, Jamaica.

4 Vision

J. N. Lythgoe

Introduction

If two people were to dive into a huge tank of pure water they might be able to see each other at a range of about 100 m.; but such clarity is not to be expected in natural waters and in the exceptionally limpid waters of the Caribbean or Mediterranean they would be lucky to see each other at a horizontal range of 40 m. Usually inshore waters are much less clean than this and off the West Coast of England or California a visual range of 25 m. would be good. Where rivers bring down suspended matter into the sea or where a sandy or muddy shore is continually beaten by waves a visible range of 1 m. is usual, whilst in the muddy rivers themselves the visible range can be less than 10 cm. As the diver sinks deeper into the sea, less and less light penetrates down to him until finally he does not receive enough to see by. This depth varies enormously and could be 1 m. in really dirty water or 1000 m. when it is very clear.

The conditions for vision under water can aptly be compared to those in a fog on land. It is not low light intensity which prevents distant objects being seen, but rather that the light scattered by the water droplets in the fog interposes a veil of light which degrades contrast below the level required for vision. There is the same 'veiling brightness' under water and it is the problems of contrast perception by the eye and contrast degradation by the water which are central to the study of underwater vision.

In a really heavy industrial fog distant objects look yellow, and this is because the shorter wavelength light has been absorbed by the dissolved industrial waste in the fog droplets. Differential absorption of the spectrum is much more marked under water and in very clear natural water the long wavelength red light is so strongly absorbed relative to the blue that at 30 m. the whole visual environment takes on a blue cast. In coastal waters where the chlorophyll contained in phytoplankton and the yellow products of vegetable decay absorb light of short wavelength, it is green not blue light which penetrates deepest into the water. Nevertheless, an optical filter can only absorb light and the presence of yellow coloration in the water reduces light penetration at all wavelengths, but the reduction is greater for blue than for red.

Once the light has reached the eye it has to be absorbed by the light-sensitive visual pigment, translated into nervous impulses, and ultimately interpreted by the brain. No problem in underwater vision can be solved without regard both to the physiology of the eye and to the optical properties of the water; the two phases of the problem can be separated but they cannot be divorced.

Light Under Water*

In the upper layers of the sea most of the natural light comes from the sun, but in deeper water, or at night, light must be provided by the animals themselves—either by bioluminescence in the case of the fishes, cephalopods, and plankton, or a torch in the case of the diver.

In Fig. 44 are shown some of the visually important aspects of light transmission through the sea. Some of the light which penetrates from the surface to the object (the fish in the diagram) is reflected in the direction of the eye. Some of this image-forming light is absorbed by the water and some is scattered out of its path by suspended particles in the water. The ambient light is also scattered into the eye and it is this scattered light which is responsible for the brightness of the water background and for the 'veiling brightness' between the object and the eye. If the range between the object and the diver is increased, the image-forming light reaching the diver gets less, but the non-image-forming light increases until finally the light from the object can no longer be distinguished from the background and the object becomes invisible.

This situation has been described mathematically by Duntley (1962) and in the simplest but important case where the path of sight is horizontal, the light reaching the eye from a target suspended in deep water is represented in equation 1. The extent that light is scattered and absorbed by the water (and by the eye) depends strongly on its wavelength. For this reason the light radiating from a source is described by its spectral radiance.†

* N. G. Jerlov (1968) and Tyler and Smith (1970) have given us accounts of the optics of natural water which cover all the aspects of optical oceanography only touched upon here.

† Radiance is defined as the radiant flux per unit solid angle per unit projected area of a surface. The spectral radiance is the distribution of radiant energy as a function of wavelength (or frequency).

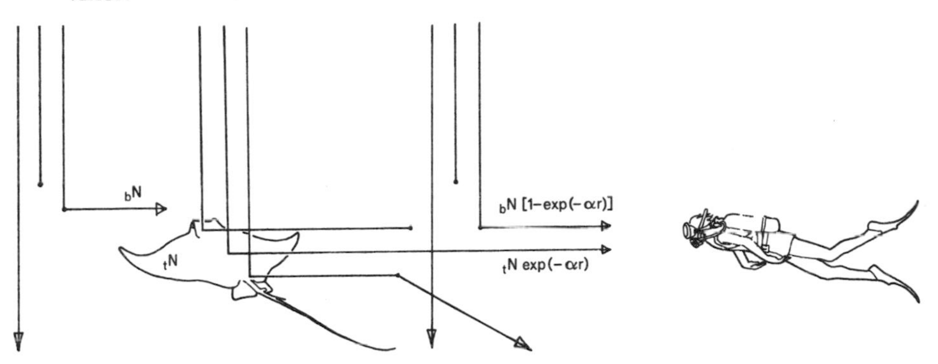

Fig. 44
A general diagram showing some of the visually important optical events which occur as light penetrates through a body of water.

Daylight may either be absorbed or scattered in its passage from the surface; the light scattered into the eye is the chief contributor to the water background spacelight, $_bN$. When the light meets an object suspended in the water (the sting ray in the diagram), some of it is reflected into the direction of the eye. This image-forming light, $_tN$, is reduced in its passage through the water by being both absorbed and scattered out of the light beam. At the same time the apparent brigh ness of the object is increased by the addition of diffuse veiling light scattered into he eye by the water between the object and the eye.

When the distance, r, between the eye and the object is large the residual image-forming light becomes small; but the contribution of non-image-forming light increases to equal that of the background spacelight. When the distance, r, is sufficiently large the eye can no longer distinguish the object from the water background.

$$_tN_r = {_tN} \exp(-\alpr) + {_bN}[1 - \exp(-\alpha r)] \qquad (1)$$

where $_tN$ is the inherent spectral radiance of the target,

$_tN_r$ is the apparent spectral radiance of the target at distance r,

$_bN$ is the spectral radiance of the background spacelight,

r is the path length or range,

α is the beam attenuation coefficient.

In equation 1 the first term on the right represents the residual image-forming light from the target and the second term represents the radiance contributed by the scattering of ambient light in the sea throughout the path of sight (the 'veiling brightness').

Almost every problem in underwater vision requires data about the apparent target radiance throughout the visible spectrum as it mig t be measured at the level of the faceplate, and $_tN_r$ cannot . calc lated unless the spectral radiance of the target, the water background and the beam attenuation coefficient are all known. In addition it is almost always necessary to know the diffuse attenuation coefficient of daylight penetrating through the water. To find all these data for the same body of water is uncommon to say the least. Tyler (1965) has made some of the relevant measurements for very clear natural water, and an estimate of these quantities can be obtained from the measurements of contrast reduction made by a diver-held device in medium clarity Mediterranean water (Lythgoe 1968).

The Colour of Water and the Attenuation of Light

When a diver swims downward into the sea, the daylight reaching him gets dimmer and dimmer and if the diver were to measure the light reaching a suitably designed light meter* he would find that the light at any depth at one particular wavelength could be described by equation 2 (Atkins *et al.* 1938).

$$I_Z = I_0 \exp. (-KZ) \tag{2}$$

where I_0 is the spectral irradiance of light at the surface,†
I_Z is the spectral irradiance of light at depth Z,
K is the diffuse attenuation coefficient.

If, on the other hand, the diver was to go down at night and shine the parallel beam of a torch on to a light meter he would find that the reduction of intensity with distance between the torch and the meter would be described by an equation similar to equation 2 except that K would be replaced by α, the beam attenuation coefficient. The attenuation of a beam of light will always be greater than that of diffuse light ($\alpha > K$) for the same body of water. This is because light is lost from the beam both by absorption of light and by light being scattered out of the beam and thus lost to the light meter. The downwelling daylight is also scattered, but its diffuse component is largely included in the measurements.

* The collecting surface must always be similarly orientated and must obey characteristics of a cosine collector:

$$J_\theta = J_0 \cos \theta$$

where J_θ = relative sensitivity at any angle θ,
J_0 = sensitivity normal to the surface.

† Irradiance is defined as the radiant flux incident on an infinitesimal element of surface containing the point under consideration, divided by the area of that element.

K, the diffuse attenuation coefficient, is the most commonly used function in photobiology and especially in studies of plant productivity, etc. (see Chapter 6) where the directionality of the daylight is less important than its energy. It has also been used extensively in zoological studies to compute the limiting depth for vision. However, *K* alone is of very little value in computing how far through the water an animal (or a television camera) can form an image.

Water does not absorb light of all wavelengths equally, and it is those wavelengths which are least absorbed which give the water its characteristic colour. As the depth increases the ambient light is restricted to a narrow band of wavelengths and the water behaves as a very efficient monochromator (Tyler 1959). In the very pure water of Crater Lake, Oregon, Smith and Tyler (1967) found that the wavelength of daylight which penetrates farthest into the water was centred at about 420 nm. in the blue (Fig. 45). Such water is quite exceptional and most measurements of the ambient light in clear ocean water shows it to be brightest in the blue-green around 470 nm. (Jerlov 1951, Kampa 1961). The rather greener colour of ocean water is partly due to the presence of chlorophyll-containing phytoplankton, and partly to the yellow products of vegetable decay (Kalle 1961). Chlorophyll uses red and blue light in the photosynthetic process, but allows the green to pass. The 'yellow substances' absorb blue light but allow the green and other wavelengths to pass. However, the water itself strongly absorbs red light and only green escapes absorption. Coastal water which contains much yellow substance washed down by the rivers and often much phytoplankton as well is usually green or yellow-green. Jerlov (1951) has traced the change in water colour from the enclosed yellow-green waters of the Baltic to the blue of the open sea (see Fig. 66 in Chapter 6).

The Radiance of the Water Background and the Distribution and Polarization of Light

If the water did not scatter light but only absorbed it, the water background would appear black and all the light would come from above. If the water did not absorb light, but only scattered it, the diver would find that as he went deeper the light reaching him would come equally from all directions. However, as the water both absorbs and scatters, light will reach him from the surrounding water in all directions, but upwelling light will be about 2 per cent of the downwelling light. The polar distribution of light varies according to the water, the position

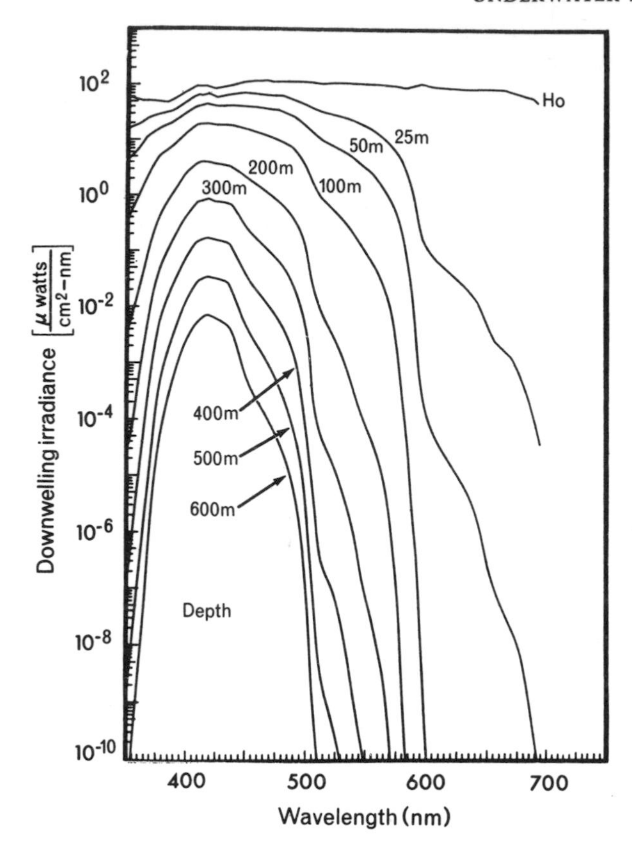

Fig. 45

The downwelling irradiance at various depths in the exceedingly clear waters of Crater Lake, Oregon. These irradiances have been calculated from K values measured in the lake, and assume the lake is homogeneous.
(From Tyler and Smith 1967.)

of the sun, and the cloudiness of the sky, but a complete set of measurements for total overcast and clear sky has been made (Tyler and Preisendorfer 1962) for one particular lake. Such measurements of a great variety of azimuth and tilt angles, sky states and sun elevations for a useful range of wavelengths would be dauntingly long, but Tyler (1965), Tyler and Smith (1967, 1970), and Smith and Tyler (1967) have measured the ratio between downwelling light arriving horizontally and from below throughout the spectrum in a variety of natural waters.

Light is scattered by particles in the sea that vary in size from the water molecules themselves to plankton which can be seen by the naked eye. The larger particles scatter all wavelengths more or less equally, but the smaller particles less than 13 μm. in diameter (Jerlov 1951) scatter the short wavelength blue light most strongly. The larger particles appear to be chiefly responsible for the forward scattering of light where the angle of scatter is close to the primary ray's direction of travel (Atkins and Poole 1952, Jerlov 1953, Gazey 1970). This forward scatter has the effect of degrading the sharp edge of the object as seen by the eye. This indistinct edge between the object and the background reduces the eye's ability to distinguish small brightness differences (p. 121) and also the eye's ability to detect objects subtending very small visual angles.

The underwater spacelight is strongly plane-polarized (Waterman 1954); and this is thought to be the result of the scattering of light rather than refraction at the water's surface. The discovery and characterization of the polarized nature of the underwater spacelight is one of the earliest and certainly one of the most successful pieces of optical research using divers that has yet been carried out. The analyser (Waterman 1960) was specially developed for use by divers. It is mounted on an astro-compass which in turn is clamped to a heavy tripod resting on the bottom. The type and orientation of the underwater polarization is deduced from sighting through an interference analyser, and the presence of polarization arises by sighting through a split-field analyser with axes at 90° to each other. Most of the underwater polarization arises from the scattering of the sun's rays by the water, and thus the plane of maximum polarization, which is at 90° to the sun's rays, tilts markedly through the day (Fig. 46). Near the surface in clear deep water there might be about 60 per cent polarization,* but this falls to an asymptotic value of about 30 per cent in deep water (Ivanoff and Waterman 1958, Tyler 1963). The degree of polarization decreases when the sky becomes overcast or the water is turbid. It also decreases near the bottom (especially a light-coloured bottom) owing to the light reflected upwards from it (see Ivanoff and Waterman 1958, for a review).

* The polarization factor, P, is

$$P = \frac{I_p}{I_n - I_p}$$

where I_n and I_p are respectively the unpolarized and linearly polarized fractions of the incident light.

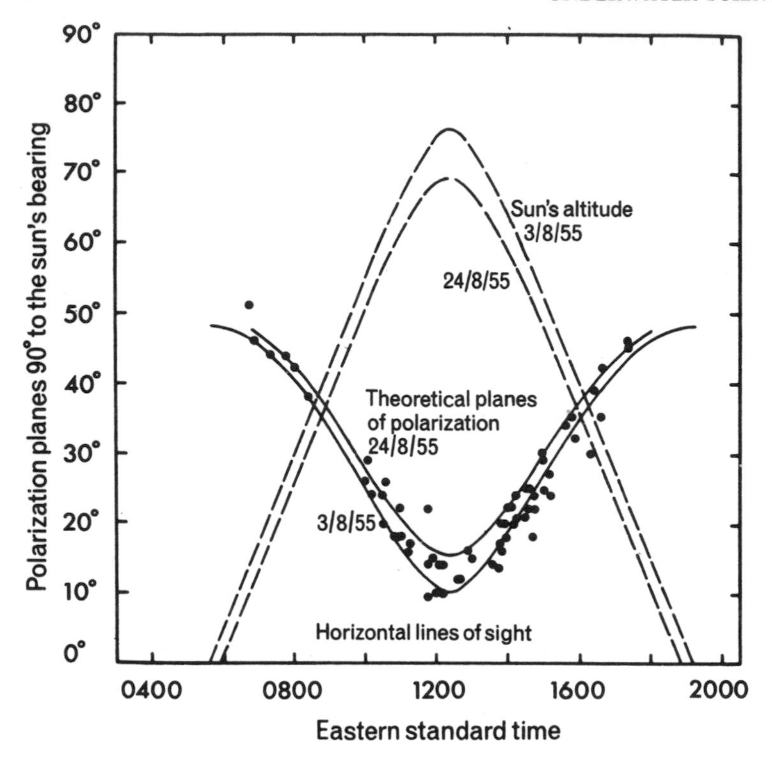

Fig. 46

Relationship between the planes of natural underwater polarization (plotted as tilt of the e-vector from the horizontal) and the sun's altitude. These polarimeter observations were made at depths between 4 and 31 m. between 3 and 25 August 1955, looking horizontally at 90° to the solar bearing and plotted (black circles) as a function of the time of day. The close correlation of the observations with the theoretical curves indicates that the orientation of the plane of polarization is basically determined by the angle of sunlight refraction in the water (Waterman 1961).

The Reduction of Contrast

A diver does not usually have the time or the inclination to decide why some objects under water are visible and some are not; yet the question has to be answered if the limitations of underwater vision are to be explained and improvements suggested. Briefly, an object is seen if it appears of a slightly different brightness or a slightly different colour to its background. The distinction has to be made by the eye and the brain, and whether an object can be seen or not depends ultimately on them. But the raw data which reaches the eye can be described (and measured) by purely physical means. Speaking very generally the human eye can detect brightness differences of between 1 per cent and 2 per cent if the light is fairly bright. Brightness contrast is defined as follows:

$$C_r = \frac{{}_tN_r - {}_bN}{{}_bN} \tag{3}$$

where C_r is called the 'apparent contrast' (Duntley 1960). However, light from the object and background must first be absorbed by the light-sensitive visual pigments before it has any relevance to vision, and it is the proportional differences in the number of photons actually absorbed from the object and background which constitute the visual contrast (${}_{\text{vis}}C_r$). Nevertheless for any particular wavelength the apparent contrast and visual contrast are equal and equation 3 can be used.

The apparent contrast between an object and its background decreases as the visual range gets longer and by substituting the defining function for ${}_tN_r$ (equation 1) for ${}_tN_r$ in equation 3, it can be shown (Duntley 1962) that:

$$C_r = C_0 \,.\, \exp.(-\alpha r) \tag{4}$$

It will be remembered that this equation is derived for horizontal paths of sight, and in this case the rate at which contrast is degraded depends only upon α and r. r is a length and can be measured, and it becomes possible to measure α if the apparent contrast of an object against an infinite water background is known. There is one particular circumstance where this can easily be done and that is when a black target is used (le Grande 1939, Duntley 1960). It is evident from equation 3 that if the target is black (${}_tN = 0$) then C will always equal -1. Further, a black object will become invisible, like any other object, when C_r falls below about 0·02 (it is immaterial whether C_r is positive or negative). Thus if the range at which a black object just becomes invisible is measured (two divers can stretch a tape between them) the beam attenuation coefficient, α, of the water can be estimated (Hemmings 1966).

When the visibility is poor, for example at night or in turbid water, a trombone-like device (Fig. 47) consisting of a measured rod with a target attached to the visible end and which slides through a shoulder-supported holder is useful for measuring visible ranges (Hemmings and

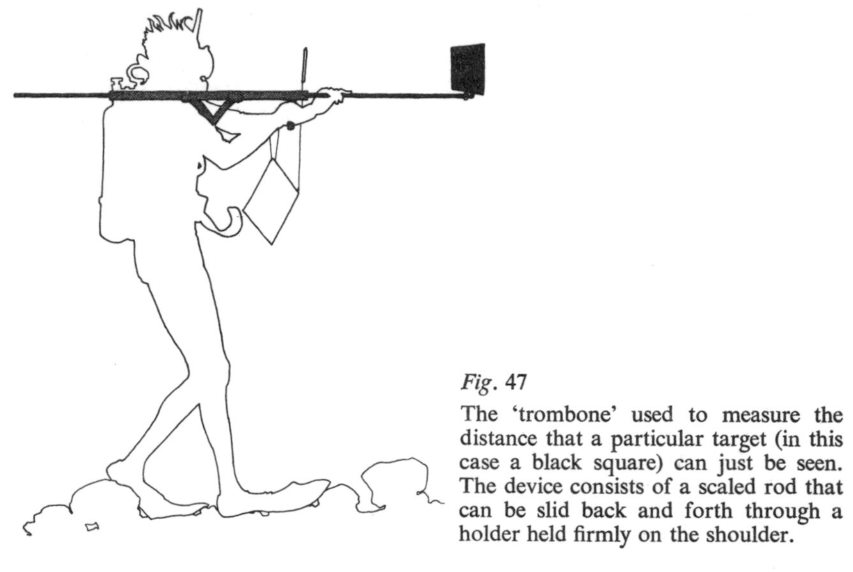

Fig. 47
The 'trombone' used to measure the distance that a particular target (in this case a black square) can just be seen. The device consists of a scaled rod that can be slid back and forth through a holder held firmly on the shoulder.

Lythgoe 1965). This 'black body distance' is an extremely valuable quantity because it can be interpreted in physical terms. It is far superior to the Secchi disc in all save convenience, for the Secchi disc measurements cannot be interpreted in physical terms (Tyler 1968).

A diver-held device that can be used to estimate α as a function of wavelength has been described by Lythgoe (1968). It is shown in Fig. 48 and can either be mounted on the 'trombone' or can be held to the eye like a telescope (Fig. 49). In essence a field of light derived from a target at a known distance is compared to one from the background spacelight in a split-field fashion. The split field is separated from the eye by a perspex rod of sufficient length to allow the eye to focus on the split field, care being taken that the target completely fills the relevant half of the field. A neutral density filter is fixed (by magnets) so as to reduce the intensity of either the target or the background field, whichever is the brighter. The waveband of observation is then restricted by interposing one of a series of interference filters between the split field and the eye. The range between the target and the observer is then adjusted until the brightness of the background and target fields are equal. A knowledge of the absorption of light by the mirrored surfaces and the neutral filter allows the spectral contrast between target and background to be calculated. If a series of neutral density filters are used the reduction of C_r with r, and from thence α can be

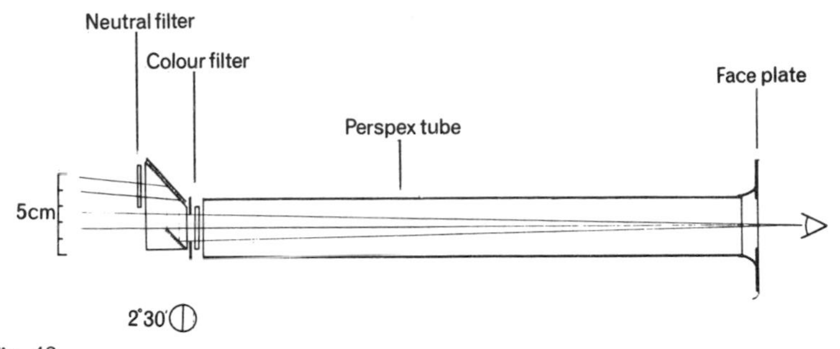

Fig. 48
Plan of a diver-held device that is used for comparing the spectral radiance of a flat target and the water background spacelight. The device can either be held up to the eye like a telescope (Fig. 49) or it can be mounted on the 'trombone' illustrated in Fig. 47. (Lythgoe 1968.)

Fig. 49
A method for measuring the relationship between visual range and the contrast between a target and its water background. The diver is looking through the device shown in Fig. 48 and aligning it so that the black target completely fills the right-hand field of the instrument; the brighter water background is seen through the left-hand field whose intensity is reduced by the neutral-density filter. The diver then moves back and forth along the measuring tape he is holding in his left hand until the two halves of the split-field appear to be equal in brightness.

The tape and the targets are held up by inverted air-filled buckets. The complicated mooring lines for the targets are designed to minimize movement in the swell. Measurements using this arrangement can only be made when the sea is calm and the water is clear with a silt-free bottom.

calculated for selected wavelengths (Fig. 50). These measurements are too protracted to measure the variation of α through the spectrum, but if a black target is used C_0 approximated to -1 at all wavelengths and, theoretically at least, r need only be measured at one value of C_r.

The Eye and the Faceplate

We have now traced the light which is ultimately to be translated into visual messages, as far as the eye. This light has to be focused by the cornea and lens on to the light-sensitive retina if any image is to be

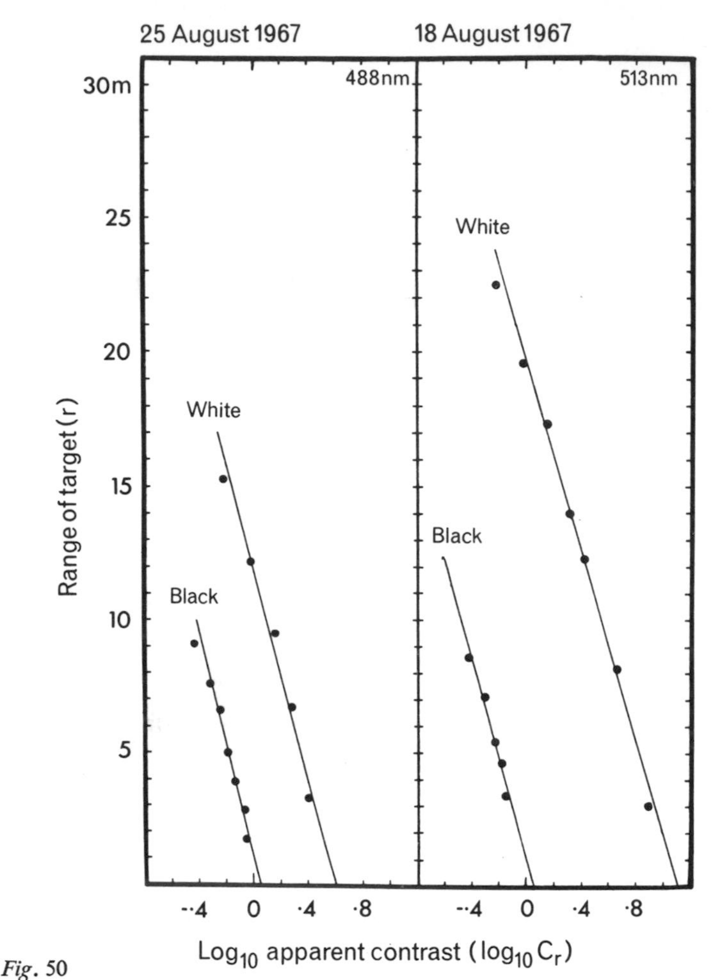

Fig. 50

The results of two experiments in Malta to show how the contrast, C_r, between a black or a white target and its background is reduced as the horizontal range, r, increases. The measurements were made at a depth of 7·5 m. using the instrument shown in Fig. 48 and the sighting range illustrated in Fig. 49. On 18 August the 513 nm. interference filter was used and the measurements were made between 14.10 14.30 local time (target slightly down-sun); on 25 August a 488 nm. interference filter was used and the measurements were made between 11.55 and 12.20 local time (overhead sun).

According to equation 4 a plot of $\log_{10} C_r$ v. r should yield a family of parallel straight lines irrespective of the target radiance, provided that the optical properties of the water remain constant (the lines drawn through the experimental points are calculated regression lines). For black targets when $C_0 = -1$ ($\log -1 = 0$), $\log C_r$ should equal 0 when $r = 0$. For other targets the value of C_0 varies with the angular distribution of the incident light.

formed. The human eye is designed to work in air with most of its focusing power lying at the curved cornea, and the lens is used only for fine focus. When the eye is immersed in water most of the focusing power of the cornea is lost because the refractive indexes of the water and the aqueous humour are nearly equal and the eye is unable to focus on any object whatever. The face-mask allows the eye to be surrounded by air in the normal way and the eye can focus normally. But the use of the flat face-plate raises some interesting problems which are described in Chapters 2 and 3.

The Eye and Vision*

The light which has reached the eye has then to pass through the cornea, the aqueous humour, the lens, the vitreous humour, and the retina, there to be absorbed by the visual pigment contained in the rods and cones. Optical filters are often present in different eyes, either in the lens, the cornea, or as coloured oil droplets in the cones (Walls 1963). In man the lens becomes progressively yellower from birth (see Wyszecki and Stiles 1967) and its absorption has to be taken into account. Amongst shallow water fishes the lens may absorb blue light relatively strongly (Denton 1956) and in other fishes the cornea may act as strong yellow filter (Walls and Judd 1933, Moreland and Lythgoe 1968, Muntz, in press).

Once the light has reached the rods and cones it must be absorbed by the visual pigment they contain. The visual pigments belong to a very homogeneous group of chromoproteins which all have very similarly shaped single-peak absorption curves, but the position of the absorption peak can vary through the spectrum (Dartnall 1957). In man and in most other animals there are two quasi-independent visual systems, one adapted for seeing in dim light at night and the other for vision in the daytime. The night vision system is mediated by the rods which contain a single visual pigment with its maximum absorption at 502 nm. in the blue-green. The day vision system is mediated by the cones which are of three types and separately contain visual pigments of maximum absorption at about 440, 540, 567 nm. (see Marks 1965b, for a review). The night vision system, containing only one visual pigment, is incapable of detecting differences in colour, but only differences in brightness. The day-vision system, having three types of visual pigment, is capable of detecting both colour and brightness

* Wyszecki and Stiles (1967) have published a valuable compendium of psychophysical data for the human eye.

differences. The spectral sensitivity of the eye when dark- and light-adapted is somewhat different—the eye being relatively more sensitive to blue-green light at night and yellow-green light in the daytime (Fig. 51).

When the light gets dimmer the sensitivity of the eye becomes correspondingly greater. However, this increase in sensitivity is accompanied by a decline in visual acuity, and at low illuminations by contrast sensitivity (see Pirenne 1962, for a review). Colour vision is also lost when the eye switches from the cones to the rods. (An unobscured three-quarters moon gives just enough light for colour vision.) The influence of low light intensities on the ability to detect the contrast presented by objects of different size is shown in Fig. 52. It will be noticed that for targets whose size exceeds 55 minutes of arc and when the background luminance exceeds about 1 mL., the contrast perception threshold remains almost constant, and it is partly for these reasons that the 'Black Body Distance' mentioned on page 112 gives such reliable results if the underwater light is good and the target exceeds 900 cm.^{2} in area. It should be noted, however, that even partial misting of the face-plate will considerably reduce the eye's ability to detect contrast differences.

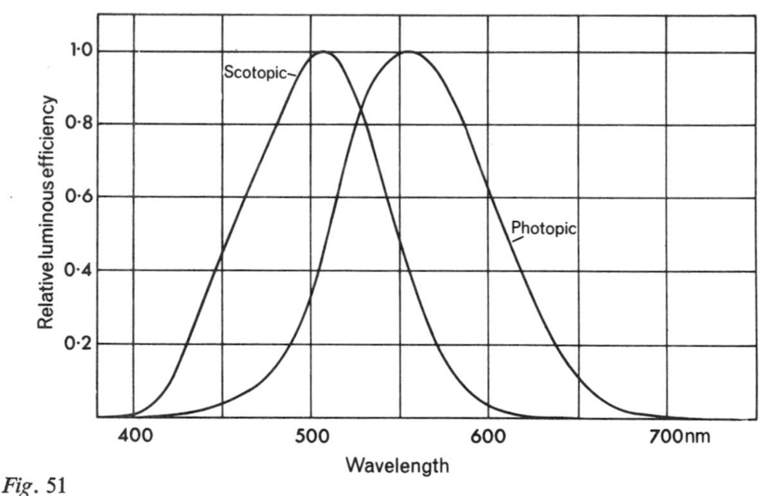

Fig. 51
The spectral sensitivity of the eye to different wavelengths at low light intensities (scotopic, or night vision) and at higher light intensities (photopic or day vision). Note that at 650 nm. the photopic system alone is stimulated.
(Wyszecki and Stiles 1967.)

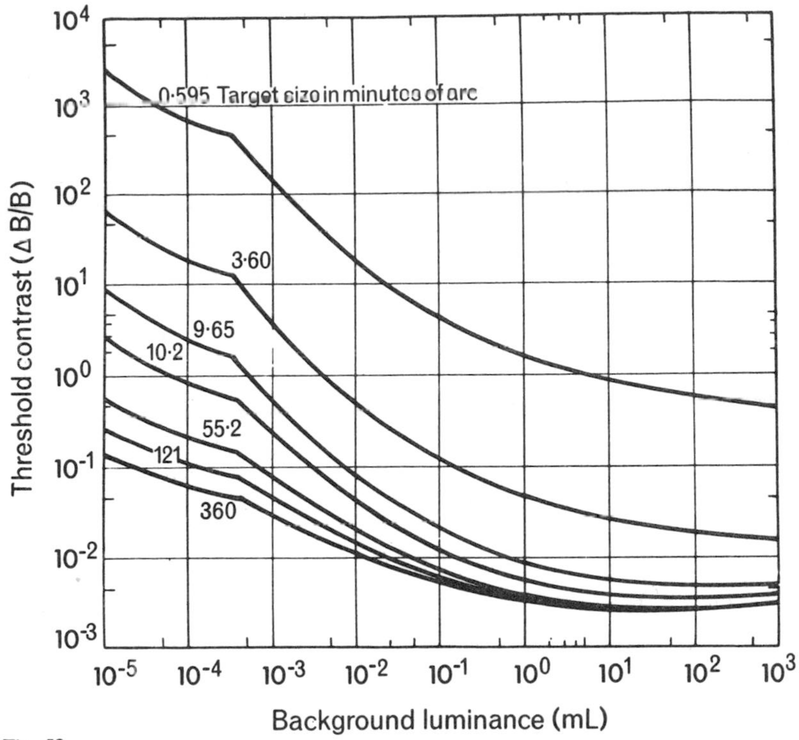

Fig. 52

The relationship between the threshold for contrast perception, the brightness of the background and the angle subtended by the target at the eye. In general the eye's ability to detect contrasts falls as the target gets smaller and the light gets dimmer. However, when the target subtends more than 55 minutes of arc and the general illumination is good, the contrast perception of the eye remains almost constant.

The kink in all the curves between 10^{-4} and 10^{-2} mL. marks the transition between the photopic and scotopic visual system in the eye. (From White 1964, after Blackwell 1946.)

Dark Adaptation in Dark Water

It is a foolish diver who believes the evidence of his own eyes. Dr. Ross has brought this out well in Chapter 3, and the eye is also well able to provide misleading information as a result of the processes of dark adaptation. Thus it is quite common for a diver to complain that the water changed from a blue-green colour near the surface to grey near the bottom, or he may wish that he had dived 30 minutes later because the light, barely adequate for vision at the beginning of the dive, had begun to improve by the time he had been forced to come up.

At the surface on a bright sunny day there may be an illumination of about 10^4 mL., but in fairly dirty water this may have fallen to 10^{-5} mL. at 30 m. The diver is well able to reach a depth of 30 m. in 60 seconds, but the eye requires at least 25 minutes (Fig. 53) to make the 1,000,000,000-fold increase in sensitivity required to make out objects on the bottom in very dim light in the last few minutes of the dive. The unwary diver usually interprets this late improvement in 'visibility' as an improvement in water clarity.

In the course of the eye's adaptation to the increasingly dim light as he swims downwards there is a switch from the day-vision system where there is colour vision to the night-vision system where there is not. In the dive discussed above this switch would come at about 20 m. The water above that would appear coloured (green probably in this kind of water), but below it would be seen in shades of grey.

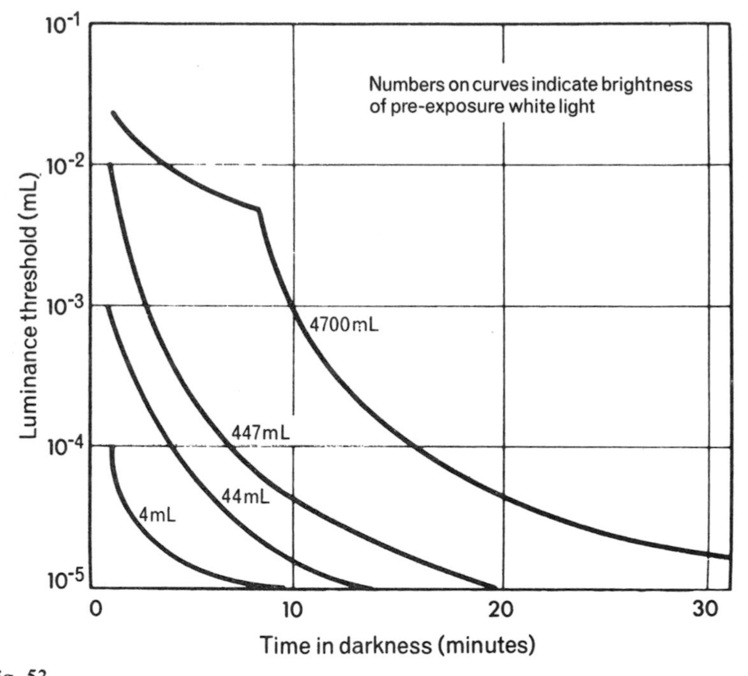

Fig. 53

The rate of dark adaptation after exposure to light of various brightnesses. The diver will normally have been exposed to at least 10^3 mL. before diving and it will take him more than 20 minutes to become fully adapted to the dim light at depth. (From White 1964, after Haig 1942.)

The difficulty that the diver's dark adaptation lags behind his rate of descent can in fact be overcome by pre-adaptation using a red-pass filter before diving (Hemmings and Lythgoe 1964). It will be seen from Fig. 51 that the night vision system is relatively insensitive to red light compared to the day-vision system. If a red face-plate is worn for about 30 minutes before diving the night-vision system reacts as though it were dark and adapts accordingly. On the bottom, the red face-plate is exchanged for a clear one; the eye thus reaches its full sensitivity at the beginning of the dive rather than at the end. The effects can be dramatic with a 30 per cent increase in working depth (when no torch is used) and bioluminescent plankton can be seen where its presence was never suspected. Light adaptation is very much faster and is completed in a minute or two. Thus a very brief exposure to sunlight or an unshielded torch on the bottom can instantly wreck a carefully nurtured dark-adapted state.

The rate of dark-adaptation depends to a large extent on the level of light to which the eye had previously been exposed (Haig 1941) and a day spent on a bright beach can measurably injure dark-adaptation (Hecht *et al.* 1948) for several hours afterwards. It is only sensible to wear dark sun-glasses during the day when a task such as diving (or, incidentally, night driving) is in prospect.

There is no doubt that it is an advantage to those fishes which perpetually live in very dim light to be as light-sensitive as possible. One way that this can be achieved is to increase the amount of visual pigment in the retinal rods. Many fishes do indeed have a greater optical density of visual pigment in their retinas than does man (Denton and Nicol 1964). And Denton and Warren (1957) have calculated that deep-sea fishes may be 10 to 100 times more sensitive to dim light than is man. Thus Clarke and Denton (1962) estimate that in the Brownson Deep, fifty miles north of Puerto Rico, man should just be able to detect daylight at 880 m., whilst a deep-sea fish should be able to detect it at 1,000 m. They point out, however, that these calculations ignore the presence of bioluminescence which must certainly be present and extends the downward range of vision.

Brightness Contrast and Visual Range

If a length of fishing net is dyed in sections in a series of colours ranging from black through grey to white it is certain that for any particular condition of viewing either the black or the white will be seen at the greatest distance for it is evident from equation 3 that the inherent brightness contrast presented by a target will be greatest if the target brightness is very high (white) or very low (black). The rate at which the inherent contrast is reduced to just detectable levels does not depend on the inherent contrast itself but only on the optical properties of the intervening water, and upon the contrast sensitivity of the eye; thus it follows that either black or white targets can be separated from the observer by a greater distance before their apparent contrast becomes subliminal than can grey targets of intermediate brightness.

The relationship between brightness and visual range has been investigated for horizontal paths of sight (Hemmings and Lythgoe 1965, Lythgoe and Hemmings 1967), the sighting range being set up on the bottom in the way shown in Fig. 54. It is important that the observer should always view the target from the same angle and in this case this was done by means of a plumb-line of known length, the leaded end bouncing on the bottom and the free end held up to the level of the face-plate. In this case a surveyor's tape was laid over the bottom, but the arrangement shown in Fig. 49 has strong advantages, especially when the bottom is irregular. The targets were fixed to a perspex board which, being transparent and of an almost identical refractive index to water, is practically invisible (Fig. 56). Care was taken that the targets were seen against an unobscured water background and that the target board was held as immovable as possible. The mooring of the target board did cause problems in shallow water, and an entirely rigid structure should be used for preference.

The targets themselves are rectangular and measure 15 cm. × 20 cm. They are fastened in groups of four to the perspex board and are rotated at random. The diver gradually approaches the target and judges it to be just visible when he can just make out its orientation. At this point the diver records the position of his plumb along the tape and proceeds to the next most visible target. The results of one such experiment are shown in Fig. 55. In both these cases it is the white targets which are the most visible. The theoretically expected results (equations 3 and 4) assuming that the reflectance of the targets (measured in air) is proportional to the target radiance and that the relative radiance of the water background is equal to that of the least visible target, is shown by the black line. It is perhaps not surprising that no tile could be camouflaged sufficiently to reduce its visible range to zero. This is due in large part to slight imperfections on the surface and the glint of bright light reflected from the top edge. In addition the grey targets appeared very slightly redder in colour to the water background

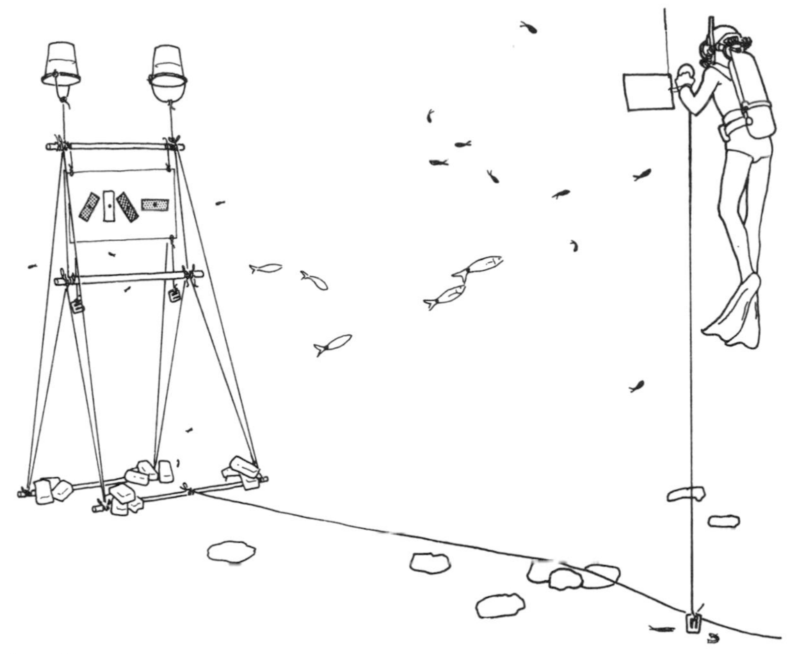

Fig. 54

A sighting range that has been used to measure the visible range of targets in clear water. The targets are fixed to perspex sheet by means of a central bolt in a random orientation. Perspex is used because, being transparent, colourless, and of the same refractive index as water, it is almost invisible under water (see Fig. 56).

The diver swims away from the targets until none are visible; he then swims back along the measuring tape stretched along the bottom until he can just make out the orientation of each target. He keeps himself at a constant height above the bottom by dangling a lead weight on the end of a measured length of cord and allowing the weight to just touch the tape. If the diver is just buoyant when the weight rests upon the bottom and just heavy when it is unsupported, the task is quite simple. The range, *r*, is the position along the tape adjacent to the lead weight.

(see pp. 130–3) and thus the target was detected through colour rather than brightness contrast.

More puzzling is the shorter than expected range at which the most visible light grey and white targets can be seen. The targets are rather too big for the effect to be due to the fall off in contrast perception at small visual angles (Blackwell 1946). A possible explanation may lie in the blurring of the edge of the target image due to forward scattering in the water. This edge-degradation is known to occur in natural water (Replogle and Steiner 1965) and forward scattering is probably greater in water which contains large suspended particles such as plankton

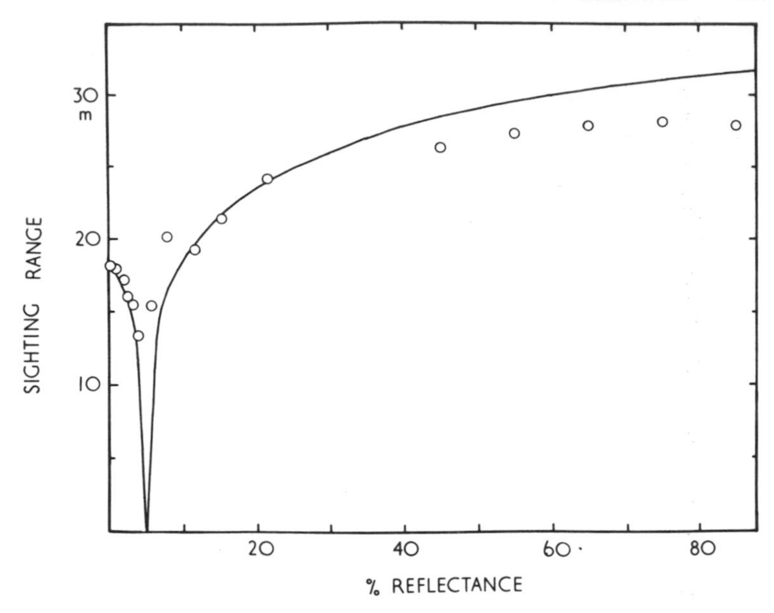

Fig. 55

The visible range of sixteen 15 × 20 cm. matt grey targets. The targets varied in reflectance (at 500 nm.) from 0·5 per cent (black) to 86 per cent (white). These measurements were made over a sandy bottom at a depth of 40 m. in Malta. The sighting range shown in Fig. 54 was used.

Under these conditions the white targets are the most visible. The targets which most closely match the water background in brightness are the least visible.

The continuous line shows the theoretically expected results (equations 3 and 4) and is calculated from the visible range of a target of reflectance 0 per cent (black); ${}_tN_o$ is taken as proportional to the diffuse reflectance of the target measured in air; ${}_bN$ is taken as equal to the reflectance of the target that would be least visible; the liminal level for contrast perception for a diver wearing a face mask is taken as 0·02. The visible range for the white and light grey targets is less than predicted and the three brightest targets are all seen at about the same range. A possible explanation is discussed on page 121.

(Atkins and Poole 1952, Jerlov 1953, Gazey 1970). The efficient perception of contrast by the eye depends upon a sharp well-defined edge between the target and background, and where the edge is slightly blurred contrast perception (and hence visible range) is impaired (Thomas and Kovar 1965). This effect has been discussed by Middleton (1941, pp. 56–7). Although this book does not deal with vision through the atmosphere many of the problems are fundamentally the same and his book is most valuable in a study of vision under water.

Kinney *et al.* (1967) have reported that although white was a very visible colour in a wide variety of different natural waters, black was often very difficult to see. At first sight this contradicts Hemmings and Lythgoe's (1966) finding that black shows up rather well. One explanation may be that Kinney's 'black' had a reflectance of 3·7 per cent whilst Hemmings and Lythgoe's had a reflectance of 1·5 per cent. But the most likely explanation lies in the position of the target relative to the observer. In Hemmings and Lythgoe's experiments the path of sight was horizontal, but in Kinney *et al.*'s experiments, where black proved so difficult to see, the target was vertically below the observer. Under these conditions about 98 per cent of the ambient light would be falling upon the target from above and there would only be 2 per cent upwelling from below. In this situation a target of 3·7 per cent reflectance would appear almost identical in brightness to the background and would indeed be very hard to see. This effect, together with the related one that white objects will be hard to see when viewed directly from below was in fact predicted by Denton and Nicol (1962, 1966) in their studies of the camouflage of silvery-sided fishes.

The general principle that either black or white objects can be seen at the greater distance through the water can be extended with confidence to the study of coloration in fishes and the visibility of one fish to another. Usually the assessment of relative brightness requires a knowledge of the apparent spectral radiance of the object and background at the eye, plus a knowledge of the physiological and optical properties of the eye itself. However, for white and grey objects with their rather flat spectral reflectance curves, the brighter objects reflect more light than the darker at all wavelengths, and thus black or white objects always present the greatest brightness contrast irrespective of the spectral sensitivity of the eye viewing them.

Polarized Light and Vision

For all practical purposes the unaided human eye is unable to detect that light is polarized, but in a recent review Waterman and Horch (1966) state that more than ninety species of animals, mostly arthropods, but also some cephalopods, have been shown to possess the ability to detect polarized light and to determine its plane of polarization. In invertebrates the analysing mechanism is intra-retinal, but in some fishes there is an adipose eyelid which is birefringent and may act as a polarizing filter (Stewart 1962, Blaxter and Jones 1967).

The function of these polarized light analysers has chiefly been discussed in relation to orientation and phototaxis but differential sensitivity to light vibrating in different planes also has significance in underwater visibility (Lythgoe and Hemmings 1967). This is because the underwater spacelight is plane-polarized as a result of light scatter in the water (see p. 109). Yet most natural objects tend to depolarize the light reflected from them. Thus in equations 1 and 3, ${}_tN_r$ is unpolarized and ${}_bN$ is plane-polarized, and if an object is viewed horizontally under water through a rotating plane-polarized filter the apparent brightness of the background will fluctuate more than that of the object. In other words, the value C_0 in equation 3 will fluctuate. This can be demonstrated photographically. In Fig. 56 an array of painted grey targets was photographed through a polarizing filter. The camera with its attached filter was then rotated through 90° and another photograph taken. The brightness of the background is greater in the one photograph than in the other, but the brightness of the targets is little changed. The most striking demonstration of this lies in the target which is brighter than the background with the camera horizontal, but darker when the camera is rotated through 90°. It follows from the discussion in earlier sections (p. 110) that when the inherent contrast, C_0 as seen through a polarizing filter is at a maximum, C_r will fall to the perception limen (the object will become invisible) at a greater range than when C_0 is at a minimum.

The results of an experiment to demonstrate the visual effects of wearing plane-polarizing filters in the face-plate is shown in Fig. 57. This experiment was carried out in moderately clear water in Malta at a depth of 7 m. The sighting range of the kind described on page 121 was set up so that the targets were down-sun from the observer and for horizontal paths of sight the plane of maximum polarization was nearly horizontal. The visible range of sixteen painted grey targets was measured, first with polarizing filters orientated to exclude the maximum spacelight, and then with the filters removed. It is quite clear that the visual range of the light grey targets was significantly increased with the filters in place. But, as is to be expected in theory, there is an indication that the visual range of the darker grey targets is reduced.

Unfortunately the use of polarizing filters in the face-plate is rather limited in usefulness for three main reasons. Firstly, the plane of maximum spacelight polarization tilts through 90° or more as the sun rises, passes overhead and sets (Waterman and Westell 1956) (Fig. 46). Thus provision would have to be made to rotate the filters in the face-plate. Secondly, the filter orientation required to see bright objects best

Fig. 56

Four matt-grey painted targets photographed at a depth of 3 m. through a plane-polarizing filter. In the upper photograph the camera was held horizontally and the polarizing filter was orientated to allow the maximum background spacelight to pass. In the lower photograph the camera (with attached filter) was rotated through 90°. There is a considerable difference in the brightness of the water background in the two photographs, but the brightness of the grey targets has scarcely varied. The contrast between the brightest target and the water background is much higher in the lower photograph. Note also that the brighter of the two middle targets is darker than the water background in one photograph but lighter in the other. (Lythgoe and Hemmings 1967.)

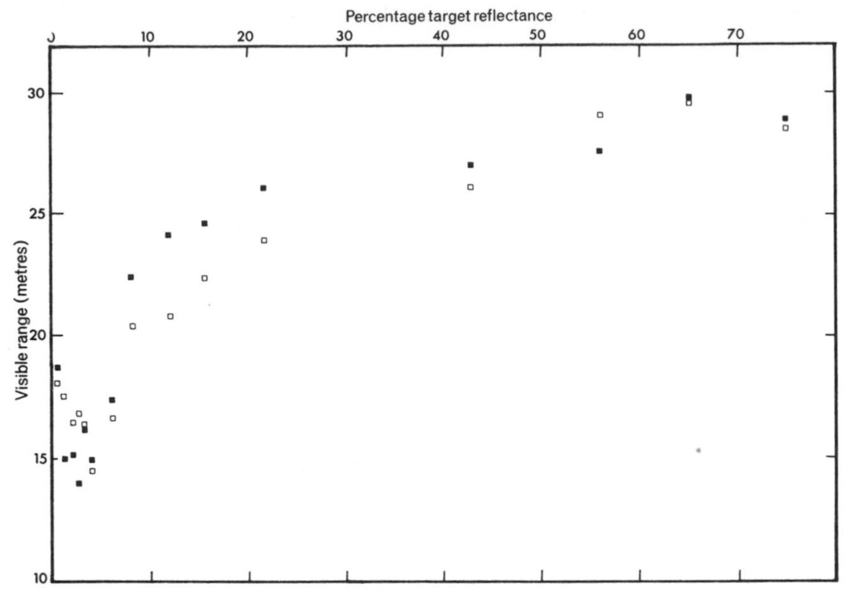

Fig. 57

The visible range of sixteen 15 × 20 cm. matt-painted grey targets. Open symbols: no filter in face-mask. Closed symbols: plane-polarizing filter in face-mask fitted to exclude the maximum amount of background spacelight. Depth of targets: 3·5 m. Depth of bottom (sea-grass covered): 7 m. Measurements made between 13.30 and 14.30, 25 August 1966. Target down sun from observer. (After Lythgoe and Hemmings 1967.)

is precisely that which is worst for seeing dark objects. Thirdly, a polarizing filter reduces the light reaching the eye by about 50 per cent. But for work close to the surface where there is a bright mid-day sun the polarizing filters do take out the dancing rays of the sun which normally make vision difficult, and most fishes and suspended objects in the water are more easily seen with filters in place.

There might be some use for polarizing filters in photography when the maximum contrast between an object and its water background is required. In practice the rotation of the filter to its optimum position is a fairly delicate job and would be difficult to do when, say, photographing a moving fish.

So far it has been assumed that only natural light is used, but should there be an artificial light source the use of polarizing techniques is more promising. In a series of tank experiments, Gilbert and Pernicka (1967) placed a light source near to the camera in such a way as to illuminate a white painted target; a circularly polarizing filter was fitted over the light and another was placed over the camera lens. If there was no other light source the target could be photographed at a considerably greater distance through turbid water when both filters were in place than when one or both were absent. These authors suggest that light reflected from the small suspended particles in the water is reflected only once and the returning light is rotated in the opposite sense to the out-going light and is unable to pass the camera filter. On the other hand there is multiple reflection of light at the target's surface and thus half the returning light is rotating in a sense which allows it to pass the camera filter.

Coloured Face-plates, Spectral Sensitivity, and Visual Range

Face-masks fitted with a yellow face-plate instead of the usual clear glass are occasionally sold because they are supposed to improve underwater contrast. Also some fishes (mostly freshwater but some marine) have yellow corneas (Walls 1963, Moreland and Lythgoe 1968, Muntz in press). These yellow filters, whether natural or artificial, decrease the eye's sensitivity to blue light whilst leaving its sensitivity to the red relatively undiminished. The shift in spectral sensitivity by the use of a coloured filter is always gained at the expense of sensitivity and an eye with no filter will always be more sensitive than one with a colour filter whatever its colour.

On the other hand, a shift in the spectral sensitivity of the photo-sensitive pigment itself can result in greater sensitivity to the ambient light. The visual pigments are a good example for they all have the same shaped spectral absorption curve (when plotted on a light-frequency basis) but the wavelength of maximum absorption can be located in widely different parts of the spectrum (see Dartnall and Lythgoe 1965). Thus a visual pigment which absorbs maximally in the green will be more sensitive in green water than would a blue-absorbing visual pigment, and it is clear that the correct visual pigment can extend downwards the limits of vision in the sea in a way that a coloured filter cannot.

Nearer the surface the amount of light is not limiting, and any improvement in visual performance is likely to be found in contrast perception (and hence visual range) rather than in sensitivity. The effect on contrast perception springs from the fact that the spectral distribution curve of the relatively bright downwelling light is rather flatter than that for light arriving from other directions (see, for instance, Tyler and Smith 1967). This leads to the possibility that the spectral radiance curve of a close small grey or white target suspended in the water will be flatter than that from the water background and measurements with the device described on page 113 confirm this to be true. When the distance between the target and eye is increased the target radiance approaches that of the water background (Fig. 58) until the radiances from both sources can no longer be distinguished and the target becomes invisible. From Fig. 58 it is clear that an eye with its maximum sensitivity at, say, 550 nm. will absorb relatively more light from a near target relative to the water background than will an eye with its greatest sensitivity at 500 nm. Thus an eye with its wavelength of maximum sensitivity offset from the wavelength of maximum background radiance will enjoy enhanced contrast perception for near bright objects, but for near dark objects contrast perception will be reduced.

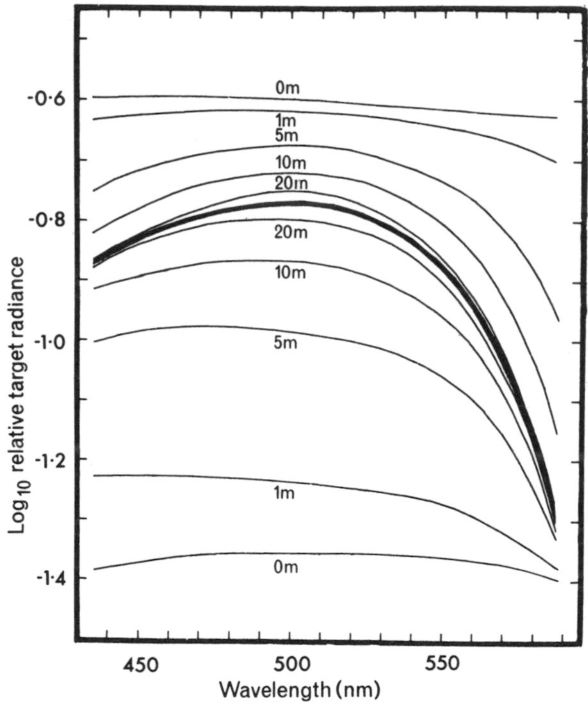

Fig. 58

Diagram to show how the apparent spectral radiances of a grey target slightly brighter than the water background (upper lines) and a grey target slightly darker than the water background (lower lines) approach the spectral radiance of the water background spacelight, as the horizontal range of the target (M = metres) increases. These curves are calculated from data obtained for Malta using the equipment shown in Figs. 48 and 49 and refer to targets just beneath the water's surface (Lythgoe 1968).

The wavelength of maximum water background radiance is, in fact, close to that where the beam attenuation coefficient (α) is at a minimum (Tyler 1965) but contrasts fall off most rapidly where α is large. And at the greater ranges of vision possible with large targets both dark and light targets present their greatest visual contrast at wavelengths where the water background radiance is brightest. Although it is not possible to specify a filter which will allow grey objects of all sizes to be seen at the maximum possible distance through the water, it might sometimes be useful to know that a filter which excludes wavelengths close to the transmission maximum of the water is best suited for seeing small bright objects; whereas a filter which passes only those wavelengths near to the transmission maximum of the water is best for seeing small dark objects and all large objects, be they dark or bright.

These arguments are potentially suitable for a quantitative analysis in the real terms important to the fish or diver, namely the range at which objects become visible. At the moment, however, the theory of radiance transfer through water and our knowledge of the visual functions outstrip the available data on light transmission through the sea. In previous sections it has been sufficiently accurate to assume that the apparent contrast (C_r in equation 3) is equal to the visual contrast as perceived by the eye, but for the analysis of the effect of filters or various visual pigments (Lythgoe 1968) this approximation becomes inadequate.

If the eye is to distinguish between the two radiances ${}_tN_r$ and ${}_bN$ from the direction of the target and background respectively, the actual visual contrast can be described as follows:

$$ {}_{\text{vis}}C_r = \frac{\int {}_tN_r \,.\, V - \int {}_bN \,.\, V}{\int {}_bN \,.\, V} \, d\lambda $$

where V is the spectral sensitivity of the eye, and λ the wavelength of light. The object will be visible when ${}_{\text{vis}}C_r$ is greater than some threshold value which depends upon the conditions of viewing (p. 110). The two radiances, ${}_tN_r$ and ${}_bN$ are usually measured in energy units but should be converted into quantum energy units for these calculations (Dartnall and Goodeve 1937). The spectral sensitivity of the eye (the V_λ curve) may either be directly measured (easy in man, but more difficult in other animals). When a coloured face-plate is considered the V_λ curve must be multiplied by the spectral transmission curve of the face-plate. The techniques for calculating the V_λ for animals is given in detail by Dartnall *et al.* (1965).

The Visibility of Colours

One of the most widely known phenomenon in underwater vision is that red objects lose their colour and cannot be distinguished from black in deep clear water. The explanation is simple; namely that objects at the surface appear red because they only reflect light at the red end of the spectrum and absorb all other wavelengths. But it is the red end of the spectrum which is most strongly absorbed by the water and deeper than about 30 m. no red light remains and all the light falling on the object is absorbed. So simple is the observation and so easy the explanation that it is forgotten that the visibility of red objects in clear blue water is only an isolated case and that the visibility of other colours in different types of water has not been explained.

A surface which reflects all wavelengths of light equally looks grey, but most surfaces reflect some wavelengths more strongly than others and it is this inequality that the eye and brain exploit to form the sensation of colour. If only a very small waveband is reflected the object will look the same colour as that waveband in the spectrum. More frequently an object will reflect one or more wide bands of wavelengths spanning more than one colour region of the spectrum.

When the wavelengths of light reflected from naturally occurring yellow, orange, and red surfaces are measured it is found that the resulting spectral reflectance curves all tend to be very similar in shape; the shorter wavelengths being absorbed and the longer wavelengths reflected. The transition between almost complete absorption and very high reflectance is usually sharp and it is the position of this 'cut-off' in the spectrum which determines whether an object will appear yellow, orange, or red.

The spectral reflectance curves for the conspicuous yellow fin of the reef-dwelling angel fish *Heniochus acuminatus* and the red tail of the perch, *Perca fluviatilis* are shown in Fig. 59. The fishes were kept alive until just before a measurement was made. The fin was then severed and spread over a white ceramic tile (an identical tile was used as a blank) and its reflectance was measured in a Beckman recording spectrophotometer; a procedure that takes about five minutes. This technique is probably satisfactory for yellows, oranges, and reds which are normally produced by the relatively stable carotenoids, but structural colours which depend on the integrity of the surface layers of the fish begin to change immediately the fish dies and could not be measured in this way. Curves of the shape shown in Fig. 59 are quite typical both for natural objects and man-made paints.

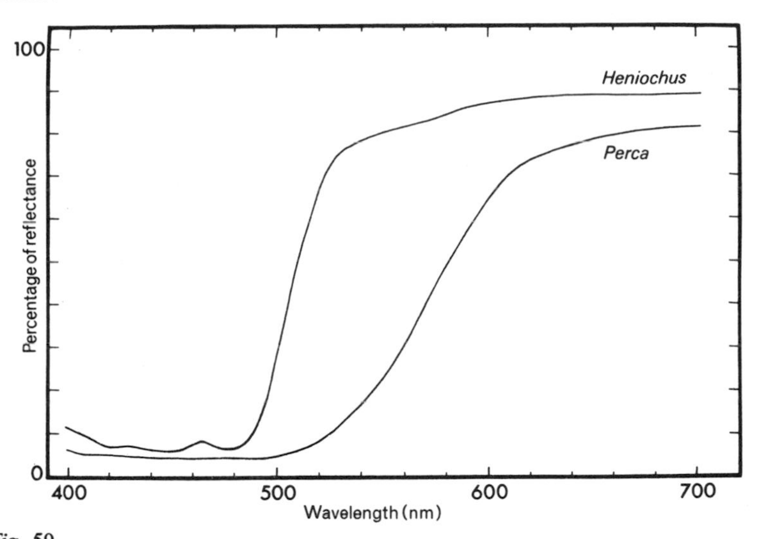

Fig. 59

The spectral reflectance curves of a yellow fin from the reef-dwelling angel fish, *Heniochus acuminatus*, and the red tail fin of the river perch, *Perca fluviatilis*. Naturally occurring reds, oranges and yellows all tend to have spectral reflectance curves of this general shape.

Figure 60 shows schematically the distribution of absorption bands in the spectrum that would produce the sensation of white, yellow, orange, red, blue, blue-green, and black. The spectral band where 90 per cent of the incident daylight energy is concentrated at the optically equivalent depths of 80 m. in Jerlov type 11 ocean water and at 18 m. in type 9 coastal water (Jerlov 1951) is also shown. In the well-known case of a red object in blue ocean water, it is evident that the object reflects no light at all in the only region of the spectrum where there is sufficient light for vision, and the object will be indistinguishable from black. By the same argument a blue object in green water will also appear black, whilst a blue-green object in ocean water and a yellow object in green water will resemble an object that was white or pale grey on the surface. Indeed it is only those surfaces that have a spectral cut-off point (the shaded areas in the diagram) in a region of the spectrum where there is light in visually adequate amounts that will appear coloured. Thus red and blue-green objects in green water and yellow and dark blue objects in clear ocean water will retain their colour to considerable depths (Lythgoe 1969).

Kinney *et al.* (1967) have provided data for the visibility of coloured

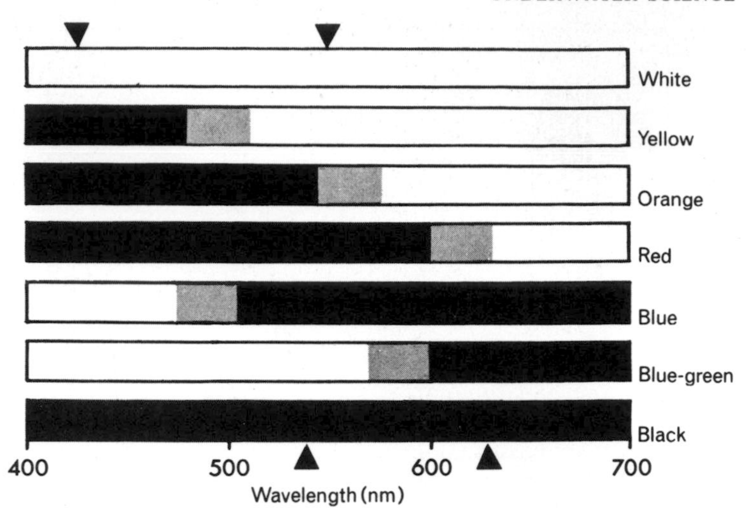

Fig. 60

The colours which result when certain regions of the spectrum are absorbed and not reflected (black areas); the stippled areas are those where the absorption varies rapidly with wavelength.

The triangular markers at the top and bottom of the diagram delimit the spectral region where 90 per cent of the daylight incident at the water's surface is absorbed at 80 m. in blue Jerlov type 11 ocean water (top markers) and yellow-green Jerlov type 9 coastal water (bottom markers). (Lythgoe 1969.)

spheres in several different types of water and their results could be explained on these lines. Furthermore, a series of coloured targets whose spectral reflectance curves are shown in Fig. 62 were viewed under water in the clear waters of the Mediterranean and in the yellow-green waters of an English lake (Fig. 61). In the Mediterranean the red targets do indeed look black, but the yellow retain their colour well, whilst in the lake the yellow resembled a light grey and it was the red that was conspicuous.

The advantage of a qualitative appraisal such as this is that it applies equally well to fishes as to man. Indeed various investigations into the colour vision of fishes show it to be essentially similar to that found in man, although there are differences in detail (e.g. Hamburger 1926, Beniuc 1933, Granit 1941, Marks 1965a and b, Liebman and Entine 1964, Liebman in press). It might be significant to note in this context the many freshwater fishes such as the perch, *Perca fluviatilis*, the stickleback, *Gasterosteus aculeatus* and the char, *Salvellinus alpinus*, all intensify their red colour around breeding time, whilst yellow fins and

Fig. 61

White, yellow, orange, red and black targets photographed at 28 m. in the clear blue water of Malta and at 0·5 m. in the yellow-green waters of an English lake (West-Tofts Mere). Each colour sample measures 2·8 × 15 cm. and they were photographed at a range of 1 m. The spectral reflectance curves of the coloured targets are shown in Fig. 61.

In the blue water red is indistinguishable from black but the yellow shows up well. In the yellow-green water the yellow closely resembles the white but it is the red that shows up well.

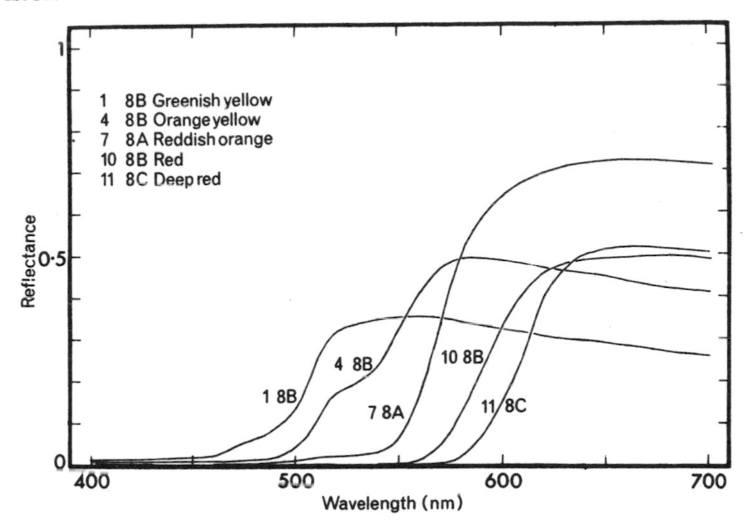

Fig. 62
The spectral reflectance curves of the yellow, orange and red perspex targets that were photographed under water (Fig. 61). The colours of the samples are white, greenish yellow (1 8B), orange yellow (4 8B), reddish orange (7 8A), red (10 8B), and deep red (11 8C). The colour names and figure references are taken from Kornerup and Wanscher (1963).

markings are common amongst fishes living in clear blue ocean water (see Chapter 5).

The 'visibility of colours' is a very broad phrase that needs more careful definition before any experiments or calculations can be attempted. There seem to be three main ways of approaching the problems of colour vision under water, namely colour recognition, colour conspicuousness, and colour differences.

1. *Colour recognition*

If a diver is presented with a collection of objects identical except in colour and asked to name the colour, first on the surface and then under water, he will give some of the objects the same colour name above and below water, but he will name others differently in the two situations. Thus at 40 m. in the Mediterranean he will correctly identify yellow, but red will be named black. Some substances with more than one peak in their spectral reflectance curve may appear quite different in colour on land from under water. Blood is a good example: at the surface the reflectance maximum in the green is swamped by the red, but at depth the water absorbs the long wavelength red light and blood appears green.

Colour recognition is important if various pieces of apparatus such as valves or pipes are painted different colours for easy recognition. In a series of experiments in waters of widely different colour and transparency Kinney *et al.* (1967) have found that the colours correctly recognized at different ranges through the water vary from place to place, but that fluorescent colours were most often correctly identified.

Fluorescent colours are particularly valuable under water (Lythgoe 1966) because of their property of absorbing the high-energy short wavelength light in the water and re-emitting it as low-energy long-wavelength green, orange, and red light. The result is that reds and oranges stand out in brilliant contrast against the predominantly blue or green of the environment, and because their colour depends more on their fluorescence than on their reflectance, their apparent colour remains 'true' at all depths and in most types of water.

The eye has a remarkable ability to adapt to the different colour of ambient light and an object which is yellow on the surface will be recognized as yellow almost to the limiting depth for colour vision in blue water. Yet the spectral radiance as measured by instruments would be quite different above and below water, and were the radiance reaching the eye from the yellow object under water to be somehow viewed on land, the object would unhesitatingly be named green.

The phenomenon of 'colour adaptation' has not received much attention in the underwater context and indeed the experiments of Kinney and Cooper (1967) under simulated underwater conditions seem to be the only ones yet described. The usual explanation of the mechanism of colour adaptation in the fundamental colour vision mechanism (p. 115) most strongly stimulated by the ambient light becomes less sensitive relative to the other mechanisms. Thus the sensation of blue from the water background, very strong on first going under water, gradually fades, whilst the sensation of red from small starfish and fishes is preserved to a rather greater depth than might have been expected. This change in the balance of sensitivity amongst the three mechanisms is thought to take rather less than five minutes to complete.

2. *Conspicuous colours*

A conspicuous colour in any particular situation is one which contrasts strongly against the general background. Thus a snorkel tube painted yellow will be more easily seen amongst seaweed than would a green one, but should the yellow snorkel fall amongst a colony of yellow sponges it might well escape detection. In general, the components of any underwater scene (weeds, rocks, encrusting animals, etc.) all tend to approach the same colour as the depth or viewing range increases and are distinguished by differences in brightness and not of colour. However, some colours (deep blue and yellow in clear blue water, or red and cyan in yellow-green water) do tend to contrast strongly in colour against the general scenery, and objects painted in these colours would be relatively easy to find. Fluorescent colours are always conspicuous because they emit light at wavelengths which are scarcely present under water. Fluorescent animals are not found in fresh water, and are restricted to coelenterates in the sea (Limbaugh and North 1956) and the only fluorescent plants are necrescent red algae. Thus, not only do fluorescent objects retain their colours at all depths where there is light enough for colour vision, but they are extremely unlikely to be displayed against a background of similarly fluorescent objects.

3. *Liminal colour differences*

The practising diver is usually interested in easily recognizable or conspicuous colour in the day-to-day tasks of identifying or finding apparatus or specimens on the sea-bed. However, these practical approaches are not at the moment easily adapted to a theoretical analysis of colour vision under water. On the other hand, Stiles's line element equation (Stiles 1946, Wyszecki and Stiles 1967) makes it possible to calculate when two radiances can just be distinguished by virtue of colour or brightness differences and thus to calculate at what range a coloured object will remain just visible. There is also a possibility that the various constants in these equations can be adapted so that the analysis can be extended to fishes as well.

References

Atkins, W. R. G., G. L. Clarke, A. Pettersson, H. H. Poole, C. L. Utterback, and A. Ångstrom (1938), 'Measurement of submarine daylight.' *J. Cons. Explor. Mer.*, **13,** 37–57.

Atkins, W. R. G. and H. H. Poole (1952), 'An experimental study of the scattering of light by natural waters.' *Proc. Roy. Soc. Lond., B,* **140,** 321–38.

Beniuc, M. (1933), 'Bewegungssehen, Vershmelzung und Moment bei Kampffischen.' *Z. Vergleich. Physiol.*, **19,** 724–46.

Blackwell, H. R. (1946), 'Contrast thresholds of the human eye.' *J. Opt. Soc. Amer.*, **36,** 623–43.

Blaxter, J. H. S. and M. P. Jones (1967), 'The development of the retina and retino-motor responses in the herring.' *J. Mar. Biol. Assoc. U.K.*, **47**, 677–97.

Clarke, G. L. and E. J. Denton (1962), 'Light and animal life.' In *The Sea*. Edited by M. N. Hill. New York and London: John Wiley & Sons.

Dartnall, H. J. A. (1957), *The Visual Pigments*. London: Methuen; New York: John Wiley.

Dartnall, H. J. A., G. B. Arden, Hisako Ikeda, C. P. Luck, M. E. Rosenberg, C. M. H. Pedler, and K. Tansley (1965), 'Anatomical, electrophysiological and pigmentary aspects of vision in the Bush Baby; an interpretative study.' *Vision Res.*, **5,** 399–424.

Dartnall, H. J. A. and C. F. Goodeve (1937), 'Scotopic luminosity curve and the absorption spectrum of visual purple.' *Nature, Lond.*, **139,** 409–11.

Dartnall, H. J. A. and J. N. Lythgoe (1965), 'The spectral clustering of visual pigments.' *Vision Res.*, **5,** 81–100.

Denton, E. J. (1956), 'Recherches sur l'absorption de la lumière par le cristallin des poissons.' *Bull. Inst. Océanogr. Monaco*, **1071,** 1–10.

Denton, E. J. and J. A. C. Nicol (1962), 'Why fishes have silvery sides, and a method of measuring reflectivity.' *J. Physiol.*, **165,** 13–15P.

Denton, E. J. and J. A. C. Nicol (1964), 'The chorioidal tapetum of some cartilagenous fishes (chondricthyes).' *J. Mar. Biol. Ass. U.K.*, **44,** 219–58.

Denton, E. J. and J. A. C. Nicol (1966), 'A survey of reflectivity in silvery Teleosts.' *J. Mar. Biol. Ass. U.K.*, **46,** 685–722.

Denton, E. J. and F. J. Warren (1957), 'The photosensitive pigments in the retinae of deep-sea fish.' *J. Mar. Biol. Assoc. U.K.*, **36,** 651–62.

Duntley, S. Q. (1960), 'Improved nomographs for calculating visibility by swimmers (natural light).' Bureau of Ships Contract NObs–72039. Rep. 5–3. Feb.

Duntley, S. Q. (1962), 'Underwater visibility.' In *The Sea*. Edited by M. N. Hill. New York and London: John Wiley & Sons.

Gazey, B. K. (1970), 'Visibility and resolution in turbid waters.' *Underwater J.*, **2,** 105–15.

Gilbert, G. D. and J. C. Pernicka (1967), 'Improvement of underwater visibility by reduction of backscatter with a circular polarizater technique.' *Appl. Optics*, **4,** 741–6.

Granit, R. (1941), 'Relation between rod and cone substances based on the scotopic and photopic spectra of *Cyprinus*, *Tinca*, *Anguilla* and *Testudo*.' *Acta. Physiol. Scand.*, **2,** 334–6.

Haig, C. (1941), 'The course of rod dark adaptation as influenced by intensity and duration of pre-adapting light.' *J. Gen. Physiol.*, **24,** 735–51.

Hamburger, V. (1926), 'Versuche über Komplementär—Farben bei Ellritzen (*Phoxinus laevis*).' *Z. Vergleich. Physiol.*, **4,** 286–304.

Hecht, S., C. D. Hendley, S. Ross, and P. M. Richmond (1948), 'The effect of exposure to sunlight on dark adaptation.' *Amer. J. Ophthalmol.*, **31,** 1573–80.

Hemmings, C. C. (1966), 'Factors affecting the visibility of objects underwater.' In *Light as an Ecological Factor*. Edited by R. Bainbridge, G. C. Evans, and O. Rackman. Oxford: Blackwell, pp. 359–74.

Hemmings, C. C. and J. N. Lythgoe (1964), 'Better visibility for divers in dark water.' *Triton*, **9,** No. 4, 28–31.

Hemmings, C. C. and J. N. Lythgoe (1966), 'The visibility of underwater objects.' *Underwater Assn. Rep.*, **1,** 23–9.

Ivanoff, A. and T. H. Waterman (1958), 'Factors, mainly depth and wavelength, affecting the degree of underwater light polarization.' *J. Mar. Res.*, **16,** 283–307.

Jerlov, N. G. (1951), 'Optical studies of ocean water.' *Rep. Swed. Deep-Sea Exped. 1947–48*, **3,** 1–59.

Jerlov, N. G. (1953), 'Particle distribution in the ocean.' *Rep. Swedish Deep-Sea Exped., 1947–48*, **31,** 71–97.

Jerlov, N. G. (1968), *Optical Oceanography*. Amsterdam, London, New York: Elsevier.

Kalle, K. (1961), 'What do we know about "Gelbstoffe"?' *Symposium on Radiant Energy in the Sea.* Edited by N. G. Jerlov. I.U.G.G., Helsinki, 1960.

Kampa, E. M. (1961), 'Daylight penetration in three Oceans.' *Symposium on Radient Energy in the Sea.* Edited by N. G. Jerlov, I.U.G.G., Helsinki, 1960.

Kinney, J. A. S. and J. C. Cooper (1967), 'Adaptation to a Homochromatic visual world.' U.S. Naval Submarine Medical Center, Groton, Conn. Report No. 499.

Kinney, J. A. S., S. M. Luria, and D. O. Weitzman (1967), 'The visibility of colours under water.' *J. Opt. Soc. Amer.*, **57,** 802–9.

Kornerup, A. and J. H. Wanscher (1963), *Colour.* London: Methuen.

Le Grand (1939), 'La pénétration de la lumière dans la mer.' *Annales Inst. Oceanogr.*, **19,** 393–436.

Liebman, P. (in press), 'Microspectrophotometry of receptors.' In *The Handbook of Sensory Physiology, VII. I. The Photochemistry of Vision.* Edited by H. J. A. Dartnall, Hamburg: Springer-Verlag.

Liebman, P. and G. Entine (1964), 'Sensitive low-light-level microspectrophotometer: detection of photosensitive pigments of retinal cones.' *J. Opt. Soc. Amer.*, **54,** 1451–9.

Limbaugh, C. and W. J. North (1956), 'Fluorescent benthic Pacific Coast coelenterates.' *Nature, Lond.*, **178,** 497–8.

Lythgoe, J. N. (1966), 'Underwater vision.' *British Sub-Aqua Club Diving Manual*, 2nd edition. London: Eaton Publications.

Lythgoe, J. N. (1968), 'Visual pigments and visual range under water.' *Vision Res.*, **8,** 997–1012.

Lythgoe, J. N. (1969), 'Red and yellow as conspicuous colours underwater.' *Underwater Assn. Rep.*, **3,** 51–4.

Lythgoe, J. N. (in press), 'The adaptation of visual pigments to the photic environment.' In *The Handbook of Sensory Physiology, VII. I. The Photochemistry of Vision.* Edited by H. J. A. Dartnall. Hamburg: Springer-Verlag.

Lythgoe, J. N. (in press), 'List of Vertebrate Visual Pigments.' *In The Handbook of Sensory Physiology, VII. I. The Photochemistry of Vision.* Edited by H. J. A. Dartnall.

Lythgoe, J. N. and C. C. Hemmings (1967), 'Polarized light and underwater vision.' *Nature, Lond.*, **213,** 893–4.

Marks, W. B. (1965a), 'Visual pigments of single goldfish cones.' *J. Physiol.*, **178,** 14–32.

Marks, W. B. (1965b), 'Visual pigments of single cones.' In *Colour Vision.* Edited by A. V. S. de Reuk and J. Knight. London: Churchill.

Middleton, W. E. K. (1941), *Visibility in Meteorology.* 2nd edition. University of Toronto Press.

Moreland, J. and J. N. Lythgoe (1968), 'Yellow corneas in fishes.' *Vision Res.*, **8,** 1377–80.

Muntz, W. R. A. (in press), 'Inert absorbing and reflecting pigments.' In *The Handbook of Sensory Physiology, VII. I. The Photochemistry of Vision.* Edited by H. J. A. Dartnall.

Pirenne, M. H. (1962), 'Dark adaptation and night vision.' In *The Eye,* **2.** Edited by H. Davson. New York, London: Academic Press.

Replogle, F. S. and I. B. Steiner (1965), 'Resolution measurements in natural water.' *J. Opt. Soc. Amer.*, **55,** 1149–51.

Smith, R. C. and J. E. Tyler (1967), 'Optical properties of clear natural water.' *J. Opt. Soc. Amer.*, **57,** 589–95.

Stewart, K. W. (1962), 'Observations on the morphology and optical properties of the adipose eyelids of fishes.' *J. Fish. Res. Bd. Can.*, **19,** 1161–2.

Stiles, W. E. (1946), 'A modified Helmholtz line element in brightness-colour space.' *Proc. Phys. Soc.* (*Lond.*), **58,** 41–65.

Thomas, J. P. and C. W. Kovar (1965), 'Effect of contour sharpness on perceived brightness.' *Vision Res.*, **5,** 559–64.

Tyler, J. E. (1959), 'Natural water on a monochromater.' *Limnol. Oceanogr.*, **4,** 102–5.

Tyler, J. E. (1963), 'Estimation of *percent* polarization in deep oceanic water.' *J. Mar. Res.*, **21,** 102–9.

Tyler, J. E. (1965), '*In situ* spectroscopy in ocean and lake waters.' *J. Opt. Soc. Amer.*, **55,** 800–5.

Tyler, J. E. (1968), 'The Secchi Disc.' *Limnol. Oceanogr.*, **13,** 1–7.

Tyler, J. E. and R. W. Preisendorfer (1962), 'Light.' In *The Sea,* **1.** Edited by M. N. Hill. New York: John Wiley & Sons, pp. 397–451.

Tyler, J. E. and R. C. Smith (1967), 'Spectroradiometric characteristics of natural light under water.' *J. Opt. Soc. Amer.*, **57,** 595–601.

Tyler, J. E. and R. C. Smith (1970), *Measurements of spectral irradiance underwater.* New York, London, Paris: Gordon and Breach.

Walls, G. L. (1963), *The Vertebrate Eye.* New York and London: Hafner Publishing Co.

Walls, G. L. and H. D. Judd (1933), 'The intra-ocular colour filters of vertebrates.' *Brit. J. Ophthalmol.*, **17,** 641–75, 705–25.

Waterman, T. H. (1954), 'Polarization patterns in submarine illumination.' *Science*, **120,** 927–32.

Waterman, T. H. (1955), 'Polarization of scattered sunlight in deep water.' *Deep Sea Res. Suppl.*, **3,** 426–34.

Waterman, T. H. (1960), 'Polarized light and plankton navigation.' In *Perspectives in Marine Biology.* Edited by A. A. Buzzati-Traverso, pp. 429–50.

Waterman, T. H. and K. W. Horch (1966), 'Mechanisms of polarized light perception.' *Science*, **154**, 467–75.
Waterman, T. H. and W. E. Westell (1956), 'Quantitative effect of the sun's position on submarine light polarization.' *J. Mar. Res.*, **15**, 149–69.
White, W. J. (1964), 'Vision.' In *Bioastronautics Data Book*. Edited by P. Webb. National Aeronautics and Space Administration, Washington D.C., pp. 307–41.
Wyszecki, G. and W. S. Stiles (1967), *Colour Science*. New York, London, Sydney: John Wiley & Sons.

5 Fish Behaviour

C. C. Hemmings

Introduction

The behaviour of fish in aquaria may differ from that of the same species in open water. Because seas, lakes, and rivers provide such a variety of habitats, the feasibility of reproducing a suitable alternative in the laboratory varies considerably. Success at this is usually inversely related to the size and activity of the species, i.e. the behaviour of minnows is easier to study in captivity than that of tuna. Therefore any technique of *in situ* observation is inherently more likely to produce records of natural behaviour. In this chapter, emphasis is placed on the underwater observation of naturally occurring behaviour, with virtually no mention of the responses of fish to unnatural stimuli of human origin, such as artificial lights or fishing gear. The one exception to this is the attraction which a number of man-made objects in the sea seem to have for some fish. This subject is briefly discussed. It should never be forgotten that the underwater observer is himself an intruder in the fish's environment, and this intrusion may sometimes affect their behaviour.

The Nature of Behaviour

The behaviour of a fish comprises all movements of the whole animal and parts of its body such as fins, and also changes of coloration. These actions are the eventual result of information received by sense organs and transmitted, usually via the nervous system, to effector organs such as chromatophores or individual muscles. This information is derived from two quite distinct sources, firstly, the fish's external physical and chemical environment, including other organisms, and secondly, the internal environment in which hormonal balance plays an important part. Examples of this internal environment are the degree of hunger, the level of sexual activity, and the degree of exhaustion of muscles. It is important to remember that although these two sources of information are distinct in origin, their effect is complementary. Thus the internal state of an animal will usually affect its reaction to external stimuli (Baerends 1957).

There are in addition rhythmic activities which do not appear to be 'switched-on' or 'switched-off' by any specific stimulus or combination of stimuli. These apparently rigid rhythms can be modified in various ways, either temporarily as a result of other stimuli or by conflict with other activities of the animal, or more permanently from evolutionary selection pressure. The regular respiratory movements of the mouth and gill region are an example of a rhythm that is thus modified. The frequency of movement is influenced by hormonal changes and by dissolved gases in the water. Alteration of the actual movement sequence occurs when the fish feeds, especially when a voracious predator captures large prey. Finally, in some species parts of the cycle of respiratory movements have been incorporated during phylogenetic development into entirely different patterns of behaviour. The gill covers or opercula of certain Cichlid species carry brightly coloured patterns which function as visual releasers in intra-specific social contact, and the spreading of the opercula which is part of the normal respiratory cycle is highly exaggerated during threat posturing; in other species, a gaping or yawning of the mouth occurs (Baerends and Baerends van Roon 1950).

When observing fish behaviour, particularly during the early stages of an investigation, every movement and orientation of the fish, and each change of colour must be considered of possible importance. Such simple actions as fin spreading or closing, movements of buccal and opercular regions, etc., can all at some time be components of more obvious behaviour such as feeding, territorial defence, or courtship. There is thus a hierarchy of simple acts which make up these more general patterns of behaviour, sometimes in quite stereotyped ways. The hierarchical organization of behaviour is discussed by Tinbergen (1950), and with special reference to fish by Baerends (1957). Some behaviour patterns can be demonstrated as sequences of individual acts each releasing the one following, or resulting in the animal being in such a position that an external stimulus will initiate the next. Such chains of activity are usually involved in appetitive behaviour, which leads on to a consummatory act or situation. This is the obvious logical conclusion of the appetitive behaviour, such as the capture of prey after a sequence of stalking actions. These sequences of activity are not necessarily rigid, but can be changed into different behaviour or be broken off altogether. Orientation to environmental stimuli is particularly important during appetitive behaviour.

The Approach to Fish Behaviour Study

Underwater and laboratory observation compared

The experimental method conducted in the laboratory involves isolating the systems under study in such a way as to allow the occurrence of certain reactions to known stimuli, but eliminating, controlling, or monitoring all extraneous variables. This method is used particularly in work on the borderline between sensory physiology and behaviour, and again in nearly all work on learning in animals. Bull (1957) reviews learning in fish as a means of determining sensory thresholds, and briefly considers the relationship between laboratory investigation and behaviour in the natural environment.

By contrast, the underwater observer has virtually no awareness of and even less control over a bewildering variety of environmental factors which affect the behaviour of the fish he is watching. Reactions to some of these will be recognized, but a large proportion of the activities of the fish will appear to the observer to be occurring quite spontaneously. This category of 'apparently spontaneous' behaviour includes reactions not only to internal states and stimuli within the animal but also to those external factors that are imperceptible to the unaided observer in an alien medium.

The success of any behavioural work depends on attention to detail plus large measures of patience and perseverance. The end result should be a complete descriptive record of the activities of a species, ideally in all parts of its normal habitat. During this stage of cataloguing the behavioural repertoire of an animal there can rarely be any critical analysis of cause and effect relationships between associated events in the field. It might be noticed, for example, that a species of fish changed its behaviour and coloration at dusk. This could be due directly to a decrease in the ambient light intensity, or it might be evidence of an endogenous circadian rhythm related to the day/night cycle. In the field one would need to wait for a total eclipse of the sun to test these alternative hypotheses, but laboratory experiments with controlled lighting would soon supply evidence indicating which alternative is more likely.

At present the laboratory is the more convenient locale for studying fish behaviour in relation to the physical and chemical parameters of the environment operating at small scale, such as the mechanisms of orientation to gradients of stimuli. The large-scale variations that are more within the sphere of ecology are better studied in the field. Perhaps the greatest scope for underwater observation is in studying the relationships of a fish with its biological surroundings. These include the more obvious intra-specific relationships such as schooling, territorial behaviour, courtship and breeding, as well as the more remarkable inter-specific ones involved in symbiosis and parasitism.

The place of underwater investigation in fish behaviour research

1. *Investigator familiarization.* It would be difficult to prove that familiarity with the underwater world is an advantage, but most diver-researchers are vaguely aware that their thinking is influenced by an intimate knowledge of the environment. It is in the framing of problems and the formation of attitudes of mind that this makes itself most felt. For example, a diver will usually consider underwater visibility to be far more relevant and meaningful than underwater light intensity. This he may bear in mind when thinking of the reactions of fish that appear to be governed by vision. There is an enormous volume of literature on underwater light in relation to behaviour, see reviews by Clarke and Denton (1962), Nicol (1963), and Woodhead (1966). There are rather fewer references to vision in relation to visibility (Duntley 1962; Hemmings and Lythgoe 1966; Hemmings 1966c; Lythgoe 1966, and Chapter 4). Similarly, it is commonly and correctly stated that fish must have a visual or tactile reference to the substrate to determine the relative strength and direction of currents in the water mass in which they are swimming. This is brought home to any diver who has swum from the surface to the bottom in murky tidal water, only to find the bottom 'moving past' him.

2. *Direct observation of behaviour.* *In situ* observation of fish behaviour can be merely one component in a much wider study conducted partly in aquaria and partly in the field, or it may on occasions be possible to perform complete studies under water. The attitude of mind of the investigator in relation to fieldwork is important and influences the emphasis placed on different phases of the work. If fieldwork is undertaken initially, this often suggests ideas for subsequent more detailed investigation which should then be based on natural behaviour. The open-minded objective attitude implied here is something of an illusion, because inevitably the underwater observer is accompanied by his own prejudices and preconceptions. The possible influences of nitrogen narcosis and stress on observing capacity and such prejudices has not yet been adequately considered by psychologists. Thus reports from divers of strange fish behaviour in very deep water should be treated with caution. The main value of knowing what behaviour might occur is that the observer will usually recognize it when it does. There are only a privileged few who discover completely 'new' behaviour. The best example of this being the phenomenon of cleaner fish that feed on the ectoparasites of 'customer fish' (Feder 1966). Underwater investigation of such behaviour patterns as those of courtship and breeding depends on the ability of the observer to recognize specific actions and sequences as components of a complete behaviour pattern (Fiedler 1964), and it is clearly an advantage to know what behaviour to expect.

A somewhat different way of regarding direct observation in the field is to consider it as merely providing confirmation of the correctness of conclusions derived from experimental work in the laboratory. An example of this is that of Lorenz (1962, 1966) seeking corroboration of a theory of the coloration of coral reef fish after laboratory study of territorial behaviour (Zumpe 1965).

3. ***Underwater experiments.*** There are a number of cases in which human activity under water is involved in behaviour study without prolonged observation of the fish. One of these is in the study of fish hearing. It is impossible to simulate acoustic free-field conditions in the laboratory for critical studies of intensity thresholds, therefore attempts have been made to work entirely in the field. The first technique is an electrophysiological investigation with the fish suspended from a boat over deep water (Enger and Anderson 1967). The second method, using diving techniques, was to set up conditioning apparatus for determining thresholds in 20 m. of water (C. J. Chapman, personal communication). It is almost certain that an increasing amount of acoustic research will be done under water to achieve the requirements for acoustic free-fields.

Commercially important species of fish are tagged and released to indicate migration tendencies and to assist in population studies. But the closed swimbladders of physoclistous fish are usually ruptured if brought to the surface. Studies of tagging under water have been made to assess the magnitude of the disturbance to fish (Hislop and Hemmings 1967).

It has been suggested that much, if not all, work on the swimming speed and endurance of fish is suspect owing to the problems of handling them during capture and maintaining them in captivity. Underwater investigation has been suggested in this field also.

Man as an observer of behaviour under water

Attention must be given to two basic problems which are actually the same irrespective of whether man himself makes the observations, or uses a remote device such as a camera or an echo-sounder. The first consideration is the effectiveness of the observer at making valid observations, and the second is the disturbing stimulus of the observer or recorder in the environment of the animal being studied. The only advantage of man being under water is to make visual observations and to control directly additional recording apparatus such as cameras. Senses other than vision are virtually useless, and various physical, physiological, and psychological factors combine under water to reduce the effectiveness of even this sense (see the chapters by Baddeley and Ross). The most relevant variable is 'visibility'—the distance at which an object of specified contrast can be seen by the eye. Assuming that light intensities are high enough, the parameter to which visibility is related is α, the beam attenuation coefficient. Means for estimating this parameter and further discussion of it are given in Chapter 4. However, it is important to remember that fish, because of counter-shading, silvery sides, or other colour patterns, are usually much less easily visible than their surface appearance would suggest. If a silvery fish turns so that it reflects top light towards the observer, it may be highly conspicuous for some distance, but to identify the fish, note its behaviour, see the position of its fins and so on, irrespective of its orientation, one must be much closer to it. To obtain a successful photographic record one must usually be closer still.

The problem of the diver as a visual stimulus is perhaps brought home by suggesting a general rule of thumb, that 'to see is to be seen'. Although the visual acuity of some fish, particularly small specimens, may not be as good as man's, various devices are used to enhance contrast, and in any case the average diver is a much more conspicuous target than the average fish. The use of hides is almost universal in the study of bird behaviour in the field, but they do not seem to have been tried as a means of watching fish. The observation of fish from underwater houses certainly constitutes observing from a hide in some degree, but the size of the house is sufficiently large for the fish to be reacting to its presence (Clarke *et al.* 1967). The results from a comparative study of one species with and without the use of a simple 'wet' hide might be very interesting.

It is obvious to any underwater observer that the fish can see him,

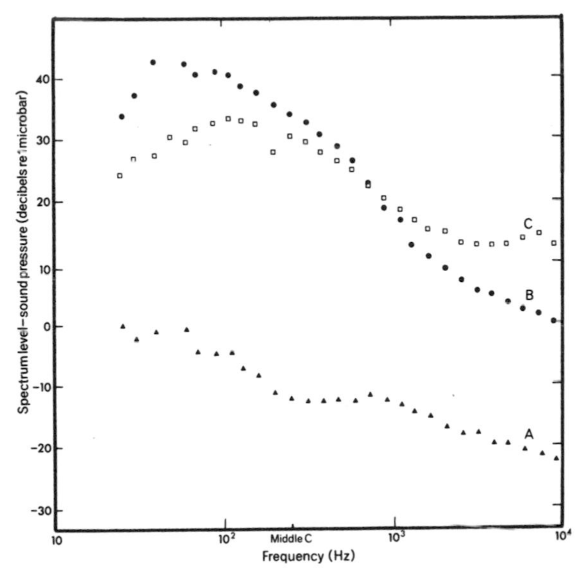

Fig. 63

Inhalent and exhalent demand valve noise against ambient sea noise.

A Ambient sea noise in Loch Erisort (sea state 2).
B Exhalent noise from demand valve.
C Inhalent noise from demand valve. This curve also contains low frequency sound from previously exhaled bubbles ascending through the water column.

Diver 1 m. distant from the hydrophone.

(Data from Chapman, unpublished Marine Laboratory Report.)

and that sometimes they react to his presence, but there is a second important factor that is occasionally forgotten. Although the aqualung diver relies entirely upon vision, he is himself not only visible but also audible to the fish, against a background of sea and biological noise. The compressed-air aqualung is an effective generator of sound of a wide range of frequencies (see Fig. 63), and not only is the sound carried with very low attenuation (e.g. approximately 10^{-6} dB per metre for 256 Hz, middle C), but most fish are very sensitive to a wide frequency range (see Fig. 64). The dual nature of the low frequency part of the curve A for fish is interesting. Wodinsky and Tavolga (1964) suggests that A_1 represents hearing via the swimbladder and inner ear, and that A_2 represents 'hearing' by the lateral line of near-field displacement effects. Variations of turbidity of the water scarcely affect the transmission of sound, and therefore the intensity of a diver's noise received by fish is proportional to distance in waters of very different

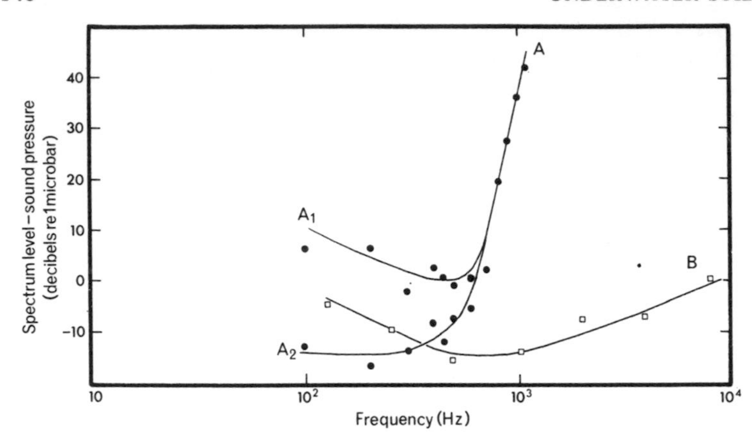

Fig. 64

Underwater hearing thresholds of a teleost fish and man for comparison with demand valve noise in Fig. 63. (See text for further details.)
(Data redrawn from Wodinsky and Tavolga 1964, and Brandt and Hollien 1967.)

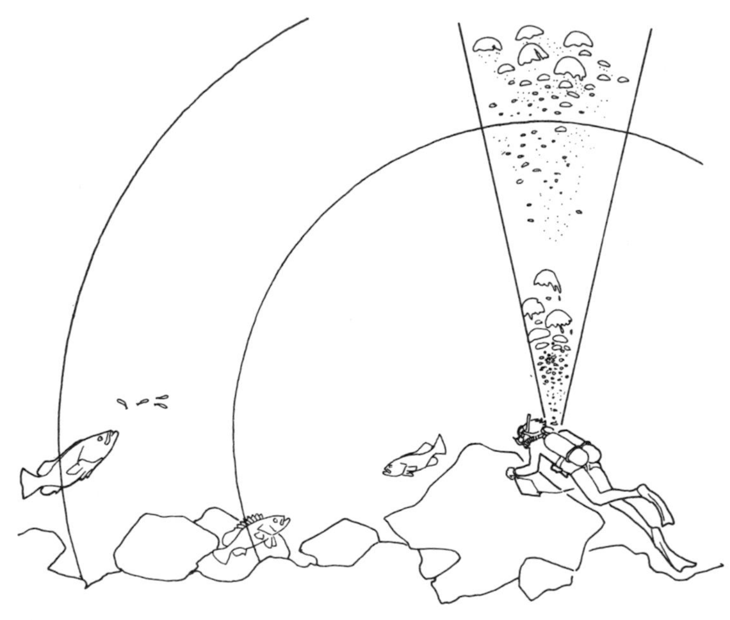

Fig. 65

Indication of the 'shell' of useful observation range around an underwater observer.

optical turbidities. Thus the relative importance of sight and sound as indicators of the diver's presence is not constant, but varies with water clarity.

It is theoretically possible to investigate the sensory physiology of the fish visual and acoustic-lateralis system in order to determine the distance at which any species would be able to detect the presence of a human observer in its vicinity, and this work is being done. However, the more important and more intractable problem is to ascertain the distance at which the presence of the diver so affects the behaviour of the fish that it is no longer to be regarded by the observer as natural. This unknown and almost unknowable distance can be related to that at which the observer can make useful records of behaviour as follows: The observer theoretically has surrounding him a shell (being the space between two concentric spheres) within which useful observations can be made (see Fig. 65). The outer boundary of the shell is set by the visibility of the fish to the observer at the centre; the inner boundary is set by the influence of the observer on the fish being watched. Both dimensions of the shell will vary with species, with their size and to some extent with the behaviour pattern being watched. Potts (1968) describes simple experiments to determine these dimensions with the Mediterranean wrasse *Crenilabrus melanocercus*. In turbid water the shell may even be negative thickness, i.e. the observer may never see the fish without their being considerably influenced by his presence, because, as has been noted, sound is virtually unaffected by turbidity. The infrequency with which some species of fish are seen in low turbidity water appears to be due not just to the encounter probability, but to the fact that they are more readily scared if they can hear but not see a diver than if they can both see and hear him. The perfect shell never exists in practice, being most nearly developed in the pelagic environment away from the bottom. Most observation is undertaken near a substrate, which cuts the available volume of observation to about half. In addition the shell is perforated above a diver by his exhalent bubble column.

Subjects of Underwater Investigation

Orientation

It is usual to recognize two uses of the word orientation, which Fraenkel and Gunn (1940) refer to as primary and secondary orientations. The primary is purely postural and refers to the normal attitude of the animal, related to the vertical as defined by gravity. The secondary orientation involves movement of the animal in response to the direction and intensity of external stimulation.

Primary orientation. The primary orientation of a typical fish is with the dorsal surface uppermost and the body length more or less horizontal. Exceptions to this occur in teleost flatfishes which from an anatomical point of view lie on one side; in *Synodontis batensoda*, the aptly-named 'Upside-down catfish'; and the remora which when 'hitching lifts' on sharks or turtles may do so in any position.

Gravity is the only force that does not vary from one environment to another, and it is natural to take its direction as that from which primary orientations are measured. Gravity also plays a part in the normal maintenance of equilibrium of fish, which have gravity receptors in the labyrinth of the middle ear. The gravitational sense is usually reinforced by a tendency of the fish to keep its dorsal region towards the maximum light intensity, which is referred to as the dorsal light reaction. Von Holst (1935), using the wrasse *Crenilabrus rostratus*, showed that the response of the gravity receptors in the middle ear was stronger than the dorsal light reaction. Unoperated fish exposed to a horizontal light source tilted over to one side to an angle substantially less than a right angle, whereas fish that were exposed to light from underneath showed no change in the normal orientation. A complete response was found only in those animals in which the labyrinth had been experimentally removed. They showed a perfect response to the light direction, even to swimming upside-down. One should notice that this result was obtained in a laboratory apparatus in which the situation of the fish had been simplified to its essentials.

Abel (1954) first observed an unrelated species, *Anthias anthias* in the Blue Grotto on the island of Capri showing quite different behaviour. In this case the fish were swimming upside-down in certain parts of the cave, apparently in response to the scattered blue light from the large underwater entrance that gives the grotto its name. Clearly there was here no question of the fish being deprived of their gravitational sense. Most divers will have noticed that fish swimming near a slope tend to tilt their bodies in such a way that they are orientated as if the slope were horizontal. In the case of fish swimming near the roof of underwater caves and ledges, the fish are to be found upside-down. Abel (1962) discusses further the relationship between fish and their immediate environment, and proposes a general definition of this behaviour, the 'ventral substrate reaction' (*Bauch-Substratreaktion*). The behaviour of *Anthias* in the Blue Grotto becomes then a special case, the 'ventral shade reaction' (*Bauch-Schattenreaktion*), as opposed to the dorsal light reflex (*Licht-Rücken-Reflex*). In this situation the fish are swimming some distance away from the substrate and are responding more to its relative darkness. Thus it appears that not only the light direction but also the visual environment of the fish is an important part of primary orientation. Further critical laboratory work taking into account the important visual environment, is essential in order to resolve this discrepancy between laboratory conclusions and field observations.

Secondary orientations. The secondary orientation involves not changes in attitude but changes of position in space. Some movements of an animal appear to be the direct result of perception by the sense organs of some external factor such as temperature, light intensity, or odour. It is important to realize that some of these factors have both intensity and directional information, e.g. sound and light, whereas others such as temperature, salinity, and odour have no directional parameter and to determine an intensity gradient, spatial and/or temporal comparisons must be made by the organism. These reactions to external stimuli are the basic components of such apparently simple behaviour patterns as habitat preference or vertical migration. These often turn out to be very subtle and complex. It is this level of orientation behaviour that results in the obviously discontinuous distribution of animals that forms the subjects of ecology and zoogeography.

Fraenkel and Gunn (1940) give an elaborate classification of secondary orientations divisible into taxes or directed responses, and kineses or undirected responses. Although vertebrates, particularly fish, are included in Fraenkel and Gunn's review, the study is based largely on research into invertebrate behaviour, but Harden Jones (1960) considers the scheme in relation to fish alone. Some doubt has been cast on the validity of applying some of Fraenkel and Gunn's categories to fish (Hemmings 1966b). It is to be expected that the considerable elaboration of the senses and central nervous development of even the lower vertebrates results in a much greater degree of central co-ordination and integration of information from different modalities than occurs in insects. Hence it is rarely possible in the field to observe responses of fish to simple environmental gradients or boundaries. For this reason, and because of the almost total restriction of the underwater observer to visual observation, most of the simple orientations are related to variations in ambient light intensity or light/shade boundaries. It is possible to deduce orientations involving senses other than vision in the field, and to investigate these experimentally in the laboratory. The determination of the senses involved in the feeding of nocturnal moray eels by Bardach *et al.* (1959) was based in part upon direct field observations in this way.

There is a considerable need for research into the difference between behaviour in relation to light intensity and to visual features of the environment. This is just as important in the case of secondary orientations as for the primary ones discussed above. The occurrence of *Anthias anthias* in caves in shallow water was mentioned above, and reported also by Abel (1962), but this species is only to be found in open water at depths of around 100 m. This is evidence of either a negative phototaxis or a low light intensity preferendum. In an extensive study of a small bay in the Gulf of Naples, Abel (1962) divides the fish population into open-water pelagic, coastal pelagic, supra-demersal, and substrate-living types. The behaviour of each of the nearly fifty species discussed is closely related to its habitat. In theory all of the subtle ecological relationships that occur are explicable in behavioural terms, but it is a much easier task to describe the distribution and habitat preference of a particular species than it is to discover what are the exact orientation mechanisms that bring them about. The magnificent monograph by Riedl (1966) on the biology of underwater holes indicates just how much behavioural work remains to be done in explaining the behavioural basis of ecological balance.

Abel (1962) touches on one of the most interesting problems that underwater observation has done much to reveal. This is the apparent attraction that certain objects and situations have for fish. The phenomenon occurs in bottom-living and pelagic fish alike, although it would be dangerous to assert that this was the same behaviour or that the same stimuli were involved. Abel called the response an 'optotaxis' although perhaps a better word would be 'optothigmotaxis', to indicate an orientation involving visual contact with the stimulus. In a relatively uniform visual field, fish tend to move towards prominent visual features, which are of course often topographically prominent also. This behaviour may account for the way in which fish seem to be attracted to artificial reefs, quays, and oil rigs. Many pelagic fisheries are based on the attraction of fish to floating and submerged objects. The 'Kannizzati' fishery for *Coryphaena hippurus*, the dolphin fish, in Malta, uses floating slabs of cork (Galea 1961), and the Java sea fishery for species of *Decapterus*, mackerel scad, uses anchored rows of coconut fibres (Soemarto 1960). This behaviour of fish in the open ocean has been studied in a simple but effective way by the use of an observation chamber projecting below the surface of a floating raft. The raft was allowed to drift in the central Pacific for up to nine days with continual observation of the attracted fish during the hours of daylight (Gooding 1965). The results of experiments with different types of shelter have recently been published by Hunter and Mitchell (1968). Divers counted the numbers of fish accumulating under each experimental object. It was found that a tent-shaped lure was more effective than flat ones orientated either vertically or horizontally, although the fish did not actually swim in the space under the tent.

A number of reasons have been advanced for the attraction of fish to floating objects, and these have been reviewed by Gooding and Magnuson (1967). In perhaps the earliest reported underwater observation of the accumulation of fish under boats, Damant (1921) suggested an explanation based on the advantage to fish of seeing plankton prey rendered more conspicuous when seen against the dark under-surface of an object, as in dark-ground illumination in microscopy. Although Damant observed mackerel feeding in this situation, it seems unlikely that this is any more than an advantage consequent on the fish showing the optothigmotactic behaviour first. Similarly another consequent advantage is Gooding and Magnuson's (1967) suggestion that floating objects act as oceanic cleaning stations. Other possibilities involve the energy-saving advantage in hiding in slack water behind an object if there is any current flow.

Station keeping by fish in a flow of water depends largely on visual fixation of features of the environment (Dijkgraaf 1933). In an otherwise featureless visual field, there may be some tendency for fish to keep station with prominent objects as if they were swimming in a flow of water irrespective of whether there actually is a current. Usually tidal streams and currents are inconvenient or even dangerous for underwater observers. But because of the significance of visual cues for the observer as for the fish, water movements are easily observed near the sea-bed or any other relatively fixed object, and consequently the behaviour of fish in relation to currents can be studied. Salmonids have been watched by Keenleyside (1962) in rapidly flowing rivers in Canada. Of particular interest was the way in which the fish seemed to gain some hydrodynamic advantage by resting against the bottom. Personal observation of the distribution of several species of Gadidae and Labridae off the Orkney islands suggests that the distribution of fish is permanently influenced by tidal flow, even though there may be long periods of slack water. During these slack water periods it is impossible to assess which gullies remain quiet and which have the full force of the tide running through them during the tidal stream, except that the fish tend to occur in those remaining slack at all stages of the tide. There may, however, be a secondary effect due to slight differences in the distribution of the sessile invertebrate fauna.

Diel variation in behaviour

Perhaps the most significant regularly fluctuating parameter of the external environment is the radiation intensity change between day and night. It now seems that the majority of fish species show some variation of activity, habitat preference or coloration during some period of the twenty-four hours. Many species in inshore habitats such as coral reefs show all three during a quiescent period of 'sleep' in which they may roost in crevices during the hours of darkness. Owing to the thermal inertia of water and the low penetration of infra-red, variation in temperature of water can be disregarded in this short-term cycle.

The earliest observations of changes of behaviour and coloration by night were made by that most important underwater pioneer, W. H. Longley, using a simple diving helmet (Longley and Hildebrand 1941). It is not surprising that many additional observations have been made of fish in the larger public aquaria, in which a variety of species is normally kept (Boulenger 1929, Weber 1961). As a result of a long study of one Caribbean coral reef area, Starck and Davis (1966) published a comprehensive catalogue of the nocturnal behaviour of reef inhabitants. They were able to divide up the population of fish into herbivores, omnivores, and plankton-feeders. The herbivores and omnivores were predominantly active by day, but the plankton-feeders fed either by night and day or only at night. Many of the fish feeding on small invertebrates fed both by day and by night, the only strictly diurnal ones being those that excavated for their prey in the sand. The larger piscivorous predators tended to have dawn and dusk peaks of feeding activity when the change-over in the rest of the fish population occurred. Starck and Davis suggest that unless account is taken of the complete difference in feeding activity and behaviour in the coral reef environment by day and by night, an entirely misleading picture of the trophic structure of the community may result.

In a general study of the behaviour of sand-eel schools in the Baltic,

Kühlmann and Karst (1967) describe the circadian rhythm of mixed schools of *Hyperoplus lanceolatus* and *Ammodytes lancea*. These fish rest overnight in the sand inshore and at dawn undertake a feeding migration offshore in large schools. During the afternoon and evening, small groups return to the inshore area prior to roosting overnight. It is interesting to note that the period of greatest nervousness of the sand-eels was during the dusk and dawn periods when large eels, their main predator, were active. This predator activity peak is exactly as described for the coral reef.

Diving techniques have been sometimes used merely as confirmation of conclusions drawn from entirely different evidence. Hasler and Villemonte (1953) in studying the daily movements of perch, *Perca flavescens*, in a lake with an echo-sounder, reached the conclusion that the fish approached the bottom at dusk, and then apparently disappeared from the echo-trace. Simple diver-observations showed that the fish had in fact settled and were resting overnight on the bottom. Kiselev (1960, personal communication) using a deepwater observation chamber has described many instances of fish resting on the bottom when of course they become difficult to spot on echo-sounders. There is considerable scope for the combination of visual techniques of underwater observation with the use of echo-sounders because the identification of echo-traces is always somewhat uncertain.

Non-diel movements of fish

Migration is often used as a synonym for fish movements of more than a few metres, being equally applied to small-scale horizontal and vertical migration as to the journeys of several hundred miles undertaken by pelagic tuna. Harden-Jones (1968) reviews fish migration and discusses the mechanisms of fish movement on both scales. He concentrates on the behaviour of fish species of commercially important fish. Underwater techniques are only valuable in studying the small-scale movements in which the full potential of direct observation has certainly not yet been realized.

In this type of study it is usually necessary for individual animals to be recognized. Exceptions to this occur only when the whole population moves, as did the sand-eels described by Kühlmann and Karst (1967). Problems are raised by the necessity for tagging in that the physiological stress involved in capture, handling, narcotizing, and tagging may affect the behaviour of the fish for an unknown period after the process.

This is more relevant in the case of fish that have been brought from some depth to the surface for tagging.

Strictly diurnal species may be captured at night. The author has assisted with the tagging of a common tropical species, *Dascyllus trimaculatus* using a technique devised by Proper (personal communication). The dascyllus roost during the night in coral, and those specimens being studied roosted regularly in a large mass of stagshorn coral, *Acropora*. This coral mass was covered with a large polythene sheet, and after it had been weighted down at the edges, anaesthetic was introduced underneath. The slightly anaesthetized fish were taken a few yards to the shore for tagging, and immediately afterwards were returned to the coral roosting place. This method involved very little disturbance to the fish, and might be equally applicable to nocturnal fish which hide during the day, provided that they can be extracted from their hiding places when stunned.

In shallow water the tidal cycle is probably second only to light intensity in order of importance as a regularly fluctuating environmental parameter. Few observations seem to have been made on the variation of behaviour in relation to tides, except for Carlisle's (1961) brief report of observations of feeding and territorial behaviour in *Morone labrax*, the bass, and *Mugil labrosus*, the grey mullet. Two specimens of the first species and seven of the latter were individually recognizable and were seen to advance over the inter-tidal zone in 15–20 cm. of water with the rising tide. The fish then fed and defended quite distinct territories, but an hour after high water they swam out to sea. The long investigation by Abel (1962) of the fish population of a small bay was similarly conducted without the use of breathing apparatus, as the water was quite shallow. In this case individual fish were marked both by glass beads and fin clipping, so that the position of individual fish could be plotted on a grid. Bardach's (1958) study of fish movements in a patch of coral reef at Bermuda involved a still larger area, in which trap/tag/release/recapture methods were more important than the direct underwater observation which was restricted to the occasional watching of tagged fish. This really represents work in which diving is a useful ancillary technique but is not crucial.

Additional evidence of the home range of some herbivorous fish can be gained in rather the same way as the home range of limpets is revealed by the browsed area around the scar in which they sit at low tide. In the Caribbean, bare patches of sand occur around rocky reef areas, gradually giving way to a richer growth of plants away from the reef.

A simple method was used by Randall (1965) to investigate this; he built out the reef artificially into an area of good plant growth and transplanted some of the good plant growth back into the bare sandy area. The result was somewhat as expected, the weed was quickly reduced in both cases by the browsing of *Scaridae*, parrotfish, and by *Acanthuridae*, surgeonfish. Randall suggested that this behaviour was related to the avoidance of predators by hiding in holes in the reef. Therefore the distance to which the fish would feed away from it was related to their escape path. This interesting behaviour could be studied further by quite simple experimental techniques, to investigate, for example, whether the simple territorial feeding area described by Carlisle is held by these fish.

Social behaviour

There is unfortunately no agreed definition of the term 'Social Behaviour'. It is used in this context to cover behaviour in which there is interaction between individuals whether of the same or different species. It is convenient to divide social behaviour into intra-specific behaviour such as schooling, territorial defence, and breeding, and inter-specific behaviour such as prey-predator reactions, commensalism, and symbiosis. The part that direct observation can play in the different types of social behaviour varies. Although it is relatively easy to observe the breeding behaviour of some species in captivity it is more difficult to contain enough fish in a volume of water sufficient to allow the full range of schooling to occur. Similarly, without foreknowledge one would not expect to observe cleaning symbiosis in aquaria unless the 'right' species were associated by chance. Close commensal or symbiotic relationships in which some anatomical modification is involved are usually more obvious.

1. *Schooling*. The study of schooling behaviour like social behaviour suffers from a shortage of good definitions and a surfeit of bad ones. There is a basic agreement that schooling is brought about by mutual attraction between individuals of the same species, and it is most highly developed in pelagic species. The subject is reviewed by Morrow (1948) and most recently by Shaw (1967).

Owing, perhaps, to the difficulty of establishing schooling fish in conditions suitable for experimentation, laboratory and field investigations have tended to concentrate on rather different aspects of the subject. The laboratory work is concerned more with the senses involved, and thus with the orientation component or the appetitive behaviour of individuals (Keenleyside 1955; Hemmings 1966a). On the other hand, the open-water observations have been more concerned with the behaviour of the whole school as a unit in which all the fish are in a consummatory situation (Ulrich 1951, Eibl-Eibesfeldt 1962, Kühlmann and Karst 1967).

Eibl-Eibesfeldt's survey gives simple descriptions of a wide range of tropical species in different habitats. This ecological variation leads to his division of schooling behaviour into six categories, depending entirely on the natural environment of the species:

(a) Pelagic wandering fish;
(b) Semi-pelagic wandering fish;
(c) Pelagic predators;
(d) Wandering demersal fish;
(e) Schooling fish with fixed refuges on the reef;
(f) Facultative schooling fish.

This separation into categories has an important bearing on one of the central arguments about the definition of schooling. This concerns whether or not parallel orientation between individuals is an integral part of schooling. Clearly parallel orientation must occur when a group of fish is in motion if they are to remain together, but Eibl-Eibesfeldt includes a number of species which whilst showing mutual attraction do not always maintain parallel orientation. Techniques of measuring the relative positions of fish in a school have been developed for use in aquaria and shallow water (Cullen *et al.* 1965, Hunter 1966). It would be valuable to extend these methods for use in open water and to relate orientation in a school to activity.

Schooling is considered by Eibl-Eibesfeldt as protective behaviour, particularly valuable in the case of peaceable and young fish. On a number of occasions the behaviour of predators and schooling prey was watched. Only individual stragglers were taken by a predator from a school of plankton-feeding fish which gave a very disorientated appearance, with each fish darting and snapping in a different direction. The typical activity of schooling fish, when disturbed, of streaming backwards and forwards is thought to confuse the predator, as is some of the coloration shown by schooling fish. This is clearly an example of 'protean' behaviour as first described by Chance and Russell (1959) for small mammals, and more extensively by Driver and Humphries (1966) and Humphries and Driver (1967). In open water a school of fish tends to form a 'vacuole' around a hunting predator but they approach rather closer when it shows no intention of feeding. The shape change of sand-eel schools was shown by Kühlmann and Karst (1967) to be also very similar to vacuoles forming around possible hiding places of predators. These authors investigated the structure of a school under stress by themselves 'attacking' it from different directions. They observed a strong tendency for a split school to reform, but without the individual fish approaching at all closely to them. However Ulrich (1951) making similar observations on young herring found that fish would swim within a few centimetres of him to regain contact with the other fragment of the school.

Eibl-Eibesfeldt suggests that the protective value of schooling has allowed fish to extend into the pelagic environment, but it is here, where schooling is particularly well developed that the additional benefit suggested by Brock and Riffenburgh (1960) would occur. In this theoretical extension of convoy theory to prey/predator relationships, they suggest the particular advantage of schooling in clear water. The larger the number of fish gathered in the school, the lower is the probability of any one fish encountering and being eaten by a predator. This same effect operates against the human observer and makes the observation of pelagic species a particularly chancy business. Underwater duration is clearly important, and perhaps the most valuable tool is the large submarine capable of extensive submergence and equipped with acoustic location devices as well as direct observation facilities. The Russian submarine *Severyanka* is the best-known example of such a machine, and has been used by Radakov (1960, Radakov and Solovyev 1959) for observing large concentrations of herring in the open ocean.

2. *Territorial behaviour*. Some aspects of territory were discussed in the section dealing with behaviour in relation to the environment. In most cases these are no more than the expression of a habitat preference. This leads to the concept of a home range through which other fish of both the same and different species are permitted to swim. Wynne-Edwards (1962) gives a very full discussion of territory and home range in relation to social behaviour, with examples taken to a large extent from the behaviour of birds and mammals. These can be paralleled with a number of examples from fish where the same sort of territory tenure seems to happen.

Lorenz (1966) points out that on coral reefs, which are probably the richest known marine environment for variety of species of invertebrates and fish, highly specialized methods of feeding and restriction to certain types of food have evolved. Thus some fish have become so adapted to their ecological niche, that competition exists only from their own species, and consequently the territory is only defended against their own species. A number of examples are given of 'poster coloration' of territorial fish which advertise their presence at the maximum possible distance to others of their own species. In addition they only show intra-specific aggressive behaviour. The more drab coloration is associated with fish of non-territorial habit which are correspondingly less aggressive. A short publication by Fricke (1966b) suggests that there may be exceptions to this apparently neat division. Fricke used poster-coloured models under water in the Red Sea, and found that one of the poster-coloured fish showed both inter- and intra-specific aggression, and that another showed no territorial defence at all. Cases of exclusion of all other species from a territory are usually associated with breeding behaviour, when the territory is often much smaller in extent.

Keenleyside (1962) observed the behaviour of trout and salmon in a Canadian river system. Although juvenile salmon were completely territorial, trout varied according to habitat. In fast-moving water over gravel and stone bottom, territories were closely maintained. In pools and quiet weedy backwaters, large numbers of fish associated together. No suggestion was given of possible explanations for such differences of behaviour.

There is evidence that learning plays a significant part in some aspects of spatial orientation in fish (Aronoson 1951, Hoar 1958). It would be possible and useful to investigate this experimentally under water. Stones and rocks as well as completely artificial objects could be added to, and removed from territories of marked fish to determine the key stimuli to orientation, in the same way that van Iersel (1964) describes for the orientation of *Bembix rostrat*, the digger wasp. Hole-living species such as *Bleniidae* would be suitable as they tend to show constant courses (Abel 1962).

3. ***Reproductive behaviour***. The popularity of home aquaria is based commercially on the ease with which many species of fish can be bred in captivity. The result is that fuller study has been made of those fish that are popular and successful inmates of aquaria. The reproductive behaviour of *Gasterosteus aculeatus*, the stickleback, has probably been more intensively studied than any other species, whilst the studies of various species of *Cichlidae* best exemplify truly comparative ethology.

Aronoson (1957) reviews the literature with particular respect to physiology and the effect of environmental variables on behaviour. Breder and Rosen (1966) provide a taxonomic review of reproduction in fish which excludes much of the European ethological work. Small fish, particularly those that build nests, seem to carry out all the courtship and reproductive behaviour in quite a small volume of water, and as a result the behaviour in aquaria may not differ greatly from that in the field. Problems certainly arise in those species which normally defend a territory larger than the usual study tank. Fiedler (1964) reports that the nest-building male wrasse, *Crenilabrus ocellatus*, which does not grow to more than 10 cm. length, requires an area of about 15 m. radius for the collection of building material in the natural environment.

Relatively few studies have been conducted partly or wholly in the field, and for obvious reasons these have been restricted to fairly shallow water. Abel (1961) published a description of the reproduction of the Pomacentrid *Chromis chromis* in the Bay of Naples. The so-called 'signal jumps' of the male serve to attract an individual female from the loose school in which they swam down to the nest-site under rocks; after spawning, the male defends the eggs until they hatch. As this reproductive behaviour occurs in fairly shallow water during the summer, and as *Chromis chromis* is one of the commonest fish in the Mediterranean, it is interesting to think how many divers have actually seen this breeding behaviour without realizing it. The reproduction of a Californian species *Chromis punctipinnis*, was described by Turner and Ebert (1962). In this species no signal jump was described, the male had to force the female into the previously cleaned nest-site. During defence of the eggs a pronounced change in colour of the males occurred. If they were driven away from the nest, this colour pattern was lost and they reverted to the normal coloration. Myrberg, Brahy, and Emery (1967) describe the behaviour of *Chromis multilineatus* in the Caribbean. This species shows certain interesting points of similarity to and differences from the congeneric species which occur at such great distances from it. Only in the outer parts of the breeding area does there occur a signal jump which may thus serve to attract females from a greater distance. There was no evidence of preparation or cleaning of a nest-site which was right in the open. Changes of coloration were reported in both males and females. The increasing numbers of studies of Pomacentrid behaviour are particularly interesting in view of the comparisons that may be made between this entirely marine family and the closely related freshwater *Cichlidae* (Wickler 1967). It seems unlikely that many underwater studies will be undertaken of Cichlid behaviour owing to the dangers from Schistasomiasis in the tropical fresh water in which they occur.

A comparative study is published by Abel (1964) of two Bleniid species, *Blennius canevae*, and *Blennius inaequalis*. Both species live in holes in the shallow water of the Mediterranean, and were studied by marking individuals and the use of models and fish contained in glass tubes. The distance at which courtship of females by males commences in the two species differed considerably. Abel related this to the differing degree of sexual dimorphism between the two species.

The most detailed and valuable study to date, conducted to a large extent under water, is that of Fiedler (1964) on eight Mediterranean species of wrasse of the genus *Crenilabrus*. Ciné-photography was extensively used as a means of recording the sequences of behavioural acts in agonistic and courtship behaviour. Fiedler compares the value of ethological and morphological characters in the taxonomy of the eight species, and concludes that the behavioural are more valuable in view of the variability of such anatomical features as fin ray counts and scale counts.

4. *Inter-specific social behaviour.* Some of the most bizarre results of evolution occur in the inter-specific relationships involved in commensalism, symbiosis, parasitism, and mimicry. So wide is the range of associations that with the exception of the last no perfect definitions are agreed for the associations. The range extends from cases of animals of different species living in close proximity without any obvious advantage to either, to cases where there is either a high degree of mutual advantage, or where the advantage is very one-sided leading directly to parasitism. Nicol (1960) considers only the occurrence of algae or bacteria living in animal tissue as symbiosis and all other non-parasitic relationships as commensal. Davenport (1955) broadens symbiosis to include all cases that many others would consider merely commensal. In few of these cases have detailed behaviour studies been made and much work remains to be done along the lines suggested by Davenport (1966). Dales (1966) provides the latest review of symbiosis in marine organisms including fish.

Mimicry describes the result of convergent evolution of two unrelated species that closely resemble each other, when one or both gains a protective advantage. In the known terrestrial cases, mimicry is divided into two categories, Batesian and Müllerian, but both are based on the unpalatable nature of either the model, or both the model and the mimic. It is not possible to fit the cases of fish mimicry into either of these categories with complete satisfaction because the mimicry is not based on the unpalatable nature of either model or mimic.

Symbiotic or commensal associations are likely to be discovered without direct observation only if the animals remain in contact during capture with dredge or trawl, or, if separated, are fortuitously placed in the same aquarium. Alternatively, marked anatomical modification such as the dorsal sucker of the remora will suggest a specialized mode of life. Underwater observation has added a number of examples to the previously recorded cases, and all that is possible here is to draw attention to some of them.

A burrowing shrimp, *Alpheus djiboutensis*, and a Gobiid fish were observed in the Red Sea living in the same burrow excavated in sand, by Luther (1958). Related species of *Alpheus* were studied under water in greater detail by Magnus (1967), prior to a laboratory study. He showed that there was quite a degree of mutual benefit, the fish being provided with a burrow and the nearly blind shrimp being warned of the approach of a potentially dangerous object by feeling the fish's movements. Another invertebrate group commonly associated with fish is the Echinodermata. Eibl-Eibesfeldt (1961) reported on an apparently advantageous association in which small flatfish in the Mediterranean followed large starfish moving over the sand, to feed on small invertebrates disturbed by the starfish. This is the same behaviour as that reported by Abel (1962) when red mullet, *Mullus* sp. rooting in the sand were followed by other fish. This behaviour has been observed by the author, who as a result of simple experiments in which the sand was disturbed, concluded that the orientation component was towards the slight sediment cloud that resulted. Fricke (1966c) noticed in the Red Sea a nocturnal association between a species of *Apogon* and a basket star, a filter-feeding ophiuroid. No suggestion was given of the possible reasons for this behaviour.

Almost certainly the best known symbiosis between fish and invertebrates is the habit of many Pomacentrid fish to live in and near a sea anemone. The first detailed descriptions based on observation partly in the field and partly in large aquarium tanks is given by Verwey (1930), who describes the basic elements of the behaviour. The fish are apparently immune to the nematocysts of the anemones and consequently are protected by them. They also to some extent feed and clean the anemones. Experimental studies by Davenport and Norris (1958) showed that the attraction to the anemones was visual and that a process of gradual acclimation reduced its stinging. There is an interesting problem in how this behaviour has evolved. Being stung would normally be regarded as a negatively reinforcing stimulus with the effect of conditioning the fish not to approach. Here the reverse appears to be the case, with habituation to a self-inflicted unpleasant stimulus which wanes with successive trials. It may be that the waning of the stinging effect forms the rewarding stimulus. It seems an interesting field for closer behavioural analysis.

Eibl-Eibesfeldt (1960) describes the defence of territory around the sea anemone and compares the agonistic behaviour of *Amphiprion percula* and *Amphiprion akallopisus* with that of the closely related Cichlidae. Fischelson (1964) gives an account of the symbiotic association in one of the Red Sea species, *A. bicinctus*. By moving the anemone, Fischelson was able to show that the 'territory' was transferred with it up to 20 m. distance. Graefe (1964) working in the same area as Fischelson, moved *Amphiprion bicinctus* from one spot to another. He showed that juvenile fish which are much more common inshore hold sub-territories within those of adult fish. He also studied the fish's immunity from stinging by the anemone, and concluded that one species *Antheopsis koseirensis* becomes conditioned to the degree and extent of stimulation from the fluttering movements of the fish. This is considered to be an important component of the protection received by the fish. Fricke (1966) made some limited experiments with painted models which released the aggressive behaviour of *Amphiprion*. Finally it should be pointed out that Abel (1960) has reported on the occurrence of a facultative association between a Gobiid, *Gobius bucchichii*, and the snakelocks anemone, *Anemonia sulcata*, in the Mediterranean. This is the only example known to the author of an association that does not involve the Pomacentridae. With this lead one should be encouraged to look for this type of symbiosis in temperate as well as tropical waters.

Turning to associations between fish, these are well documented by Eibl-Eibesfeldt (1955). In order to indicate the range that exists, one cannot do better than quote from the summary to that paper: 'A fish can play various roles in the life of another species, not only that of prey, predator, or competitor, but also that of a symbiont removing parasites, of a true parasite, of a protector, a means of transport, a scrubbing tool and a camouflage.' Attention will be restricted here to one subject in which underwater observation has played a most important part, namely cleaning symbiosis. This has been recently reviewed by Feder (1966b). Cleaning symbiosis was first observed by William Beebe but the name usually associated with this topic is that of Conrad Limbaugh whose death in 1960 was a great loss to underwater biological investigation. It is valuable that his work including that on cleaning behaviour (Limbaugh 1961a, Feder 1966) and on the life histories of some species that he studied (Limbaugh 1961b) has been published posthumously. Cleaning behaviour involves two different species of fish one of which picks external parasites from the body surface and sometimes also the buccal and opercular cavities of the other. This behaviour often occurs at special cleaning reefs, which although usually on the bottom may apparently occur also in the open ocean under drifting objects (Gooding and Magnuson 1967). Eibl-Eibesfeldt (1955) describes a number of cleaners most of which are strikingly coloured with areas of yellow (see p. 130), which he considers to be a guild sign. One of the most specialized cleaners in the Caribbean, *Elecatinus oceanops*, is not coloured yellow, but is blue with black stripes. Eibl-Eibesfeldt (1955) also describes the parasitic behaviour of *Runula albolinea* which takes bites out of passing fish. Randall (1955) described the behaviour of *Labroides dimidiatus* (Labridae) which is a highly specialized blue and black striped cleaner in the Indo-Pacific region, and commented on the similarity between this species and the parasitic Bleniid *Aspidontus taeniatus*, a close relative of *Runula* and with very similar behaviour. Randall actually suggested that mimicry might be involved. This was confirmed by Eibl-Eibesfeldt (1959) who observed cleaning by *Labroides dimidiatus* in the Indian Ocean, and noticed some rather unusual behaviour. The customer fish waiting to be cleaned would occasionally swim rapidly away at the approach of a cleaner. On catching one of these 'cleaners' it was discovered that it was not a cleaner, but *Aspidontus taeniatus* not only mimicking the genuine Labrid cleaner in coloration and morphology, but also swimming in a wrasse-like fashion with rigidly held body using the pectoral fins for propulsion. Wickler (1960, 1961) investigated the behaviour of these parasitic Blenniids in aquaria.

Potts (1968) suggests that the similarity between the species that he investigated, *Crenilabrus melanocercus* and the Californian *Oxyjulis californica* supports Eibl-Eibesfeldt's 'guild mark' theory. It would not appear from Feder's (1966) review that the theory of a universal guild mark is tenable unless yellow coloration, longitudinal stripes, and dark tail spots are all to be included. However, there is still the remarkable situation of two cleaners in different oceans resembling each other very closely, and one of them being mimicked by a third unrelated fish.

Fricke (1966a) made some very simple observations on the behaviour of cleaner and customer, using models. He concluded that although *Labroides* carries out a special 'cleaner dance' this is not essential for the release of the reciprocal response of the customer fish.

Mimicry has been introduced in relation to cleaning behaviour, and in an extensive review Randall and Randall (1960) discuss the *Labroides/Aspidontus* case at length, and quote another case of mimicry of a cleaner. This fish appears merely to be protected from predation by sharing the coloration of the real cleaner. Two examples are given in which there is no obvious biological advantage in the resemblance, but the authors point out that these might be cases of convergent evolution.

In conclusion, it seems highly unlikely that all cases of commensalism, symbiosis, and mimicry have now been discovered. One of the interesting gaps is the apparent restriction of these highly evolved patterns of social behaviour to tropical waters. External parasites are by no means restricted to them, and it seems unusual that there are as yet no cases of for example Gadoids being cleaned.

Techniques and Potential of Underwater Observation in Fish Behaviour Studies

The basic essential for visual observation under water is some means of enabling the eye to focus in a medium of high refractive index. The simplest way of doing this is with a glass-bottomed bucket or box as used by Verwey (1930). The floating raft with an observation chamber (Gooding 1965, Gooding and Magnuson 1967), and observation chambers incorporated into ships (Strasburg and Yuen 1960), represent a logical extrapolation by providing a bucket or box big enough to take the observer. The schooling study of Kühlmann and Karst (1967) and the detailed work of Abel (1962) fall into the same category in that they also relied upon the free atmosphere. In both these cases, observations were made largely without the use of breathing apparatus but just mask, fins, and snorkel, with the obvious advantage of cheapness in the field.

Extension of underwater observation from the surface to greater depths proceeds in two ways, either from the surface-swimmer to the aqualung diver, or from the surface observation chamber to sub-surface observation chambers, submersibles and underwater houses. These two groups of techniques have somewhat opposed advantages and disadvantages; diving techniques are cheap and easy, but are of very limited duration especially in depths greater than 20 m. Observation chambers, submersibles, and underwater houses are relatively very expensive but have the advantage that prolonged observation is possible.

Riedl (1967) gives a comprehensive review and appraisal of diving techniques in biological and geological studies. He points out that diving is but one approach, and sometimes one phase of a study, and that in many cases a remote-operating device may take a diver's place. It seems, though, that for such an instrument package to be correctly designed, it will usually be necessary to use a flexible human observer first to decide what to measure and with how much accuracy.

It is unfortunately the case that no really detailed behavioural studies have yet been undertaken from the more sophisticated underwater houses and submersibles. Kiselev's (1960) was the first and in some ways the most important work because of the fact that it was undertaken in the arctic. It provided knowledge on the behaviour of cold-water species of fish about which very little is known and which are commercially significant. Strasburg, Jones, and Iverson's (1968) observations from only fifty dives with the General Dynamics submersible *Asherah* are indicative of the information that waits to be collected. Small submersibles can be handled sufficiently easily to observe the behaviour of quite small pelagic fish as described by Barham (1966), and this has interesting possibilities for the investigation of deep-sea fish and scattering layers in the ocean. Clarke *et al.* (1967) made a number of biological observations from *Sealab II*, but the attraction of fish to the structure and its artificial lights means that the behaviour was not entirely natural. MacInnis (1966) reviews underwater dwellings, almost any one of which could be used in behavioural research were it not for the expense. However Heath (1967) describes a 'cheap' underwater house for biological research.

No mention has been made of tools and techniques of observation, largely because these have to date been so simple as to be hardly worthy of mention. However, it is often the case that the simplest methods are the best and this certainly seems to apply to underwater behaviour study, the observer's memory being supplemented with scribbling pad and pencil, still and ciné-cameras. Very little use has been made of tape-recorders for verbal descriptions despite the waste of time, and the necessity to look away when making a written record (but see Chapter 1, p. 19). The same applies to portable event recorders which could be very useful in those fields of research in which behavioural acts could be recognized. On the borderline between behaviour and ecology, there is a need for the use of pH, temperature, oxygen, and salinity electrode-probes with a hand-held read-out which could indicate the occurrence of some of the secondary orientations that at present must be deduced from other behaviour of the fish.

Conclusions

If the work discussed in this chapter is considered, one main fact emerges: the quality of the work done is inversely proportional to the depth (more accurately, perhaps, to the log of the depth). The following may be some of the reasons for this:

(1) Behavioural biologists are not good divers, or are not interested in diving.
(2) The short time available per dive does not allow detailed work.
(3) Research money is not available for behavioural work requiring deep submergence.
(4) The attitude of mind is surface-orientated and deep-water problems have not been thought of at all.

The real explanation is possibly a combination of all these. Certainly there is no evidence that biologists make any better or worse divers than other scientists, but they are sometimes inclined to regard underwater research as irrelevant, perhaps because marine biology has always been done quite successfully from over the side of a research ship. Even the biologists who do dive are somewhat surface-orientated which shows itself in their reluctance to use proper recording tools. Underwater research is still regarded by them as being 'different' and not always requiring the same degree of rigour in approach. It is pleasant to be able to quote one exception to this, the ecological work of Riedl (1966), which is in a class by itself.

The question of financial backing is connected with the other fact to emerge from this chapter, the restriction of much of the detailed work to tropical and warm temperate areas. There is an obvious reason for this, but it is worth remembering that it is in the cold waters of the world where fisheries are most important that the least is known of the behaviour of the fish. Simple diving techniques are not going to answer questions about the behaviour of cod, haddock or hake in 100 m. depth in the way they might about coral reef fish in 10 m. Therefore large amounts of money must be involved in a deep-water behaviour research programme if one is to use the acknowledged flexibility that direct underwater observation provides in the difficult environment of high latitude seas.

References

Abel, E. (1954), 'Lichtrückenreflex eines Fisches in der Blauen Grotte.' *Österr. Zool. Zeitschr.*, **4,** 397–401.

Abel, E. (1960), 'Liaison facultative d'un Poisson (*Gobius bucchichii* Steindachner) et d'une Anemone (*Anemonia sulcata* Pennant).' *Vie et Milieu*, **11,** 517–31.

Abel, E. (1961), 'Freiwasserstudien über das Fortpflanzungsverhalten des Mönchfisches *Chromis chromis* einen Vertreter der Pomacentriden im Mittelmeer.' *Z. Tierpsychol.*, **18,** 441–9.

Abel, E. (1962), 'Freiwasserbeobachtungen an Fischen im Golf von Neapel als Beitrag zur Kenntnis ihrer Ökologie und ihres Verhaltens.' *Int. Rev. Ges. Hydrobiol. Hydrogr.*, **47,** 219–90.

Abel, E. (1964), 'Freiwasserstudien zur Fortpflanzungsethologie zweier Mittelmeerfische, *Blennius canevae* Vinc. und *Blennius inaequalis* C.V.' *Z. Tierpsychol.*, **21,** 205–22.

Aronoson, L. R. (1951), 'Orientation and jumping behaviour in the Gobiid *Bathygobius soporator*.' *Am. Mus. Novit.*, **1486.**

Aronoson, L. R. (1957), 'Reproductive and parental behaviour.' In *The Physiology of Fishes*. Edited by M. E. Brown. New York: Academic Press.

Baerends, G. P. (1957), 'The ethological analysis of fish behavior.' In *The Physiology of Fishes*. Edited by M. E. Brown.

Baerends, G. P. and J. M. Baerends van Roon (1950), 'An introduction to the study of the ethology of cichlid fishes.' *Behaviour*, Suppl. 1.

Bardach, J. E. (1958), 'On the movements of certain reef fishes.' *Ecology*, **39,** 139–46.

Bardach, J. E., H. E. Winn, and D. W. Menzel (1959), 'The role of the senses in the feeding of nocturnal reef predators *Gymnothorax* spp.' *Copeia*, 1959 (2), 133–9.

Barham, E. G. (1966), 'An unusual flatfish observed and photographed from a diving saucer.' *Copeia*, 1966 (4), 865–7.

Boulenger, E. G. (1929), 'Observations on the nocturnal behaviour of certain inhabitants of the society's aquarium.' *Proc. Zool. Soc. London.*, Pt. I, 359–62.

Brandt, J. F. and H. Hollien (1967), 'Underwater hearing thresholds in man.' *J. Acoust. Soc. Amer.*, **42,** 966–71.

Breder, C. M., Jr., and D. E. Rosen (1966), *Modes of Reproduction in Fishes*. Published for The American Museum of Natural History by Natural History Press, New York.

Brock, V. E. and R. H. Riffenburgh (1960), 'Fish schooling: a possible factor in reducing predation.' *J. Cons. Perm. Int. Explor. Mer.*, **25,** 307–17.

Bull, H. O. (1957), 'Conditioned responses.' In *The Physiology of Fishes*. Edited by M. E. Brown.

Carlisle, D. B. (1961), 'Intertidal territory in fish.' *Anim. Behav.*, **9,** 106–7.

Chance, M. R. A. and W. M. S. Russell (1959), 'Protean displays: a form of allaesthetic behaviour.' *Proc. Zool. Soc. Lond.*, **132,** 65–70.

Clarke, G. L. and E. J. Denton (1962), 'Light and animal life.' In *The Sea*, **1.** Edited by M. N. Hill. New York and London: John Wiley & Sons.

Clarke, T. A., A. O. Flechsig, and R. W. Grigg (1967), 'Ecological studies during project Sealab II.' *Science*, **157** (3795), 1381–9.

Cullen, J. M., E. Shaw, and H. A. Baldwin (1965), 'Methods for measuring the three-dimensional structure of fish schools.' *Anim. Behav.*, **13**, 534–43.

Dales, R. P. (1966), 'Symbiosis in marine organisms.' *Symbiosis* **1.** Edited by S. Mark Henry. New York and London: Academic Press.

Damant, G. C. C. (1921), 'Illumination of plankton.' *Nature*, **108**, 42–3.

Davenport, D. (1955), 'Specificity and behavior in symbioses.' *Q. Rev. Biol.*, **30**, 29–46.

Davenport, D. (1966), 'The experimental analysis of behavior in symbioses.' In *Symbiosis*, **1**. Edited by S. Mark Henry.

Davenport, D. and K. S. Norris (1958), 'Observations on the symbiosis of the sea-anemone *Stoichactis* and the Pomacentrid fish *Amphiprion percula*.' *Biol. Bull. Woods Hole*, **115**, 397–410.

Dijkgraaf, S. (1933), 'Untersuchungen über die Funktion der Seitenorgane an Fische.' *Z. vergl. Physiol.*, **20**, 162–214.

Driver, P. M. and D. A. Humphries (1966), 'Protean behaviour: systematic unpredictability in interspecific encounters.' *Mental Health Res. Inst. Rep. No.*, **197**, Univ. Michigan.

Duntley, S. Q. (1962), 'Underwater visibility.' In *The Sea*, **1**. Edited by M. N. Hill.

Eibl-Eibesfeldt, I. (1955), 'Über Symbiosen Parasitismus und andere besondere zwischenartliche Beziehungen tropischer Meeresfische.' *Z. Tierpsychol.*, **12**, 203–19.

Eibl-Eibesfeldt, I. (1959), 'Der Fisch *Aspidontus taeniatus* als Nachahmer des Putzers *Labroides dimidiatus*.' *Z. Tierpsychol.*, **16**, 19–25.

Eibl-Eibesfeldt, I. (1960), 'Beobachtungen und Versuche an Anemonenfischen *Amphiprion* der Maldiven und der Nicobaren.' *Z. Tierpsychol.*, **17**, 1–10.

Eibl-Eibesfeldt, I. (1961), 'Eine Symbiose zwischen Fischen *Siphamia versicolor* und Seeigeln.' *Z. Tierpsychol.*, **18**, 56–9.

Eibl-Eibesfeldt, I. (1962), 'Freiwasserbeobachtungen zur Deutung des Schwarmverhaltens verschiedener Fische.' *Z. Tierpsychol.*, **19**, 165–82.

Enger, P. S. and R. Anderson (1967), 'An electrophysiological field study of hearing in fish.' *Comp. Biochem. Physiol.*, **22**, 517–25.

Feder, H. M. (1966), 'Cleaning symbiosis in the marine environment.' In *Symbiosis*, **1.** Edited by S. Mark Henry.

Fiedler, K. (1964), 'Verhaltenstudien an Lippfischen der Gattung *Crenilabrus*, (Labridae, Perciformes).' *Z. Tierpsychol.*, **21**, 521–91.

Fischelson, L. (1965), 'Observations and experiments on the Red Sea anemones and their symbiotic fish *Amphiprion bicinctus*.' *Sea Fish. Res. Stat. Haifa*, Bull. No. **39**, 3–16.

Fraenkel, G. S. and D. L. Gunn (1940), *The Orientation of Animals*. Oxford University Press.

Fricke, H-W. (1966a), 'Zum Verhalten der Putzerfisches *Labroides dimidiatus*.' *Z. Tierpsychol.*, **23**, 1–3.

Fricke, H-W. (1966b), 'Attrapenversuche mit einigen plakartfarbigen Korallfischen im Roten Meer.' *Z. Tierpsychol.*, **23**, 4–7.

Fricke, H-W. (1966c), 'Partnerschaft zwischen Kardinalfischen (*Apogon* sp.) und dem Gorgonenhaupt (*Astroboa nuda* Lyman) im Roten Meer.' *Z. Tierpsychol.*, **23,** 267–9.

Galea, J. A. (1961), 'The "Kannizatti fishery".' *Proc. Gen. Fish. Counc. Medit.*, **6,** 85–91.

Gooding, R. M. (1965), 'A raft for direct submarine observation at sea.' *Spec. sci. Rep. U.S. Fish. Wildl. Serv.-Fish.*, No. **517.**

Gooding, R. M. and J. J. Magnuson (1967), 'Ecological significance of a drifting object to pelagic fishes.' *Pacif. Sci.*, **21,** 486–97.

Graefe, G. (1964), 'Zur Anemonen-Fisch Symbiose, nach Freilanduntersuchungen bei Eilat/Rotes Meer.' *Z. Tierpsychol.*, **21,** 468–85.

Harden-Jones, F. R. (1960), 'Reactions of fish to stimuli.' *Proc. Indo-Pacif. Fish. Counc. 8th Session (III)*, 18–28.

Harden-Jones, F. R. (1968), *Fish Migration.* London: Edward Arnold (Publishers) Ltd.

Hasler, A. D. and J. R. Villemonte (1953), 'Observations on the daily movements of fishes.' *Science*, **118,** 321–2.

Heath, J. R. (1967), 'Operation of a cheap underwater house.' *Helgoländer wiss. Meeresunters*, **15,** 399–411.

Hemmings, C. C. (1966a), 'Olfaction and vision in fish schooling.' *J. Exp. Biol.*, **45,** 449–64.

Hemmings, C. C. (1966b), 'The mechanism of orientation of roach, *Rutilus rutilus* L. in an odour gradient.' *J. Exp. Biol.*, **45,** 465–74.

Hemmings, C. C. (1966c), 'Factors influencing the visibility of objects under water.' In *Light as an Ecological Factor*. Edited by R. Bainbridge, G. C. Evans, and O. Rackham. *Brit. Ecol. Soc. Symp. No. 6.* Oxford: Blackwell.

Hemmings, C. C. and J. N. Lythgoe (1966), 'The visibility of underwater objects.' (In *Malta '65*), *Underwater Assn. Rep.*, **1,** 23–9.

Hislop, J. R. G. and C. C. Hemmings (1967), 'A preliminary assessment of the survival of haddock (*Melanogrammus aeglefinus* L.) tagged under water.' *I.C.E.S. C.M. Demersal Fish Comm. F:28.*

Hoar, W. S. (1958), 'Rapid learning of a constant course by travelling schools of juvenile Pacific salmon.' *J. Fish. Res. Bd. Can.*, **15,** 251–74.

Humphries, D. A. and P. M. Driver (1967), 'Erratic display as a device against predators.' *Science*, **156,** 1767–8.

Hunter, J. R. (1966), 'Procedure for analysis of schooling behaviour.' *J. Fish. Res. Bd. Can.*, **23,** 547–62.

Hunter, J. R. and C. T. Mitchell (1968), 'Field experiments on the attraction of pelagic fish to floating objects.' *J. Cons. Perm. Int. Explor. Mer.*, **31,** 427–34.

Keenleyside, M. H. A. (1955), 'Some aspects of the schooling of fish.' *Behaviour*, **8,** 183–248.

Keenleyside, M. H. A. (1962), 'Skindiving observations of Atlantic Salmon and Brook trout in the Miramichi River, New Brunswick.' *J. Fish. Res. Bd. Can.*, **19,** 625–34.

Kiselev, O. N. (1960), 'Observations on fish behaviour by means of underwater techniques in the Barents Sea.' *Sov. Fish. Invest. N. European Seas Mosc. 1960* (Marine Lab. Trnsln No. 658).

Kühlmann, D. H. H. and H. Karst (1967), 'Freiwasserbeobachtungen zum Verhalten von Tobiasfischschwärmen (*Ammodytidae*) in der westlichen Ostsee.' *Z. Tierpsychol.*, **24,** 282–97.

Limbaugh, C. (1961a), 'Cleaning symbiosis.' *Scient. Am.*, August 1961. (Reprint No. 135.)

Limbaugh, C. (1961b), 'Life history and ecologic notes on the black croaker.' *Calif. Fish. Game*, **47,** 163–74.

Longley, W. H. and S. F. Hildebrand (1941), 'Systematic catalogue of the fishes of Tortugas, Florida.' *Pap. Tortugas Lab.*, **34,** 1–331. Carnegie Inst. Wash. Publ. 535.

Lorenz, K. (1962), 'The function of colour in coral reef fishes.' *Proc. Roy. Inst. G.B.*, **39,** 282–96.

Lorenz, K. (1966), *On Aggression*. London: Methuen.

Luther, W. (1958), 'Symbiose von Fischen (*Gobiidae*) mit einem Krebs (*Alpheus djiboutensis*), im Roten Meer.' *Z. Tierpsychol.*, **15,** 175–7.

Lythgoe, J. N. (1966), 'Visual pigments and underwater vision.' In *Light as an Ecological Factor*. Edited by R. Bainbridge, G. C. Evans, and O. Rackham. Brit. Ecol. Soc. Symp. No. 6. Oxford: Blackwell.

MacInnis, J. B. (1966), 'Living under the sea.' *Scient. Am.*, March 1966. (Reprint No. 1036.)

Magnus, D. B. E. (1967), 'Zur Ökologie sedimentbewöhnender Alpheus-Garneln (Decapoda-Natantia) des Roten Meer.' *Helgoländer wiss. Meeresunters*, **15,** 506–22.

Mansuetti, R. (1963), 'Symbiotic behaviour between small fishes and jellyfishes with new data on that between *Peprilus alepidotus* and the scyphomedusa *Chrysaora quinquecirrha*.' *Copeia*, 1963 (1), 40–80.

Morrow, J. E. (1948), 'Schooling behaviour in fishes.' *Q. Rev. Biol.*, **23,** 27–38.

Myrberg, A. A., B. D. Brahy, and A. R. Emery (1967), 'Field observations on reproduction of the damselfish, *Chromis multilineata* (Pomacentridae) with additional notes on general behaviour.' *Copeia*, 1967 (4), 819–27.

Nicol, J. A. C. (1960), *The Biology of Marine Animals*. London: Pitman.

Nicol, J. A. C. (1963), 'Some aspects of photoreception and vision in fishes.' In *Advances in Marine Biology*, **1.** Edited by F. S. Russell. London and New York: Academic Press.

Potts, G. W. (1968), 'The ethology of *Crenilabrus melanocercus*, with notes on cleaning symbiosis.' *J. Mar. Biol. Ass. U.K.*, **48,** 279–93.

Radakov, D. V. (1960), 'Observations on herring during a voyage of research submarine *Severyanka*.' *Byull. Okeanogr. Kom. Akad. Nauk. S.S.S.R.*, **6,** 39–40. (Marine Labty. Transln. No. 719.)

Radakov, D. V. and B. S. Solovyev (1959), 'A first attempt at using a submarine for observations on the behaviour of herring.' *Ryb. Khoz.*, **35,** 16–21. (J. Fish. Res. Bd. Can. Transln. No. 338.)

Randall, J. E. (1955), 'Fishes of the Gilbert Islands.' *Atoll Res. Bull.*, **47,** i–xi, 1–243.

Randall, J. E. (1965), 'Grazing effect on sea grasses by herbivorous reef fishes in the West Indies.' *Ecology*, **46,** 255–60.

Randall, J. E. and H. A. Randall (1960), 'Examples of mimicry and protective resemblance in tropical marine fishes.' *Bull. Mar. Sci. Gulf Carrib.*, **10,** 444–80.

Riedl, R. (1966), *Biologie der Meereshöhlen.* Hamburg and Berlin: Paul Parey.

Riedl, R. (1967), 'Die Tauchmethode, ihre Aufgaben und Leistungen bei der Erforschung des Littorals: eine kritische Untersuchung.' *Helgoländer wiss. Meeresunters,* **15,** 294–352.

Shaw, E. (1967), *Some New Thoughts on the Schooling of Fishes.* F.A.O. Conference on fish behaviour in relation to fishing techniques and tactics, Bergen, October 1967.

Soemarto (1960), 'Fish behaviour with special reference to pelagic shoaling species: Lajang (*Decapterus* spp.).' *Indo-Pacif. Fish. Counc. Proc.*, **8,** 89–93.

Starck, W. A. II and W. P. Davis (1966), 'Night habits of fishes of alligator reef, Florida.' *Ichthyologica, The Aquarium Journal,* **38,** 313–56.

Strasburg, D. W. and H. S. H. Yuen (1960), 'Progress in observing tuna underwater at sea.' *J. Cons. Perm. Int. Explor. Mer.*, **26,** 80–93.

Strasburg, D. W., E. C. Jones, and R. T. B. Iversen (1968), 'Use of a small submarine for biological and oceanographic research.' *J. Cons. Perm. Int. Explor. Mer.*, **31,** 410–26.

Tinbergen, N. (1950), 'The hierarchical organization of nervous mechanisms underlying instinctive behaviour.' In *Physiological Mechanisms in Animal Behaviour.* Soc. exp. Biol. Symp. No. 4. Cambridge University Press.

Turner, C. H. and E. E. Ebert (1962), 'The nesting of *Chromis punctipinnis* (Cooper) and a description of their eggs and larvae.' *Calif. Fish. Game,* **48,** 243–8.

Ulrich, B. (1951), 'Unterwasserbeobachtungen von Heringsschwärmen.' *Fischereiwelt* 3rd year, 164–5.

Van Iersel, J. J. A. and J. van dem Assem (1964), 'Aspects of orientation in the diggerwasp *Bembix rostrata.*' In *Learning and Associated Phenomena in Invertebrates.* Edited by W. H. Thorpe and D. Davenport. Anim. Behav. Suppl. 1, pp. 145–62.

Verwey, J. (1930), 'Coral reef studies: I. The symbiosis between damselfishes and sea anemones in Batavia Bay.' *Truebia,* **12,** 305–53.

Von Holst, E. (1935), 'Über den Lichtruckenreflex bei Fischen.' *Pubbl. Staz. zool. Napoli,* **15,** 143–58.

Weber, E. (1961), 'Über Ruhelagen von Fischen.' *Z. Tierpsychol.*, **18,** 517–33.

Weber, E. (1965), 'Eine fakultative Fressgeneinschaft von Fischen und Stachelhäutern.' *Z. Tierpsychol.*, **22,** 567–9.

Wickler, W. (1960), 'Aquarienbeobachtungen an *Aspidontus* (*Bleniidae*) einen ektoparasitischen Fisch.' *Z. Tierpsychol.*, **17,** 275–92.

Wickler, W. (1961), 'Uber das Verhalten der Bleniiden *Runula* und *Aspidontus* (Bleniidae).' *Z. Tierpsychol.*, **18,** 421–40.

Wickler, W. (1967), 'Vergleich der Ablaichverhaltens einiger paarbildender sowie nicht-paarbildender Pomacentriden und Cichliden (Pisces; Perciformes).' *Z. Tierpsychol.*, **24,** 457–70.

Wodinsky, J. and W. N. Tavolga (1964), 'Sound detection in teleost fishes.' In *Marine Bio-acoustics.* Edited by W. N. Tavolga. Oxford: Pergamon.

Woodhead, P. M. J. (1966), 'The behaviour of fish in relation to light in the sea.' In *Oceanography and Marine Biology: an Annual Review*, **4.** Edited by H. Barnes. London: Allen & Unwin.

Wynne-Edwards, V. C. (1962), *Animal Dispersion in Relation to Social Behaviour*. Edinburgh: Oliver & Boyd.

Zumpe, D. (1965), 'Laboratory observations on the aggressive behaviour of some butterfly fishes (*Chaetodontidae*).' *Z. Tierpsychol.*, **22,** 226–36.

6 Botany

E. A. Drew

Introduction

Three per cent of the ocean water mass and only 1 per cent of the sea floor are illuminated brightly enough for plants to grow. Nevertheless, plants in this euphotic zone are the primary producers of organic matter on which all other marine organisms in both this and the vast aphotic zone depend for energy sources and body-building materials.

Only in recent years have scientists entered this realm as divers and begun to apply modern scientific methods to elucidate the problems associated with growth in a submarine habitat. In this chapter the methods and results of botanical investigations will be described, investigations confined to the benthic marine plants. The microscopic phytoplankton organisms which grow in the entire water mass of the euphotic zone are more readily investigated from surface vessels.

The Plants Involved

The most common submarine plants are the algae, members of a vast assemblage of basically primitive plants which evolved in the sea and gave rise to the ancestral forms of modern land plants. There are a number of major subdivisions of the algae, but here we shall be concerned only with the Chlorophyta (green), Phaeophyta (brown), and Rhodophyta (red). These groups have macroscopic representatives which are ecologically important. The colour classification of the algae is dependent on the photosynthetic pigments which they contain; the ecological significance of the different coloured pigments will be discussed later.

Representatives of other types of plants also occur in the sea. In various parts of the world truly marine angiosperms are to be found, often forming extensive meadows in shallow waters. Sculthorpe (1967) lists twelve genera belonging to five different families of Monocotyledons. There are, however, no representatives of the other types of photosynthetic plants—the Bryophyta, Pteridophyta, or Gymnospermae.

Marine bacteria and fungi appear to be ubiquitous in the oceans, the former being found at all depths including the deepest trenches where bathyphilic species abound in the bottom mud. These organisms are of great importance in the nutrient cycles of the sea, and some are undoubtedly parasitic on the algae and marine angiosperms which will be considered in this chapter. However, there has so far been little detailed investigation of them, although clearly the use of diving techniques for accurate sterile sampling of infected material would be advantageous.

The Environment for Submarine Plants

Environmental factors

There are several environmental factors which must be considered in relation to the establishment and growth of submarine plant communities. Exposure to wave or swell action near the surface, the nature and slope of the substrate, water temperature, the availability of dissolved nutrients and grazing by herbivorous animals are all of considerable importance. It is often necessary to determine these factors accurately within a given plant community, and observation or measurement by divers is then indispensable.

However, no other single environmental factor approaches in importance the variation in intensity and spectral composition of light under the sea, since light is essential for plant growth. Submarine illumination varies between the extremes of intense tropical insolation near the surface of coral reefs to absolute darkness below the euphotic zone. The depth of the euphotic zone depends largely on the turbidity of the water mass, so that another factor—suspended matter of natural origin or introduced as man-made pollution—must be considered in this context. Most of the present section will be devoted to a consideration of the penetration of solar energy to different depths of the sea.

Light energy in the sea

Submarine illumination is discussed elsewhere in this book in relation to vision and similar phenomena (Chapter 4). However, in order to obtain a clear understanding of its effect on plant growth, it is essential to consider light in terms of absolute energy. Furthermore, it is incorrect and frequently misleading to quote values for submarine light energy in foot candles or any other units of luminous flux; these apply to illumination with the spectral composition of the standard candle flame and are defined in terms of the spectral sensitivity of the human eye. Submarine light must be measured in absolute energy units such as gram calories/cm.^{2}/minute (= langleys/min.) or watts/cm.^{2}, taking into account the change of spectral composition with depth. This subject is fully discussed by Strickland (1958).

In the open sea, light intensity declines approximately as a logarithmic function of depth. Data for a number of stations throughout the world are reported by Jerlov (1951), whose paper forms a useful reference in this context. Spectral transmittance and reduction of total light intensity with depth for several types of sea-water, from clear oceanic to very turbid, are shown in Figs. 66 and 67. To summarize, it is apparent that clear waters transmit maximally at the blue end of the visible spectrum, whilst increasing turbidity displaces this maximum towards the red end. The extinction rates for total energy also increase rapidly with increased turbidity so that light penetrates very much deeper in clear than in turbid water.

The data quoted above were obtained in the open sea, with hundreds of metres of water beneath the deepest points of measurement. The situation for attached marine plants is very different; they are on the sea bottom and often on steep sloping or vertical faces. The effect of these factors is not completely known, although clearly the proximity of a bright sandy bottom will increase the intensity of diffuse space lighting, whilst a vertical rock face, possibly cutting out as much as half the space lighting and maybe the major light vector itself, may cause considerable reduction in intensity. What is probably a combination of these 'bottom' and 'cliff' effects is illustrated in Fig. 68. Deviation from a logarithmic reduction in intensity with depth down a vertical submarine cliff occurs first at the base of the cliff and is accentuated 15 m. out from it, where it approximates more closely to comparable open sea data.

The cliff habitat, even when it faces towards the sun, is essentially a shade one. Extremes of shading occur in submarine caves, and a theoretical assessment of the reduction of light intensity inside a cave is given by Larkum *et al.* (1967). They conclude that light intensity and spectral quality inside corresponds to that at much greater depths in unshaded habitats and they recommend the usefulness of such caves in the study of shade communities found otherwise only beyond the depth range of the aqualung diver. Similar considerations undoubtedly hold for crevices and fissures but their usefulness is restricted by the small area of the communities present.

Light measurements taken inside a cave very similar to that described by Larkum *et al.* are shown in Fig. 68, plotted on to a line extrapolated from the data for the open cliff. It is clear that the light regime at a depth of only 22·5 metres, and a similar distance inside the cave, is similar to that outside at about five times that depth.

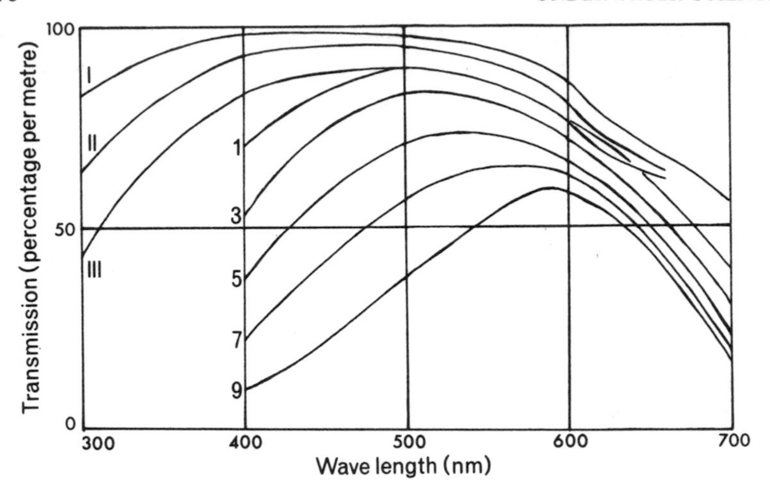

Fig. 66 Spectral transmittance of various waters. (After Jerlov 1951.)

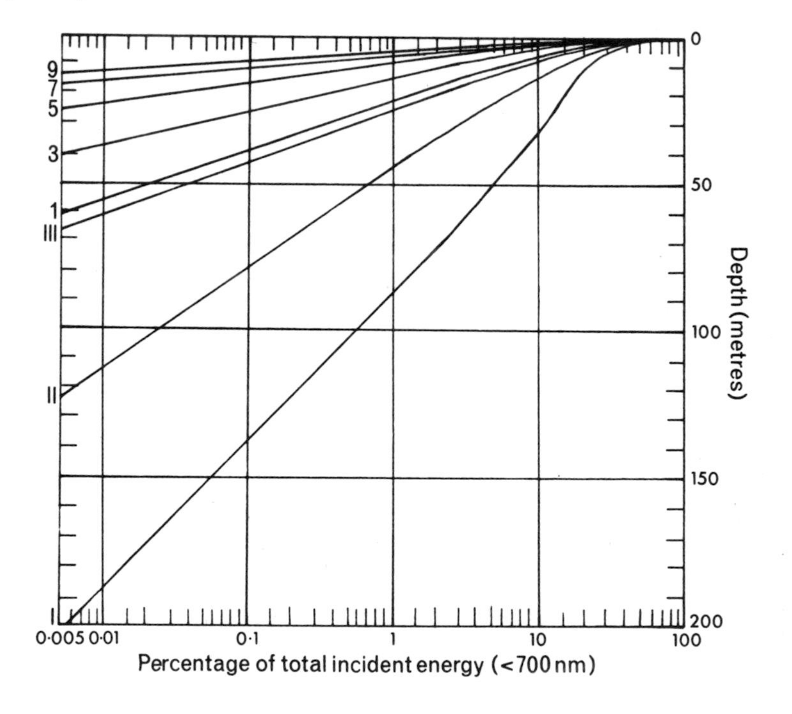

Fig. 67 Penetration of light energy into various waters. (After Jerlov 1951.)

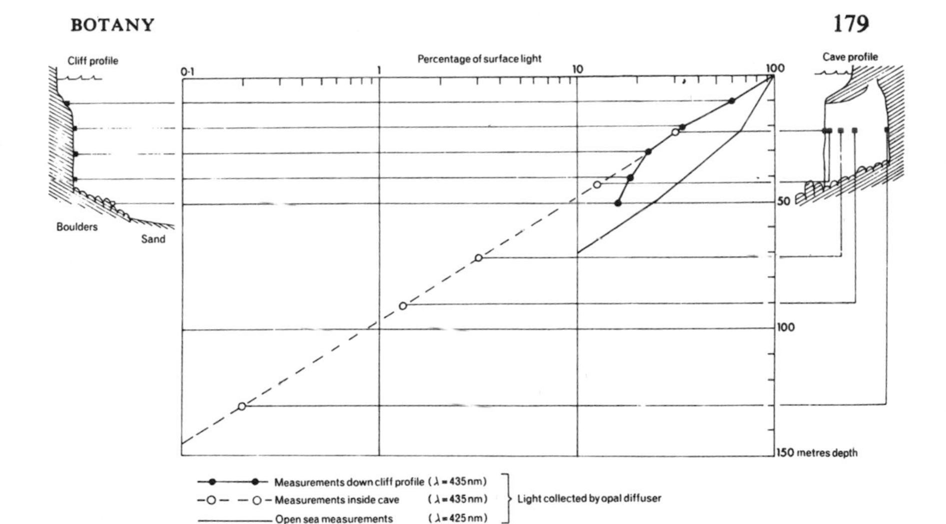

Fig. 68
Effect of shading by vertical cliff and cave on submarine light (both facing south).

Measuring light energy

In view of the factors discussed above, it is apparent that an accurate estimate of the illumination at or within a given plant community can only be obtained by actual measurement *in situ*. This requires moderately small, self-contained equipment which can be positioned and read by a diver. It is important to understand certain principles before embarking on underwater light measurement of this kind.

The use of photographic light meters, incorporating selenium photocells, is unsatisfactory unless restricted spectral regions, isolated with coloured filters, are measured. This is because a sensing system which responds differently to different wavelengths is being used to measure light which is becoming increasingly monochromatic with depth. However, the introduction of coloured filters in front of the meter greatly reduces its sensitivity; if an opal cosine collector is added to make the system absorb light more as the plant surface does, then it is of use only in shallow, brightly lit waters.

A more suitable instrument for use to greater depths is illustrated in Fig. 69. It incorporates a selenium photocell of increased surface area, thereby ensuring increased current output per unit of illumination, a system for easily changing the coloured filters, and a sensitive ammeter, the range of which can be altered by current attenuation circuitry. Such

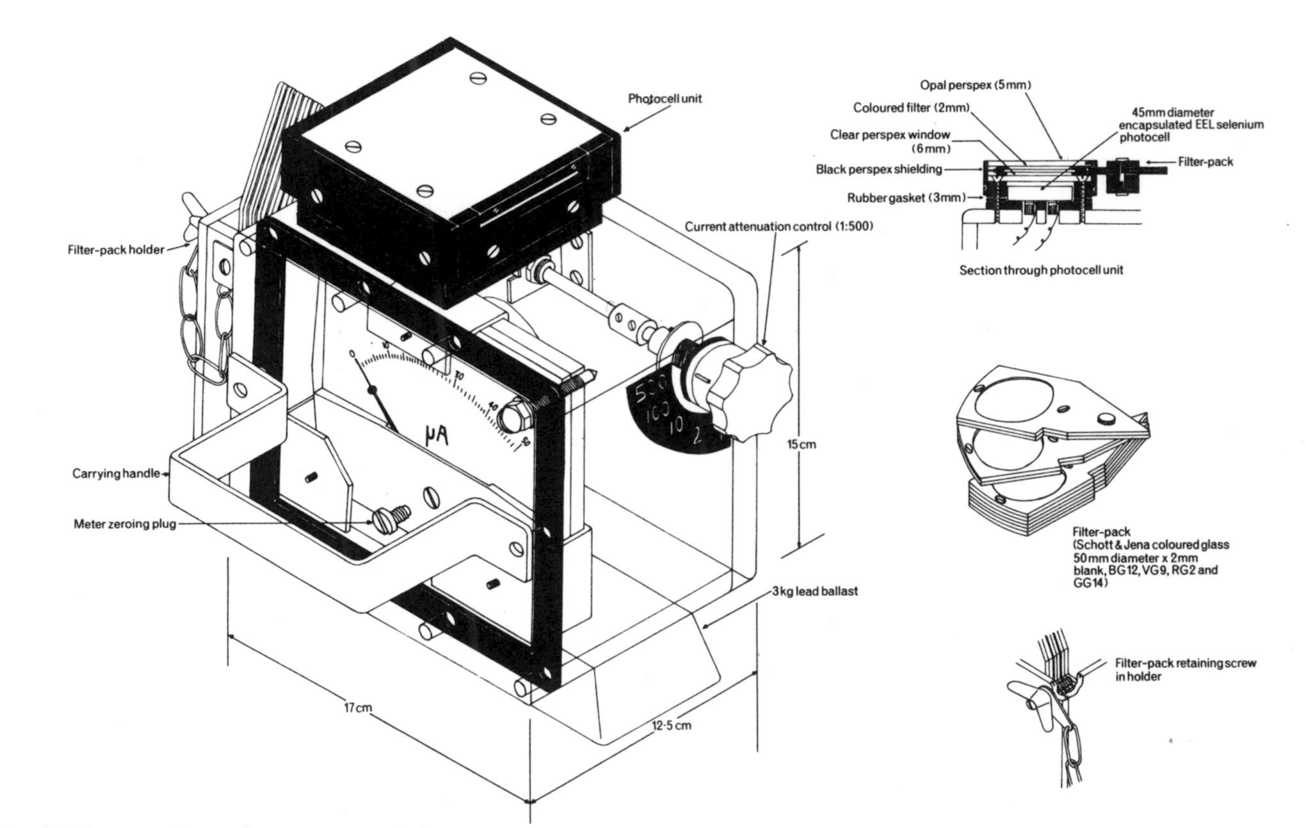

Fig. 69 Direct reading underwater spectral photometer.

an instrument was used for the light measurements shown in Fig. 68. It generates 2,450 μA when illuminated by 0·6 langleys/minute unfiltered visible solar radiation, being sensitive between 250 and 750 nm.

The first investigator to measure light *in situ* in submarine plant communities was Kitching (1941), using a selenium photocell, coloured filters, and an opal diffuser. Readings were taken at the surface but the instrument was positioned by a diver. Submarine botanists have seldom used such good equipment since. It is certainly time that modern instrumentation was used in this field, but without the complex, power-consuming electronics associated with the photomultiplier and interference filter system described by Boden *et al.* (1960).

A compromise between simplicity and sensitivity may be found in an integrating light meter, the readings of which represent the summation of light energy over a prolonged period; instantaneous output from the detector system need then only be small for a final reading of considerable magnitude. Indeed, the total light energy flux per day, month, or year is often of greater biological significance than a few supposedly representative instantaneous readings, but since sea conditions frequently prevent measurement at critical times, such as when illumination is at its lowest and ecologically most critical levels in

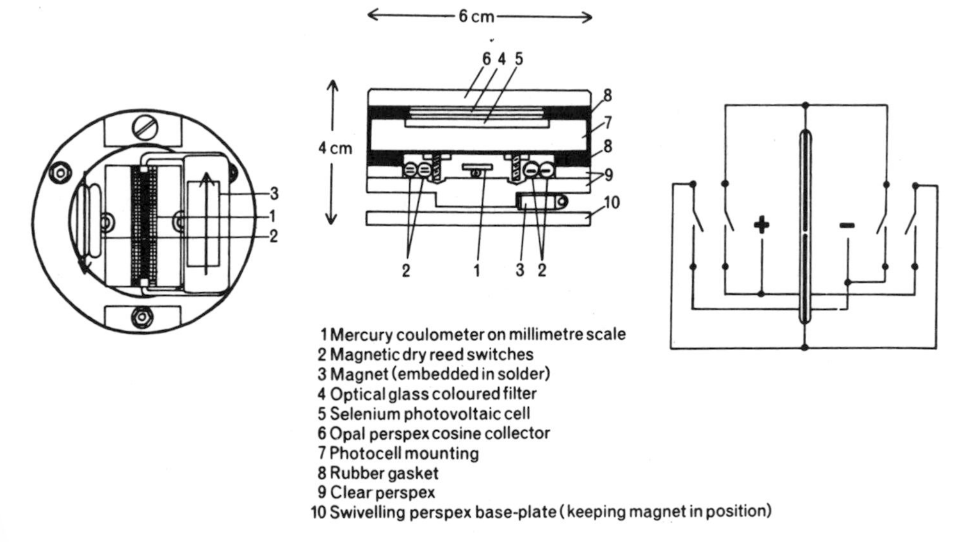

Fig. 70 Miniature integrating photometer.

winter, estimates of long-term energy flux must at present be calculated in a rather approximate manner.

Integrating light-monitoring equipment, even for use on land, is at present in an early stage of development, and most available units are complex, bulky, and unsuitable for underwater use. However, an instrument recently constructed by the author achieves the same object in a simple, inexpensive manner. This miniature integrating photometer is illustrated in Fig. 70, and will shortly be described in a detailed publication. The output from a selenium photovoltaic cell is connected to a mercury coulometer, a device in which the passage of current causes transfer of mercury across an electrolyte gap in a narrow mercury column; resultant movement of the gap is directly proportional to current flow. Using suitable coloured glass filters to isolate various spectral bands, detectable readings can be obtained, even at the maximum depths of plant growth, in the sea during exposures of only a day or so.

Subdivision of the benthic environment

The shoreline was first divided by Kjellman (1877) into the following divisions:

epilittoral — above high water spring tides
littoral — between high and low water springs
sublittoral — below low water springs
elittoral — below 36 metres depth.

Since then there have been numerous attempts to define universal subdivisions of the sublittoral, frequently based on ambient light intensity. A number of schemes, mainly applicable to the Mediterranean, have been summarized by Molinier (1960) and one particular scheme, given in detail by Pérès (1967), is discussed later in this chapter.

However, it seems unprofitable to attempt to define rigid subdivisions below the lowest tide levels, for below that the environmental factors often change but slowly. The choice of a depth of 36 m. by Kjellman as an important underwater dividing line is an example of the use of data from specific localities, obtained in this case only by dredging, as a basis for universal schemes.

In this chapter we consider the sublittoral. In any particular area there will be a unique series of sublittoral zones, each inhabited by a well-defined plant community, until the one universal sublittoral dividing line, the end of the euphotic zone, is reached.

Analysis of Submarine Plant Communities

Early techniques

The methods developed at the end of the last century and in the first quarter of this century for studying sublittoral algae by working from the surface were of necessity qualitative rather than quantitative. Thus, Darbishire's 'phycoscope' allowed direct observation of the sea bottom in calm shallow waters and was an elaboration of the glass-bottomed box still used by algologists (e.g. MacFarlane 1967). The hand dredge devised by Reinke was extensively used in that period to obtain specimens from moderately deep water without recourse to more massive ship-borne equipment, and this too still finds its uses in sublittoral surveys. However, it was not until the advent of the Petersen spring grab that quantitative collection from the sea bed became possible, although recent observations of this apparatus in operation indicate that it collects only some 30 to 65 per cent of the algae actually present (Baardseth 1954). Nevertheless, present knowledge of the kelp resources of N.W. Europe is derived mainly from studies using this implement.

In view of the non-quantitative aspects of the remote sampling procedures described above, it is not surprising that algal ecologists were amongst the first submarine scientists, and the classic study of a Scottish kelp forest by Kitching (1941) must surely be the pioneer study of this kind. Although their work was basically qualitative, using the simple diving helmet to enable them to observe and collect submarine vegetation to depths of about 12 m., the advantages of sorties into the submarine world were obvious. Since the invention of the aqualung, many of the strictly quantitative methods of the terrestrial plant ecologist have been adapted for use under water.

Transect and quadrat

Most investigators have used transect or simple quadrat methods. A description of the change in vegetation with depth or other factors can readily be obtained by subjective assessment of percentage cover along a strip transect, at the same time noting other relevant features. Modifications of the transect method for use in turbid and turbulent waters are described by Bailey *et al.* (1967) in which representative samples are collected for analysis on the surface.

More detailed descriptions of the actual plant associations present can be obtained using the phytosociological methods of the Zürich-Montpelier school of ecologists, noting the abundance and sociability

TABLE 1. *Phytosociological scales*

Abundance/Dominance	*Sociability*
+ Sparse/very sparse, little cover	
1. Plentiful but little cover	1. Growing singly
2. Very numerous or covering $> \frac{1}{20}$ of area	2. Grouped or tufted
3. Any number of individuals covering $\frac{1}{4}$ to $\frac{1}{2}$ area	3. Small patches, cushions
4. Any number of individuals covering $\frac{1}{2}$ to $\frac{3}{4}$ area	4. Small colonies, extensive patches, carpets
5. Covering more than $\frac{3}{4}$ area	5. Great crowds or pure stands

Suffixes

< reduced vitality
E epiphytic
SE upper storey
SS lower storey

of the various species within a quadrat according to a recognized system of code numbers such as those suggested by Braun-Blanquet and Pavillard (1922), with suffixes for special features. These are set out in Table 1, and well illustrated in Fig. 71, taken from Pérès and Picard (1964).

Although this is an extremely useful method for the differentiation and classification of plant communities, as well as for the analysis of changes in floristic composition over extended periods, it gives little indication of the quantities of the various plants present. Accurate quantitative data on standing crop—vital for estimates of natural resources, etc.—can best be obtained by collecting the entire vegetation from a quadrat and then sorting this into component species in the laboratory. These can then be weighed, wet or preferably dry, or other criteria determined. It is important to make allowance for the weight of calcifying material present in samples containing a considerable proportion of calcareous species, since this does not represent organic plant matter. Determination of ash-free dry weight (Bellamy and Whittick 1968b) or calorific content (Larkum *et al.* 1967) offer suitable methods for such corrections.

A suitable sampling implement, used by Larkum *et al.* (1967) for the smaller forms of submarine vegetation, is shown in Fig. 72. The plants, and sedentary animals, enclosed within the 10 cm. square were scraped

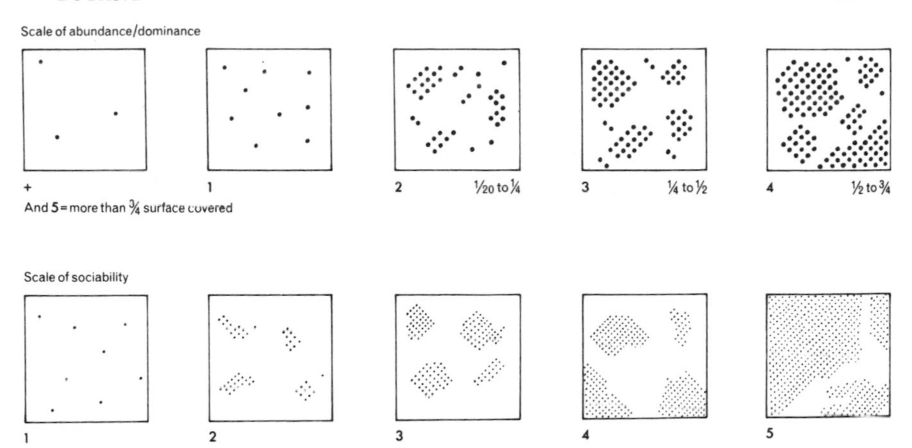

Fig. 71 Phytosociological criteria. (From Pérès and Picard 1964.)

off into the hessian bag with a chisel inserted through the slot; the bags were readily interchanged under water, and the necks closed, so that a number of quadrats could be collected on a single dive. Similar implements but with a cutting edge incorporated into the frame have been described (Kain 1960, Wilkinson *et al.* 1967), but these can be neither strictly quantitative nor effective on irregular rock surfaces.

Quadrat size

In all ecological work of this nature it is important to consider carefully the size of quadrat appropriate to the vegetation under investigation. It is obviously not practical to attempt to sample the total vegetation of a kelp forest with a 10 cm. quadrat; other important factors which are involved are adequately covered in textbooks of plant ecology (Greig-Smith 1964, Kershaw 1964, Oosting 1956). Pérès and Picard (1964) consider that a suitable quadrat size for most submarine vegetation is between $\frac{1}{4}$ and 1 square metre in area. A useful preliminary test for selection of a quadrat suitable for representative sampling of a given community is to establish the 'minimal area' which includes most

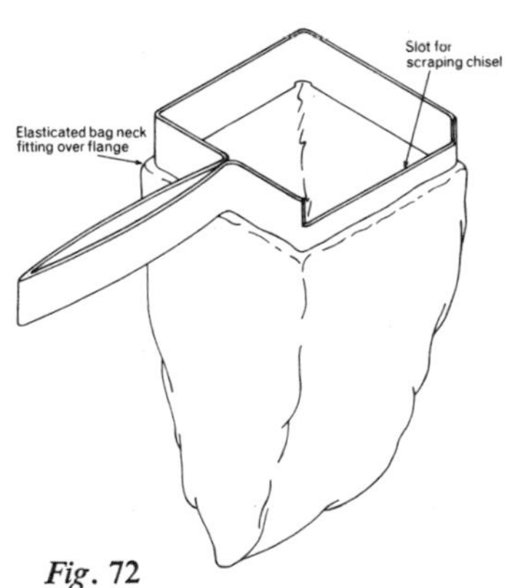

Fig. 72
Brass scraper frame and hessian bag.

of the species present. This is found by taking a number of quadrats and plotting number of species against area sampled. Such a plot for a submarine community is shown in Fig. 73. A suitable point to select as the minimal area is that at which a 100 per cent increase in area yields only 10 or perhaps 5 per cent more species.

It is advisable to use an area somewhat larger than the minimal for actual sampling. Larkum *et al.* (1967) showed that samples over 1 m. square were necessary to establish significant differences ($p < 0{\cdot}05$) between the standing crops of dominant species in similar vegetation at 30 and 45 m. depth. The minimal area of this vegetation was only 400 cm.² This difference was, however, partly attributed to the non-randomness of the quadrats taken (25 samples of area 400 cm.² within a square metre at each depth) so that the results were affected by pattern within the vegetation with an intrinsic area 500 to 1,000 cm.²

Although considered essential for statistical considerations, the value of completely random quadrats in ecology is limited, especially when distinct gradations of vegetation type occur over moderately small distances. Thus, if a number of quadrat samples are to be taken in each zone of a vertical submarine zonation, these should be arranged horizontally. Larkum *et al.* (1967) used a system of quadrat selection

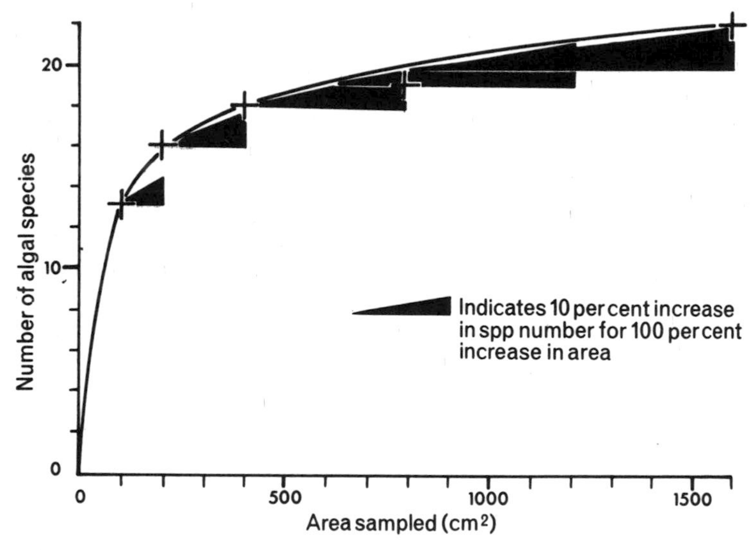

Fig. 73 Species number/quadrat area plot to determine minimal area. (*Udotea/Halimeda/Peyssonnelia* community, Malta.)

in which the positioning of the first quadrat at a specific depth, which was likely to be partially subjective, defined the position of the others which were one quadrat width to the side; in this way the four 100 cm.^{2} sub-samples which made up each sample for analysis were collected with virtually no overall subjective selection. This method can also be compared with the use of rectangular quadrats which several authors have found more efficient than square ones. Furthermore, it has been reported that taking alternate strips within a rectangular quadrat does not impair the reliability of the measurements whilst reducing the work by half.

Aids to observation and sampling

The optimal use of sampling time under water is vital in view of the severe restriction imposed by the use of the aqualung, especially if sampling at depths of 30 m. or more. For studies of low-growing stands of plants, close-up photographs of individual quadrats are rapidly executed and provide useful additional records since plant orientation and small-scale distribution is destroyed by removal of the sample. Inclusion of a calibrated three-dimensional structure in each photograph adds greatly to the information obtainable from such photographs, allowing accurate plans to be drawn from oblique, randomly oriented shots (Johnston *et al.* 1969, Morrison 1970). However, an attempt to correlate observations from photographs with actual standing crop data for the same quadrats, or even simply the number of individuals of dominant species, proved impossible (Drew and Larkum, unpublished). Even under the most advantageous conditions, photography can therefore only supplement direct sampling.

Great economy in the use of diving time can be made by using a miniature tape-recording system to allow the diver to make more copious notes than would be possible with pencil and writing board, while leaving both hands free for photography, etc. The use of the tape-cassette type of recorder greatly facilitates storage and retrieval of the data.

Recovery of sacks full of plant material can be difficult and the effort involved may reduce the productivity of a dive. The problem can be solved either by increasing diver buoyancy at the end of the dive by use of controllable compressed air life-jackets (e.g. 'Fenzy'), or by partially filling the sacks with air from the aqualung and allowing them to ascend independently for collection by a boat party. Attachment to a rope for hauling to the surface is also feasible under some conditions. It is easy to overlook such problems and valuable samples may be lost without adequate planning.

Submarine Plant Communities

Marine biologists did not venture under water until a less cumbersome diving system than the 'standard Navy diving dress' was devised. The zoologist William Beebe was the first biologist to carry out serious underwater observations, using a simple diving helmet supplied with air by small pumps on the surface. He describes in fascinating detail his underwater walks amongst coral reefs in the *Arcturus Adventure* (1926) and *Beneath Tropic Seas* (1928). Soon after this the first underwater botanical excursions were described by Kitching *et al.* (1934). Since the advent of the aqualung an extensive body of botanical data has been built up by diving investigations, data otherwise unobtainable. The methods of investigation, of data presentation and details of the communities found will now be discussed.

Geographical limitations

It is convenient to consider the investigation of submarine plant communities according to climatic zones. There are major differences in the type of vegetation present in polar, temperate, and tropical seas, and there are also limitations on the type and duration of underwater work which can be carried out in waters of various temperatures. At the same time water clarity, controlled mainly by the amount and periodicity of run-off from the land, differs according to these zones. Precipitation is considerable and more or less continuous in temperate regions and here dense human populations contribute much effluent to the already high natural turbidity. In the tropics many regions are arid except for brief rainy seasons; in the polar regions too, precipitation is low and usually bound up as ice.

The most extensive and complex underwater botanical investigations have naturally been carried out in warm clear tropical waters; here we must include the Mediterranean. The least work has been done in polar seas, and most of this of a qualitative nature. However, in temperate waters algal growth reaches a maximum luxuriance, and here the demands of the seaweed industry have led to more thorough investigation than would perhaps be expected in these cold, turbid seas. It is with this region that this survey begins.

Temperate Regions

Laminaria forests

Kitching and his co-workers followed up their first description of underwater excursions (1934) with their now classic study of *Laminaria* forests in the Sound of Jura (published 1941). Not only did they collect samples of the underwater vegetation and make direct ecological observations, they also measured ambient light intensity under water (see p. 181). By taking measurements both inside and outside the kelp canopy they demonstrated that up to 85 per cent of ambient light was cut off by the canopy so that only shade-loving species, found elsewhere only at considerably greater depths, grew within. In addition they studied the regeneration of *Laminaria* plants on an area cleared by cutting all the plants at the base of the stipe. Twelve months later they observed complete loss of the old holdfasts and a dense growth of new plants up to a metre high, half the length of the original individuals.

Forster (1954) carried out preliminary surveys of the vertical distribution of plants and animals off the south-west coast of England. Diagrammatic profiles illustrated in Fig. 74 show that the small brown algae *Dictyopteris* and *Halopteris* replace the *Laminaria hyperborea* forests in this region below 15–17 m., and penetrate to about 25 m., below which sedentary animals predominate. He also showed (1955) that *Laminaria* may only penetrate to 5–6 m. in more turbid waters with scattered small brown and red algae penetrating to about 8 m.

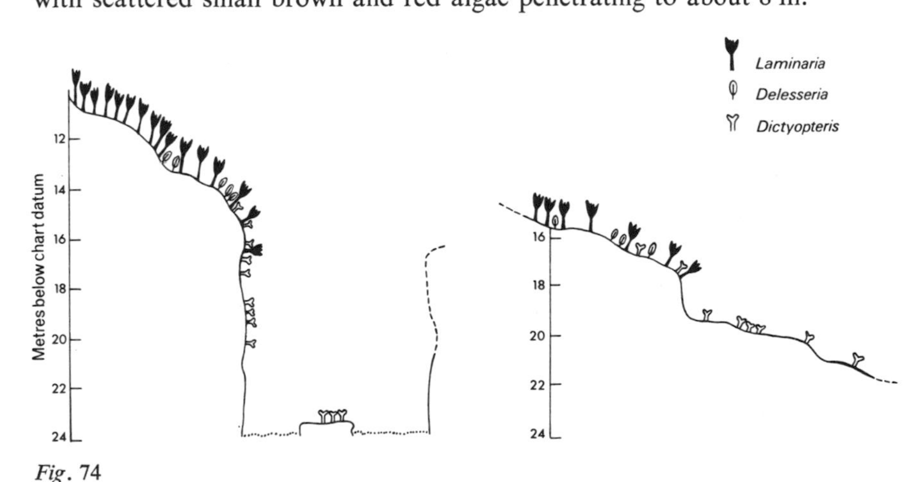

Fig. 74

Profiles of algal distribution at the lower limit of *Laminaria* forests in S.W. England. (From Forster 1954.)

The relationship between *Laminaria* growth, depth, nature of substrate, etc., have been investigated in the Isle of Man and Scotland by Kain (1960, 1962, 1963, 1964). Using transect procedures, profiles of vertical distribution and percentage cover such as those in Fig. 76 have been produced. Thus, in those areas, *Laminaria hyperborea* can grow as deep as 20 m. under favourable conditions. Apart from light intensity, Kain has shown that the nature and perhaps also slope of the substrate is important in controlling the maximum depth of growth as well as the species present. *L. digitata* predominates near the surface where its flexible stipes allow it to withstand wave action and to remain immersed even at very low tides. *L. hyperborea* forms dense forests below this until either light or moderately stable substrate runs out, whilst *L. saccharina* grows on small stones and gravel. A related plant, *Sacchariza polyschides*, is able to colonize very unstable substrates (gravel, etc.), possibly by virtue of a rather modified holdfast structure. However, the ecology of this plant is obscure and Kain has found that it can sometimes replace *L. hyperborea* even on rocky bottoms, except at the greatest depths.

Laminaria plants produce annual growth rings in the stipe (Kain 1963). The number of rings of rapid growth in a transverse section therefore indicates the age of the plant. The section, which should not be too thin if the narrow separating rings of slow winter growth are to be seen clearly, can be cut with a sharp diving knife close to the basal holdfast or hapteron (see Fig. 75) and examined in the field by holding up to a bright sky. If the section is too far up the stipe, primary tissue and even the first few years' growth rings may be lost. The oldest *Laminaria hyperborea* plants in a population are usually eight to nine years old, although individuals up to fourteen years have been reported.

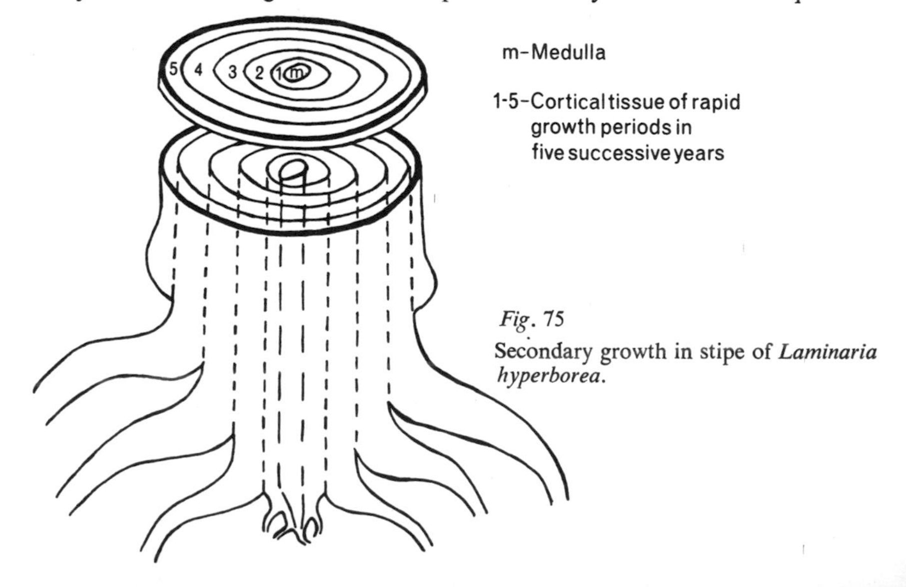

Fig. 75
Secondary growth in stipe of *Laminaria hyperborea*.

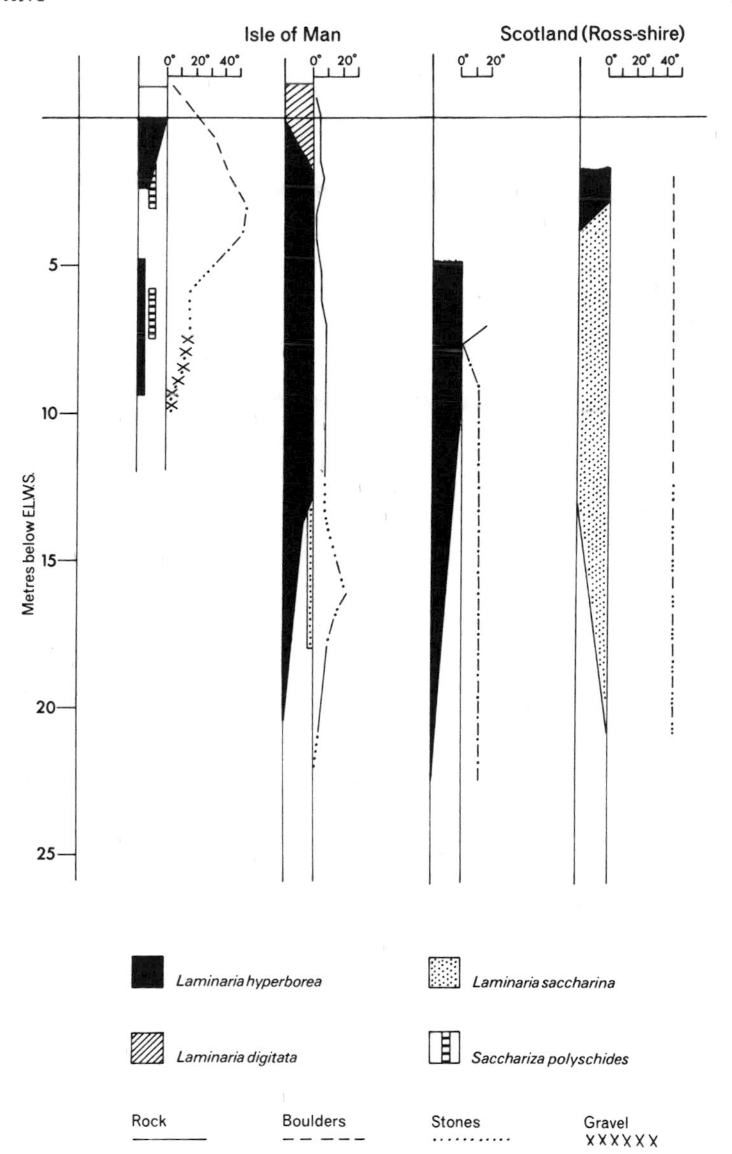

Fig. 76
Structure of some *Laminaria* forests in the British Isles.
(After Kain 1964.)

The growth rate of *Laminaria* plants can be readily determined by measuring various criteria in a number of plants of different ages, bearing in mind that the lamina is shed annually, in spring, and a new one grown. Data for plants of *Laminaria hyperborea* from Shetland in midsummer are set out in Fig. 77.

Bellamy and Whittick (in press) have compared in detail the *Laminaria hyperborea* forests in shallow (1–3 m.) and deep (10–12 m.) water at one site in south-east Scotland. They used divers to collect the entire

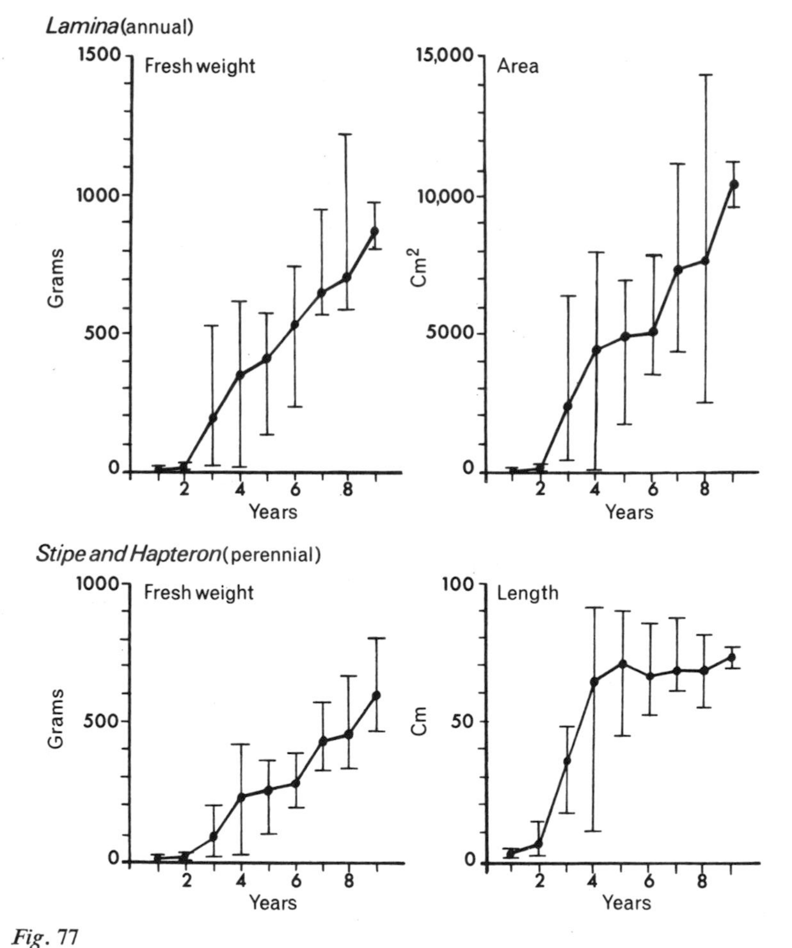

Fig. 77

Growth rates of *Laminaria hyperborea* in Shetland (vertical lines show range of valves recorded). (Drew, unpublished data.)

TABLE 2. *Cropping data and annual production of* Laminaria hyperborea *forests at two depths*
Values as grams ash-free dry weight per m.^{2} (From: Bellamy and Whittick 1968b)

	Age	Hapteron		Stipe		*Peak* Lamina *Biomass*		Total *Biomass*		Total *Increment*		*Mean No. of Plants*	Annual Production	
		Biomass	*Annual Increment*	*Biomass*	*Annual Increment*	*1966*	*1967*	*1966*	*1967*	*1966*	*1967*		*1966*	*1967*
1–3 metres depth	1	1 0·6	1	1 0·6	1	2 1·5	1 0·4	4	3	4	3	8	32	24
	2	3 2	2	3 2	2	11 4	8 4	17	14	15	12	4	60	48
	3	7 2	4	11 3	8	42 11	37 10	60	55	54	49	2	108	98
	4	13 3	6	25 6	14	65 9	49 8	103	87	85	69	4	340	276
	5	22 8	9	37 7	12	88 11	94 21	147	153	109	115	7	763	805
	6	25 9	3	45 6	8	82 31	85 17	152	155	93	96	6	558	576
	7	26 11	1	49 6	4	76 9	84 16	151	159	81	89	3	243	267
										Total			2,104	2,094
										+ Epiphytes, etc.			2,386	2,342
10–12 metres high	1	1 0·8	1	1 0·6	1	2 0·4	2 0·3	4	4	4	4	7	28	28
	2	3 2	2	4 2	3	12 4	9 3	19	16	17	14	4	68	56
	3	7 2	4	11 3	7	36 13	23 3	54	41	47	34	2	94	68
	4	12 3	5	21 5	10	36 9	37 7	69	70	51	52	2	102	104
	5	23 9	11	31 8	10	65 17	61 13	119	115	86	82	5	430	410
	6	25 9	2	37 10	6	61 13	71 13	123	133	69	79	2	138	158
	7	26 11	1	40 8	3	56 14	67 18	122	133	60	71	1	60	71
										Total			920	895
										+ Epiphytes, etc.			994	947

Annual increment = biomass − that of plants one year younger.
Annual production = total of hapteron and stipe increments and peak lamina biomass.

crop of macroscopic algae within 2-m. square quadrat frames at various times over a two-year period. When weather conditions were bad, time under water was reduced by taking only fifty representative plants from within each quadrat. Each quadrat was sited in a previously undisturbed stand of algae at the required depth. After analysis of the crops in the laboratory, where each plant was aged, the stipe measured, and then the lamina, stipe, and holdfast weighed separately after drying, growth curves and annual productivity were calculated. Values were converted to ash-free dry weight according to factors determined from ashed sub-samples. Data from which mean annual productivity was calculated are given in Table 2 together with the production values for plants of various ages and for the entire vegetation.

The productivity of *Laminaria* forests at various points along a 'pollution gradient' on the north-east coast of Britain has been estimated by the same authors (unpublished). They determined the annual productivity of a metre-wide strip of seabottom from 0 to 12 m. depth at three points, the length of each strip being about 120 m. The data shown in Table 3 indicate clearly that pollution has a very large effect on the depth of penetration and annual productivity of these algae. A considerable change in plant morphology was also noted, the plants from deeper polluted sites appearing etiolated; a similar but lesser degree of etiolation was also noted in the investigation quoted previously for *Laminaria hyperborea* growing at deeper sites in unpolluted waters.

TABLE 3. *Effect of pollution on depth penetration and productivity of* Laminaria hyperborea *forest* (Bellamy and Whittck, unpublished)

Site	*Degree of Pollution*	*Depth of Penetration*	*Production of Metre Wide Strip*	
			g. ash-free dry wt./annum	*% of max. measured*
Durham coast	Very polluted	0–3 metres	19·7	8·3
Mid-Northumberland coast	Some pollution	1–10 metres	101	43
St. Abbs Head	Unpolluted	1–>12 metres	236	100

Length of metre-wide strip from 0 to 12 metres below chart datum in all three cases approximately 120 metres (slope 1:10).

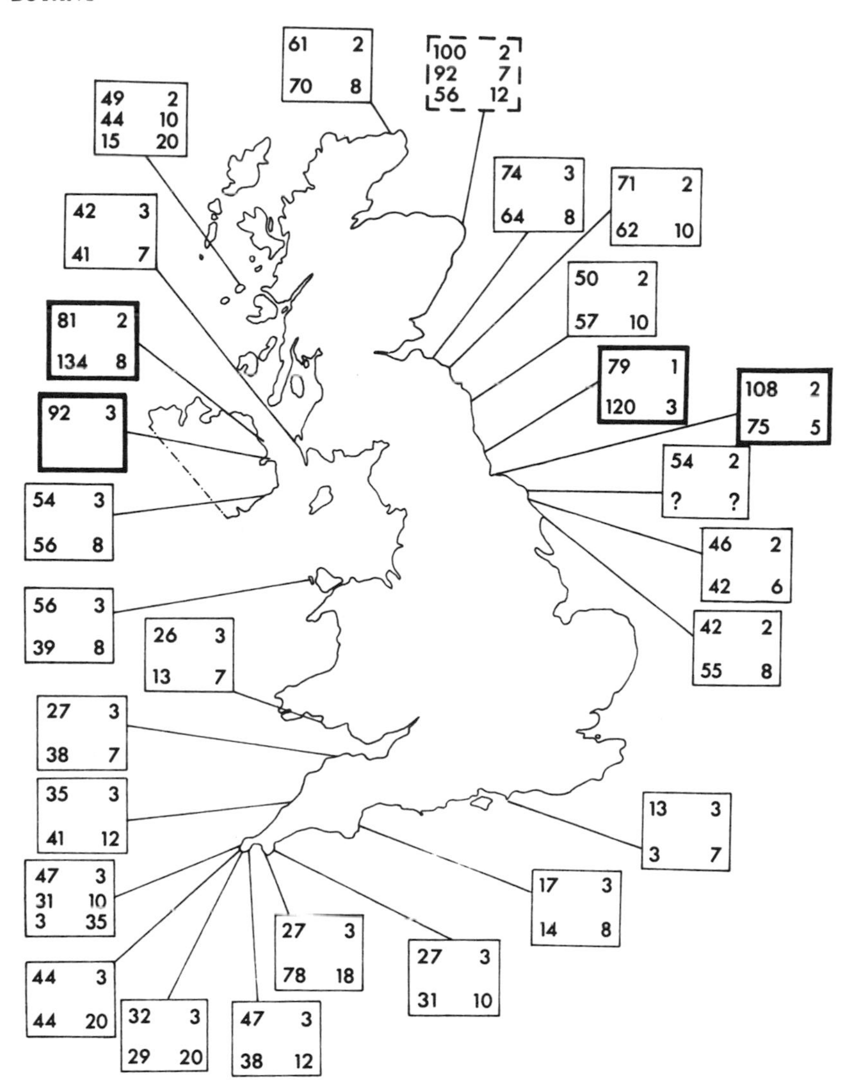

Fig. 78

Performance of *Laminaria hyperborea* at 27 stations around the British Isles. (From Bellamy and Whittick 1968a.)

Left-hand column = individual performance (stipe weight) as per cent best unpolluted site (dashed box)

Right-hand column = depth in metres

Polluted areas indicated by heavy lined boxes

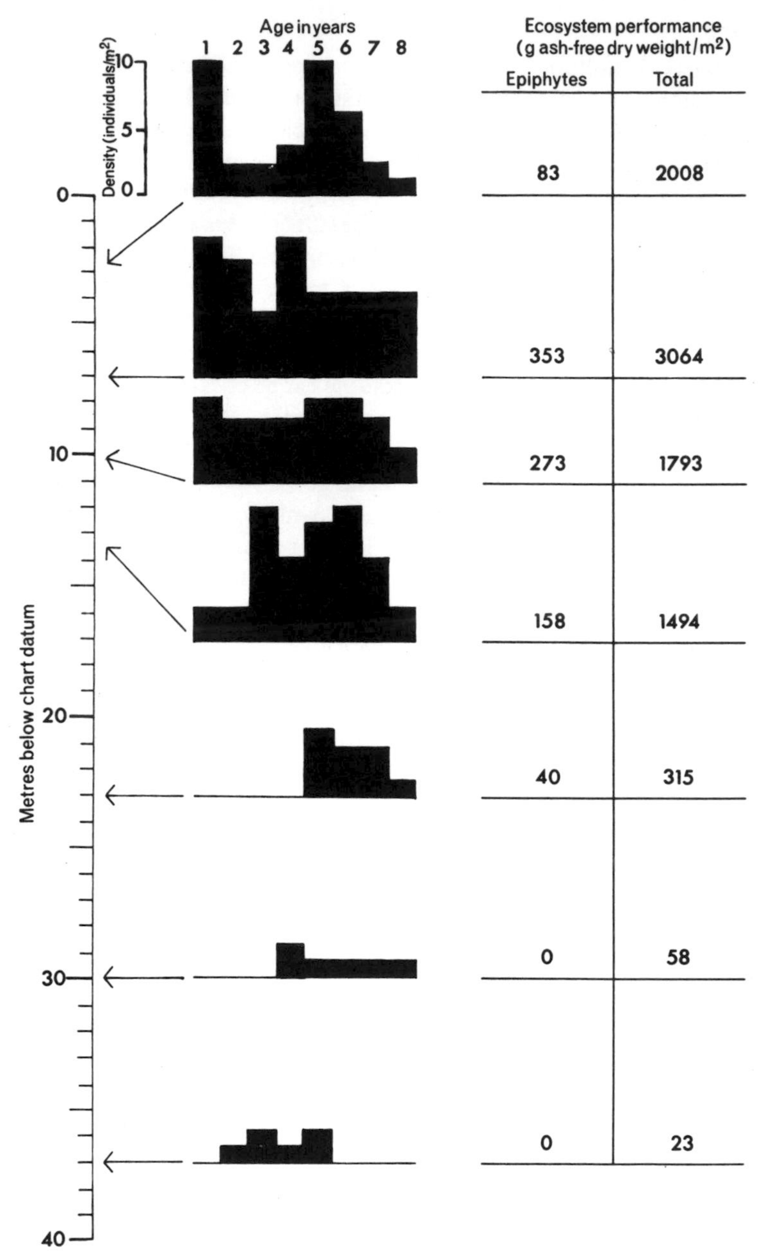

Fig. 79 Density, age structure, and standing crop of *Laminaria hyperborea* forest in S.W. England. (After Bellamy and Whittick, 1968b.)

The effects of pollution in these areas are probably due mainly to the single factor of increased water turbidity resulting in decreased light penetration. The effect of dissolved materials in the effluent has yet to be determined. However, comparison with regions of high natural turbidity at nearby sites should help clarify this point.

In order to understand more fully the effect of environmental factors on the growth of *Laminaria* forests, Bellamy and Whittick (1967, 1968a) organized 'Operation Kelp'. During a single diving season amateur divers collected, aged, measured, and weighed a total of 7,052 kelp plants; the organizers were then able to estimate performance of *Laminaria hyperborea* over a considerable range of conditions and of its actual geographic distribution. As an example, data for the best stipe performance at twenty-seven sites around Britain are shown in Fig. 78, the values being expressed as a percentage of the best unpolluted site. The trend is clearly a marked reduction from north to south, reversed only in definitely polluted areas. This operation demonstrates the potential usefulness of amateur diving organizations in submarine surveys.

Although individual plant performance has been shown to be best in northern latitudes, the number of plants per unit area is much greater farther south, resulting in greater over-all productivity. The deepest penetrating and most productive forests of *L. hyperborea* in the British Isles are undoubtedly to be found in the south-west where this plant can penetrate to a depth of 37 m. Data for age composition, plant density, and standing crop for such a forest have been given by Bellamy and Whittick (1968b) and are illustrated in Fig. 79.

Macrocystis beds

The giant kelp beds of the Pacific coast of the Americas are of considerable economic importance, much being harvested for the production of alginates for use in textiles, food, and other industries (Newton 1951). Aleem (1956) investigated *Macrocystis* beds using quadrat and line transect methods to determine their structure and the distribution of associated vegetation. He produced the diagrammatic representation shown in Fig. 80 together with estimates of the standing crop determined from samples brought up for analysis. These showed an average of 9·4 kg. fresh weight of algae per m.^{2} of sea bed, and of this some 8·0 kg. was *Macrocystis* itself. That author estimated an annual production of 10–15 metric tons/hectare per year wet weight of algae.

Since then extensive investigations of such beds have been made, involving both diving work and co-ordinated laboratory studies. Diving investigations have included further detailed description of the community, experimental transplantation, and growth measurements on tagged plants (i.e. North 1958, 1961, 1964, Neuschel 1959, Clendenning 1964). For transplanting *Macrocystis* plants it has been found necessary to wrap them in damp sacking to prevent rough handling, desiccation, or damage by bright surface light and high temperatures whilst they are out of the water.

In situ investigations of the rate of sporulation and range of spore dispersal in *Macrocystis* have recently been described by Anderson and North (1967). Sporulating sporophylls were enclosed in polythene bags whilst still attached to the plants in the sea and then the whole sporophyll removed whilst still inside the bag, the contents of which were killed and preserved with formalin. The concentration of spores in the enclosed water was then determined in the laboratory, and the results showed an average sporulation rate of $2{\cdot}5 \times 10^3$ spores/cm.^{2}/minute. This varied with water temperature, and hence with season, and the conditions for maximum rate agreed well with those appertaining at the season of maximum juvenile plant development.

To investigate the range of dispersal of spores single plants were transplanted to areas without *Macrocystis* colonization, and the appearance of juvenile plants in the vicinity noted after a period of several months. Similar experiments using a group of transplanted individuals

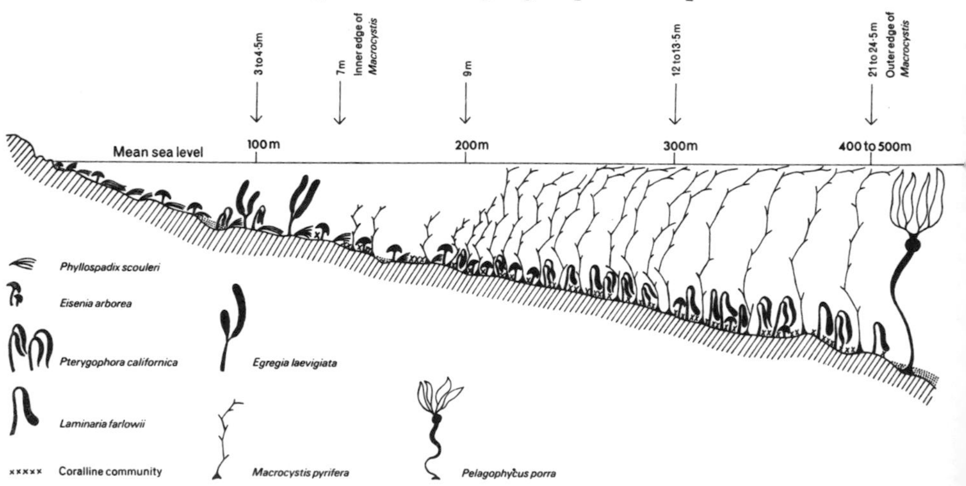

Fig. 80 Cross-section of Californian kelp bed. (From Aleem 1956.)

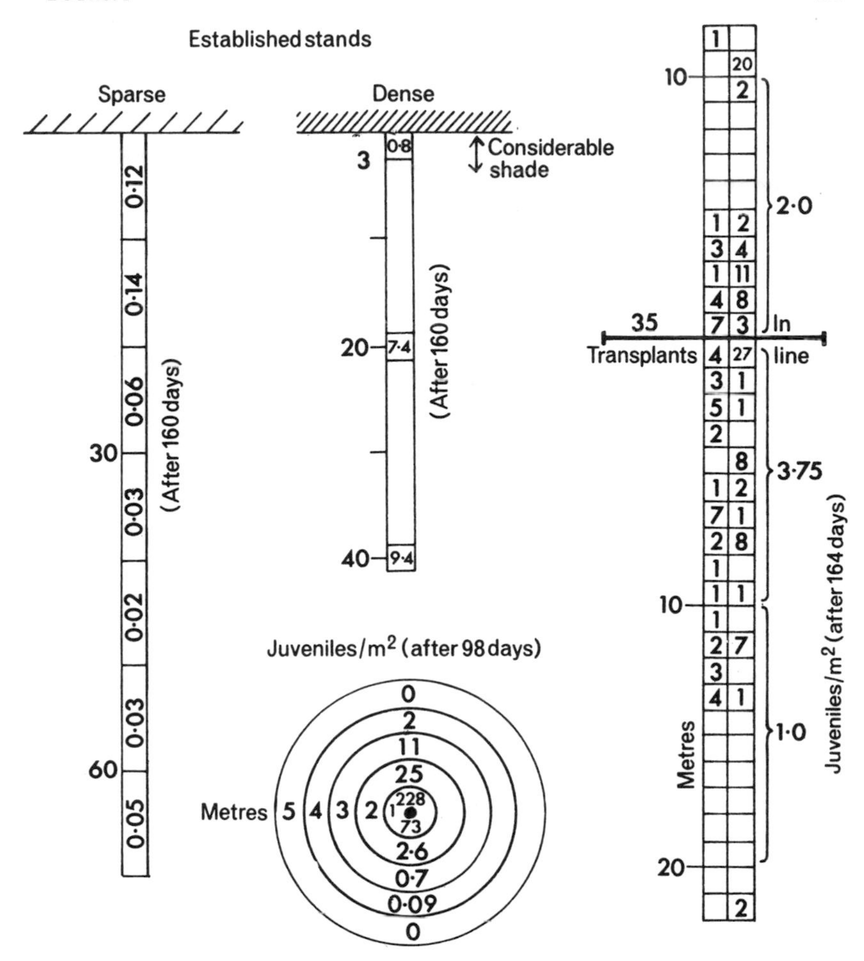

Fig. 81 Examples of spore dispersal from *Macrocystis* plants. (After Anderson and North 1967.)

and also an existing population with a clear-cut boundary were also carried out. Some of the results are shown in Fig. 81. It was important in these experiments to remove and keep out all echinoderms from the experimental areas because these graze on the plants and sporelings (see later section on 'Grazing'); the clear-cut boundary to an existing population utilized in the experiments was in fact caused by these animals in the first place.

Warmer Waters

Pacific

Gilmartin (1960) has described investigations of algal distribution on the Pacific atoll of Eniwetok. He dived at 21 stations along a transect line 28·5 km. long across the lagoon which had a maximum depth of 65 m. He found the vegetation rather patchy and had to abandon plans to collect quantitatively representative samples because he estimated an area some 100 times that initially planned would have been needed to be representative, and underwater time was too limited. Instead he resorted to subjective description of the vegetation, recorded whilst under water, and extensive collection of the algae present. This was coupled with determination of various environmental parameters along the transect line by standard oceanographic procedures from a surface vessel. He produced diagrams of the bathymetric distribution of the major algae which showed that a greater proportion of the green algae penetrated into deep waters than of the brown species, whilst an even greater proportion of the red algae were found there and none of these occurred above 20 m. depth.

In this study it was not found possible to correlate algal distribution with physical or chemical factors such as temperature, phosphate concentration or salinity, but only with light intensity. Light at 45 m. was 8 per cent of that at the surface, indicating transparency equal to the clearest oceanic water. Several changes in plant morphology with depth were noted, such as:

(a) decreased calcification and branching, coupled with elongation and lessening of green pigmentation in *Halimeda monile*;
(b) lengthening and attenuation of the thallus of *Dictyota*;
(c) decrease in the number of ramelli per unit length of stolon in *Caulerpa racemosa*, coupled with reduction in both height and diameter of the individual ramelli.

It was also noted that several species required a consolidated coral substrate for growth, whilst others, such as *Dictyota* and *Dictyopteris* could grow on large debris in sand and *Halimeda monile* and *Caulerpa serrulata* were able to grow in sand of all grades by virtue of their massive holdfasts and spreading stolons respectively.

The question of changes in degree of calcification with depth has been discussed by other authors using diving techniques to collect their specimens from accurately determined depths. Goreau (1963) noted that in Jamaica calcification in *Halimeda* increased with depth, although it was reduced in other green calcareous species. He attributed this to a considerable difference in the physiology of calcification in the former genus which, unlike many of the others, had a higher rate of calcification in the dark than in the light (see later for a discussion of the *in situ* experiments involved). Larkum *et al.* (1967) found that in Malta there was no change in degree of calcification with depth in either *Halimeda tuna* or *Udotea petiolata.* Drew and Larkum (1967) also report no change in thallus density (mg. dry weight/cm.^{2}) or chlorophyll content in *U. petiolata* between 15 and 60 m.

Mediterranean

Apart from the work of Gilmartin and of Goreau cited above, most detailed investigations of warm water algal communities have been carried out in the Mediterranean, usually in conjunction with studies of the fauna. Phytosociological methods have frequently been employed and an excellent example of their use to delimit plant communities is given by Laborel (1960); part of this is reproduced in Table 4. Here the species present on vertical rock faces are shown to be the members of the nearby community on horizontal faces which have sciaphilic (shade) tendencies.

Molinier (1960) has used these methods to demonstrate the existence of most of the characteristic Mediterranean communities around Cap Corse, recording information from quadrats ranging from 25 × 25 to 20 × 200 cm. and bringing back the entire biomass to the laboratory for detailed sorting. He also extended this work to marine angiosperm meadows which will be considered in a later section. Most of his diving investigations seem to have been confined to the very shallow sublittoral, all samples quoted being within 1 m. of the surface (except for angiosperms). Since the Mediterranean is for the most part virtually tideless, such accurate definition of the beginning of the sublittoral is possible. Sciaphilic (shade) communities such as that of *Udotea/ Peyssonnelia,* normally found much deeper than this were in fact studied in crevices near the surface.

By means of dredging, Molinier was also able to investigate the communities at depths beyond the reach of the aqualung diver. In this work he classified the communities according to a well established series which is discussed further by Pérès (1967). This uses the nature of the communities present, rather than physical factors such as depth or light intensity, as its criteria and recognizes those zones shown

Table 4. *Comparison by phytosociological methods of the plant communities on vertical and horizontal rocky substrates at different depths*

(After: Laborel 1960)

Depth	7 m.	10 m.	13 m.	15 m.	35 m.	42 m.	42 m.
Area of sample	5 m.^{2}	4 m.^{2}	10 m.^{2}	25 m.^{2}	10 m.^{2}	5 m.^{2}	2 m.^{2}
Slope	vert.	vert.	vert.	hor.	hor.	hor.	hor.
Cover	100%	100%	100%	100%	100%	100%	100%
Asparagopsis armata	.	.	.	2·1	+	+	.
Spathoglossum solierii	.	.	.	2·3	+	(+)	.
Phyllaria reniformis	.	.	.	2·3	+	(+)	.
Cystoseira opuntioides	.	.	.	1·2	+	(+)	.
Dictyopteris membranacea	2·2	1·2	.	3·3	2·2	2·2	.
Udotea petiolata	1·3	.	2·2	(+)	+	1·1	(+)
Halimeda tuna	(+)	.	2·2	(+)	+	+	(+)
Peyssonnelia spp.	1·3	4·4	2·3	3·3	1·2	1·1	3·2
P. polymorpha	.	.	.	(+)	(+)	+	(+)
Codium difforme	.	.	.	.	.	+	1·1
Zanardinia prototypus	.	.	.	(+)	+	+	1·1
Palmophyllum crassum	.	.	.	.	+	+	1·2
Mesophyllum lichenoides	.	.	.	3·4	+	+	+
Pseudolithophyllum spp.	.	.	.	.	5·5	5·5	(+)

diagrammatically in Fig. 82, with numerous facies of each. In broad outline this series is probably generally applicable both within and outside the Mediterranean.

Laminariales in the Mediterranean

Most of the algae found in the Mediterranean are of small dimensions; the presence of certain members of the massive brown Laminariales is therefore worthy of note, the more so because of the great depths at which they are able to grow. Except for the Atlantic species *Laminaria ochroleuca*, which penetrates into the Mediterranean along the coast of Algeria and into the Alboran Sea, these algae are endemic and usually confined to very deep water. Molinier (1960) discusses in detail the

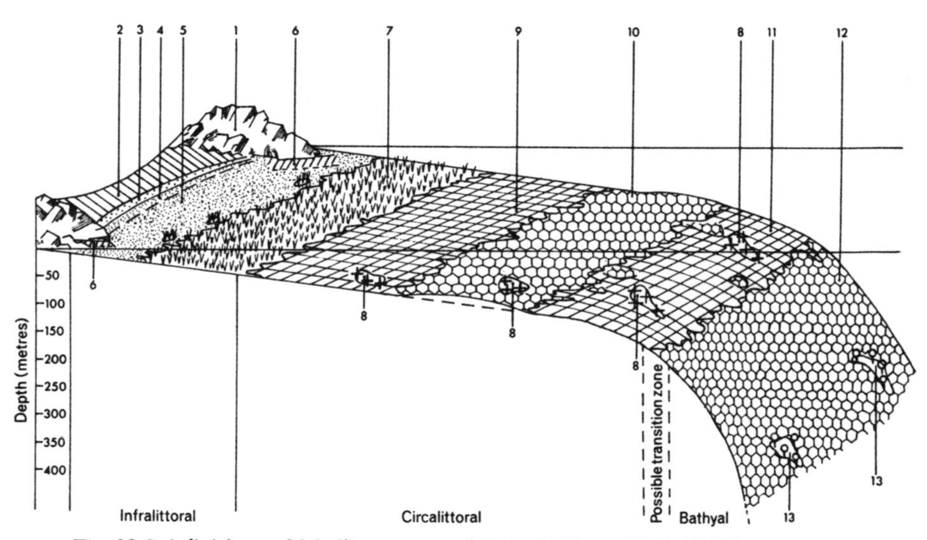

Fig. 82 Subdivisions of Mediterranean sublittoral. (From Pérès 1967.)

1. rocky points
2. alluvial area
3. high and middle beach (supralittoral and mediolittoral sandy communities)
4. SFHN — community of fine sand in very shallow waters
5. SFBC — community of well sorted sands
6. AP — community of photophilic algae on rocky substratum
7. HP — community of *Posidonia* meadow
8. C — coralligenous community
9. DC — coastal detritic community
10. VTC — community of terrigenous mud
11. DL — community of the shelf-edge detritic
12. VP — community of the bathyal mud
13. Community of the deep sea corals.

community associated with *L. rodriguezii* which he found between 90 and 120 m. depth off Cap Corse, as did Gautier and Picard (1957). This unusual *Laminaria* grows by means of stolons and its holdfasts are attached to loose 'stones' composed almost entirely of the red lithothamnoid algae—the so-called 'pralines' of which the bottom in this zone is composed. Furthermore, the blades trail along the bottom in the direction of the prevailing current. He also found off Cap Corse specimens of another member of this order—*Phyllaria reniformis*—at depths of about 60 m., associated with the shallower growing coralligeneous community. This species has also been found as shallow as 19 m. in this area, associated with *Sargassum hornsuchii, Dictyopteris membranacea*, etc. in what must therefore have been a rather shaded habitat.

Direct observation of these deep growing Laminariales is now possible from submersibles, and Giermann (personal communication) filmed extensive 'forests' of *Laminaria echroleuca* growing on the bottom of the Straits of Messina between 90 and 130 m. depth. It was clear that the plants were less dense at the greatest depth, and also appeared somewhat smaller. However, no detailed ecological work has yet been done from such vehicles which possess only crude sampling claws.

These Laminariales are almost certainly cold water species as are the massive temperate species. Cold water in the Mediterranean means considerable depth and therefore very low light intensity. It is interesting to note that both *L. rodriguezii* and the species from the Straits of Messina grow in regions of considerable bottom currents, causing their photosynthetic lamina to stream out horizontally. This gives a maximal light collecting area, a feature of obvious importance in such dimly illuminated waters.

Quantitative studies of Mediterranean algae

Quantitative data on the standing crop or productivity of marine algae in warm waters is scarce, the only two estimates being from the Mediterranean and the Canary Islands.

Bellan-Santini (1966) investigated communities very near the surface, measuring wet weight and decalcified dry weight of both plants and animals in both summer and winter. The plant data are shown in Table 5 and indicate that only one of the three communities had a large difference between summer and winter standing crop.

Larkum *et al.* (1967) have determined the quantities of the various species present on vertical submarine cliffs down to 75 m. in Malta.

They also measured ash content (dry matter remaining after determination of calorific value of important species by bomb calorimetry). Their data (for summer only) are shown diagrammatically in Fig. 83, and also in Table 5; values in the latter are directly comparable with Bellan-Santini's data as they are corrected for calcification. Larkum *et al.* did not sample at depths shallower than 5 m. It is clear from their data that the green algae *Udotea petiolata* and *Halimeda tuna* were dominant from near the surface to the deepest sites investigated, with no increase in importance of calcareous red algae with depth, although the fleshy red alga *Vidalia volubilis* occurred in the deepest quadrats. The increase in total biomass at the deepest site is probably due to its position; this was on the vertical face of a large boulder some 50 m. out from the cliff base and therefore better illuminated than the latter.

The *Udotea/Peyssonnelia* community which those authors found widespread is basically a shade community, although it was present on the south facing submarine cliffs which were illuminated only a little

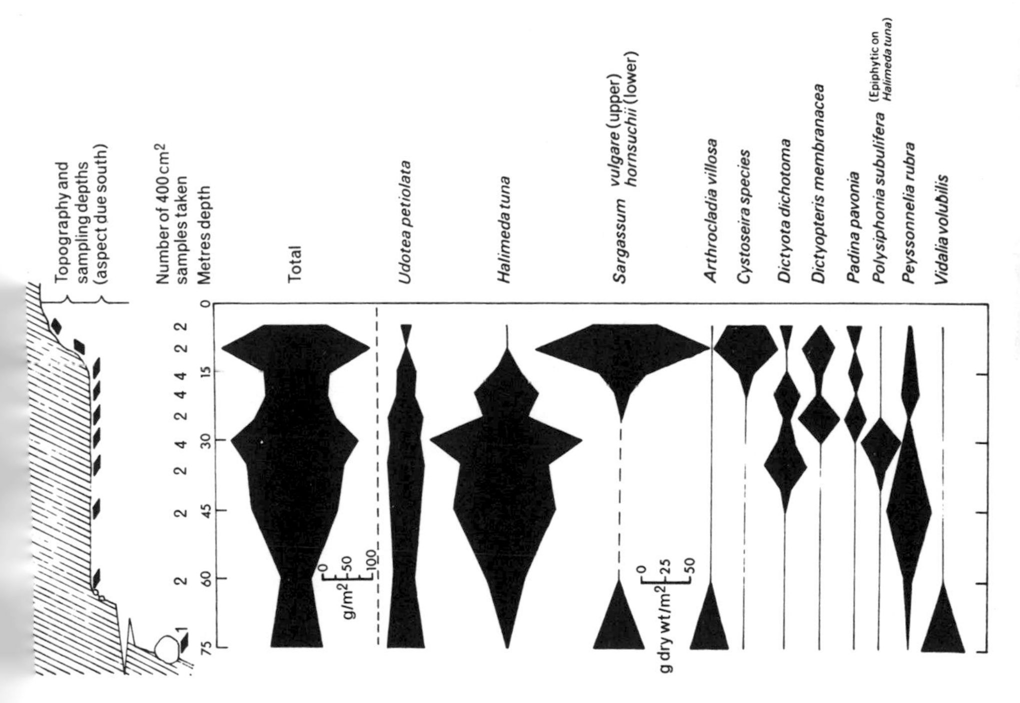

Fig. 83 Distribution of algae with depth down vertical submarine cliff face in Malta. (After Larkum *et al.* 1967.)

TABLE 5. *Standing crop of Mediterranean algal communities*

A. Comparison of three communities from very near the surface in summer and winter

(g. wet wt. or g. dry weight after decalcification/m.²)

(After: Bellan-Santini 1966)

		Cystoseira stricta	*Corallina mediterranea*	*Cystoseira crinita*
Summer	wet	13,845	2,424	8,019
	dry	2,583	186	1,397
Winter	wet	7,487	3,276	6,897
	dry	1,327	221	1,083

B. Standing crop of algae at various depths on vertical submarine cliffs in Malta —summer only

(g. dry wt./m.²)

Depth (m.)	*Total biomass*	*Biomass less calcification (approx. 50% of weight of calcareous species)*
5	137·5	127
10	325	324
20	137·5	94
30	262·5	175
40	194	111·5
50	145	85
60	60	20

less brightly than horizontal areas. However, surfaces lying at about 45° to the vertical exist both above and below the cliffs, and here that community was replaced by brown algae. *Dictyopteris* was also found growing on large debris in sand at 75 m.

Johnston (1969) reports similar maximal standing crops for the *Cystoseira* communities found down to 15 m. in the Canary Islands (1000–1600 g. dry wt./m.²). However, at most of the sites he investigated the communities were very reduced (200–400 g./m.²).

Vegetation in semi-obscure caves

Several authors have investigated the distribution of fauna within semi-obscure submarine caves in the Mediterranean (Laborel 1960,

Laubier 1966) and they find them similar to the coralligenous zone mentioned previously. Such caves are generally reported to be devoid of algae, although it would appear that attention has been focused on the roof and back of the caves with no indication of the communities on the walls. Larkum *et al.* (1967) have reported a very distinct zonation of algae on the walls of a similar cave in Malta, and the topography and vegetation cover therein are shown in Fig. 84. The green algal community graded abruptly into a zone dominated by *Peyssonnelia* which ended equally abruptly. The interior of the cave was devoid of plants except for occasional lithothamnoid incrustations which were found even at the very back near the cave floor but never on the roof itself.

Computation of the light intensities at the demarcation lines of these communities agreed well with those for the depths suggested as the lower limits of the green and red communities in unshaded habitats by Molinier (1960). They were also in good agreement with direct light measurements made in a very similar cave only a few kilometres down the coast in Malta. The values recorded are shown in Fig. 68.

It seems probable that the zonation of algae equivalent to that at much greater depths can be found within such caves where the only varying factor is light intensity. Similarly, the composition of the *Udotea/Halimeda/Peyssonnelia* community at 30 m. depth on deeply shaded north facing submarine cliffs in Malta is quantitatively similar to that at 60 m. or deeper on the south facing cliffs (see Table 6).

Much useful information can thus be obtained by studying communities under various degrees of shading, information otherwise unobtainable within the depth range of the aqualung.

TABLE 6. *Comparison of standing crop of dominant species on north and south facing submarine cliffs in Malta*
(g. dry wt./m.²)

Species	*Larkum* et al. (*1967*)		*Larkum and Crossett (1966)*	
	30 metres		*30 metres*	*60 metres*
	South	*North*	*South*	*South*
Udotea petiolata	57·0	23·5	35·0	22·5
Halimeda tuna	205·0	trace	161·0	33·5

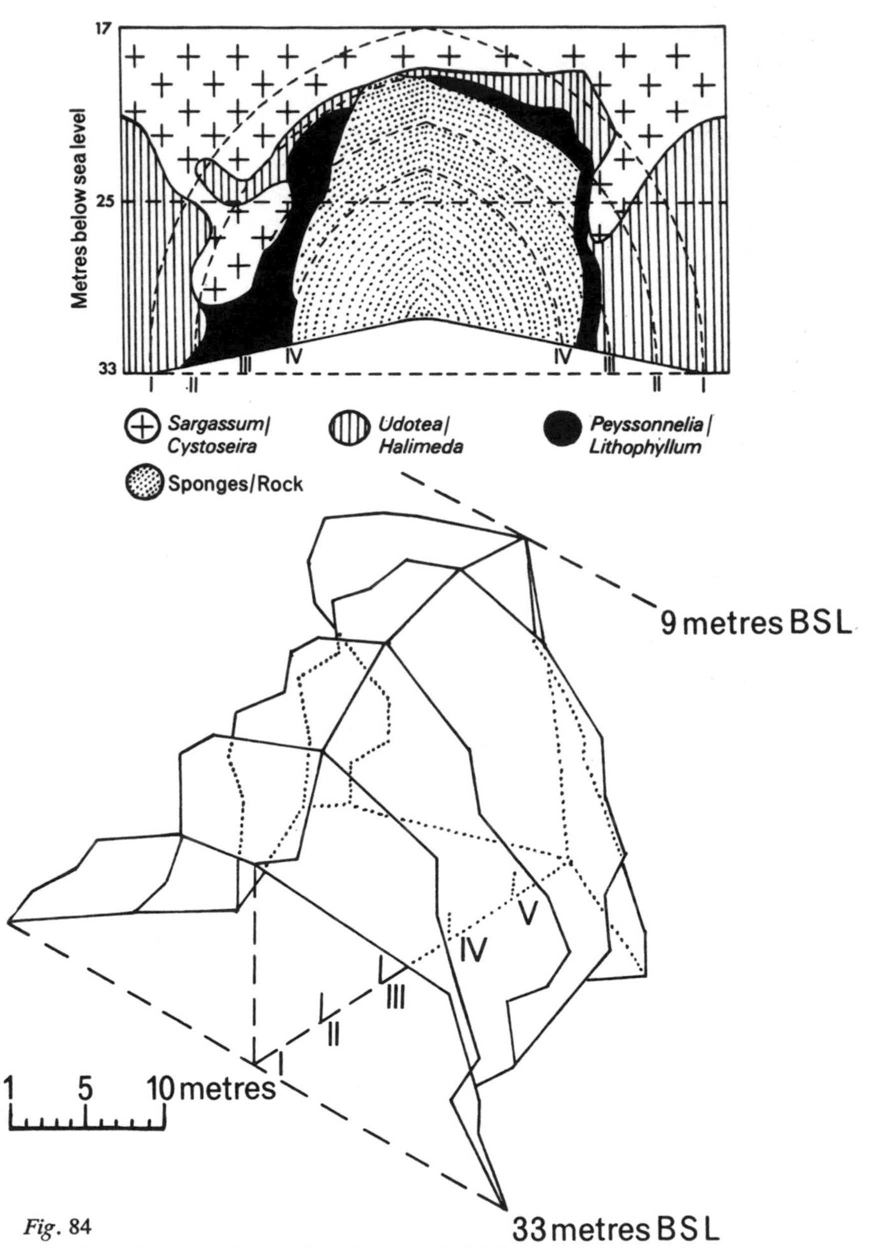

Fig. 84
Zonation of algae inside a submarine cave in Malta.
(From Larkum *et al*. 1967.)

The Antarctic

In recent years several diving biologists have worked in polar seas. Neuschel (1961) described the methods he used during his pioneering work in the Antarctic in 1957–8. He recommended the use of $\frac{3}{8}$ inch foam neoprene wet suits with trousers covering the chest and very long overlaps on bootees and mittens, together with a hood covering the entire face and with a small hole for snorkel or mouth-piece. Both Fane (1959) and Peckham (1964) recommended the use of dry suits and the latter reported that this, with a wet suit inside, allowed comfortable dives under sea ice for periods of up to an hour. However, depth was limited by the compression and loss of insulation in the wet suit and Neuschel indicated that 20 m. was the maximum depth tolerable with his equipment. Peckham describes diving through a hole in the ice under a small heated hut, using submerged lighting for winter work. All these workers found unmodified aqualung equipment suitable for use in waters even below 0°C.

Although littoral and immediately sublittoral zones in the Antarctic are usually devoid of algae as a result of the abrasive action of ice, there is frequently prolific algal growth deeper down. Neuschel (1965) described the distribution of important species at a number of stations in the South Shetland Islands and the Palmer Peninsula and some of his data are shown in the profiles reproduced in Fig. 85. A considerable number of Antarctic algae are endemic (25 per cent of the species as estimated by Zinova 1958) and Neuschel (1963a) has described important morphological details of three endemic brown algae collected by diving. One of these—*Phyllogigas*—is particularly interesting since it only becomes important at very southerly latitudes.

Delephine *et al.* (1967) have investigated benthic algae down to 42 m. in the same area and found a rich vegetation which apparently descended undiminished even deeper. Healthy material, possibly growing attached, was brought up from over 100 m. and other authors have reported certain red and brown algae obtained by dredging at depths in excess of 300 m. Although algae were found at considerable depths, water clarity was apparently not very good, normal visibility being reported as 6 m. with considerable diminution during diatom blooms. Those authors also reported the first truly sublittoral lichen—*Verrucaria serpuloides*—which was found growing abundantly at 10 m. depth.

By using the air holes of Wedell seals as entry points, Zaneveld (1967) was able to dive under fast sea ice 3 to 6 m. thick where it was present for nine to ten months of the year. That investigation revealed extensive

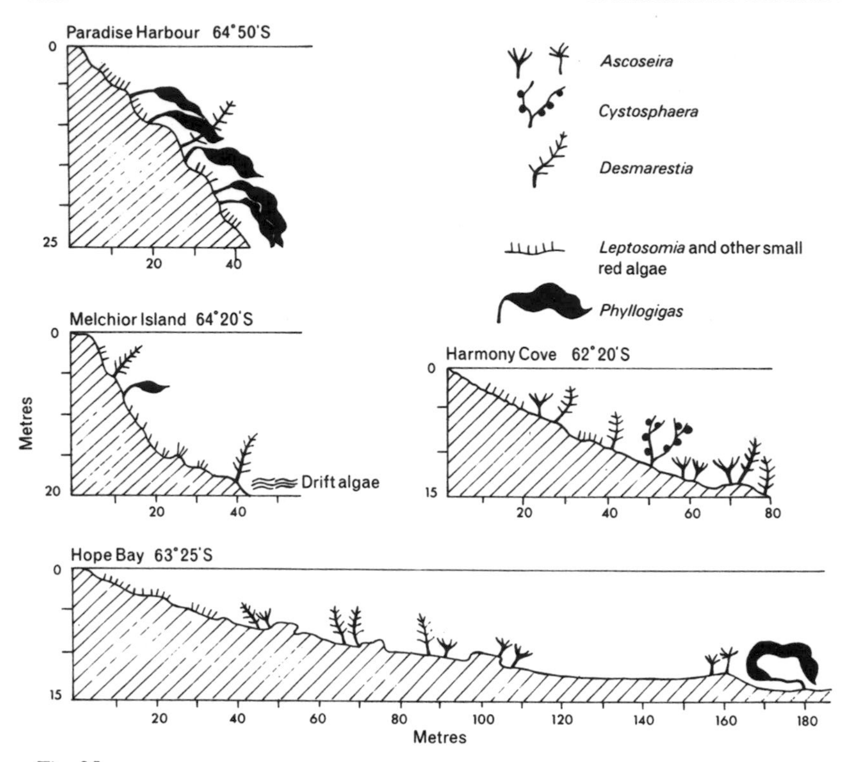

Fig. 85
Diagrammatic transects of Antarctic sublittoral algae beds. (From Neuschel 1965.)

beds of algae, dominated by red species, between 6 and 35 m. These were probably present throughout the year although inactive during the winter months without light. However, Zaneveld considered that growth began very early in the season, soon after midwinter (21 June). Light measurements taken beneath 2 m. of ice in November and December indicated an average illumination at 18 m. which was 1·39 per cent of that at the surface, this being approximately uniform over the entire day. In view of the equipment used, the absolute significance of the data must be suspect; nevertheless, there is obviously considerable illumination under sea ice and the habitat is not aphotic as suggested by Neuschel (1963b).

Extensive growth of algae at very low light intensities with extended dark winters and often turbid water is presumably possible in these cold waters because respiratory rates are correspondingly low. However,

Wilce (1967) quotes data on the penetration of light energy below sea ice in the Arctic and points out that the maximum illumination of the lower ice surface coincides with maximum melt run-off and hence turbid water so that benthic algae benefit little from the increased surface illumination in summer. His data are shown in Fig. 86 and the variation over the year of the depth illuminated by 0·001 langleys/minute (the lowest red algal compensation point according to Haxo and Blinks

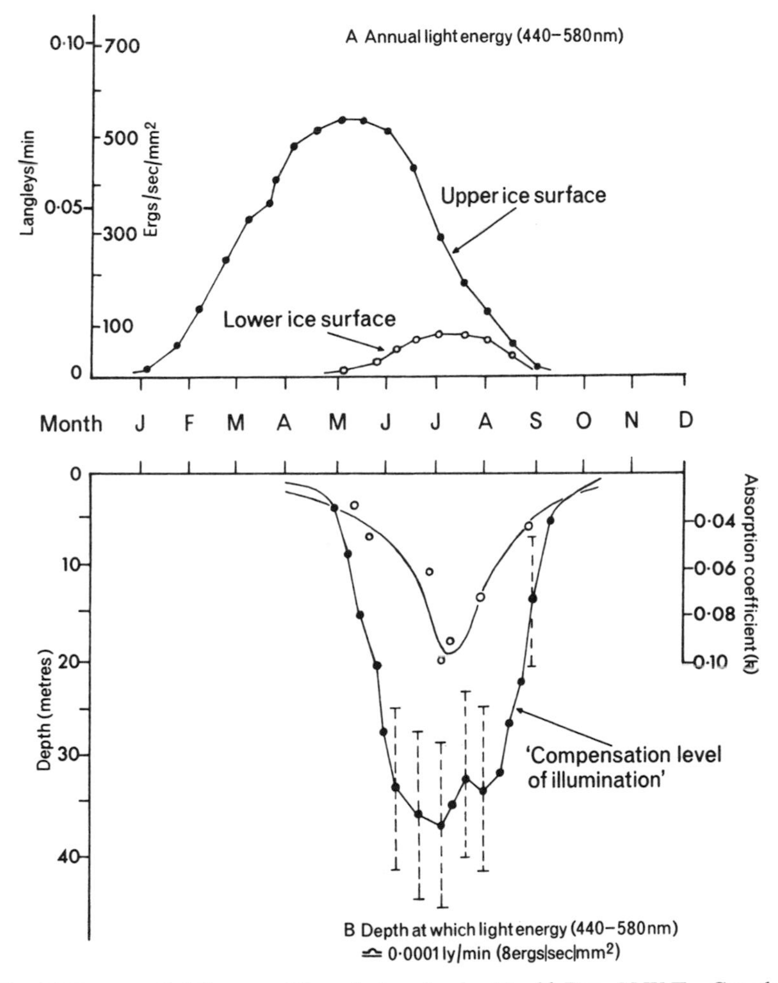

Fig. 86 Summer lighting conditions below ice in Mould Bay, N.W.T., Canada. (76° 14′ N, 119° 20′ W) (From Wilce 1967.)

1950) is also shown. Wilce suggested that, in view of the very restricted illumination which plants at 100 m. or more must receive, considerable heterotrophic nutrition may occur, possibly by utilization of dissolved organic matter released by phytoplankton blooms. Unfortunately Wilce and other workers in the Arctic have not yet published details of their ecological work, but considerable benthic populations below 100 m. have been reported.

Estimates of rates of productivity in polar seas must await quantitative investigations of the benthic plant communities described above. Investigations to date have been observational, usually supported by photographic records.

Marine Angiosperms

It has already been pointed out that there are a number of truly marine angiosperms which live entirely submerged in shallow to moderately deep water. The different species and genera are, with few exceptions, geographically segregated between tropical and temperate zones and between the New and Old Worlds (Sculthorpe 1967).

There have been few ecological investigations of the sublittoral marine angiosperms. Only one has been extensively investigated; this is *Posidonia oceanica*, a plant confined to the Mediterranean.

An important biological peculiarity of this plant is that the rhizomes, which are substantial and densely clothed with old leaf bases, grow both horizontally and vertically. Such vertical growth allows the plant to escape burial by the constant deposition of sediment trapped amongst the leaves. Molinier and Picard (1952) suggest that the submarine terraces or rhizome 'mattes' are built upwards at a rate of about 1 m. per hundred years. This figure agrees well with an observed rhizome elongation of about 1 cm. per year (Drew, unpublished), although actual upward development may be somewhat slower than this since a considerable proportion of the rhizomes at any one time grow horizontally and do not contribute directly. Thus, where erosion exposes underwater 'cliffs' of carbonized rhizomes several metres high, the lower layers are of considerable antiquity; in fact they sometimes contain archaeological material in an extremely good state of preservation.

The development of these mattes upwards to the level of swell or wave action has been described by several authors (Molinier 1960, Pérès and Picard 1964). When this level is reached erosion occurs, especially where there are small boulders which wear pot-holes, several of these joining to form channels or 'intermattes'. In sheltered bays the

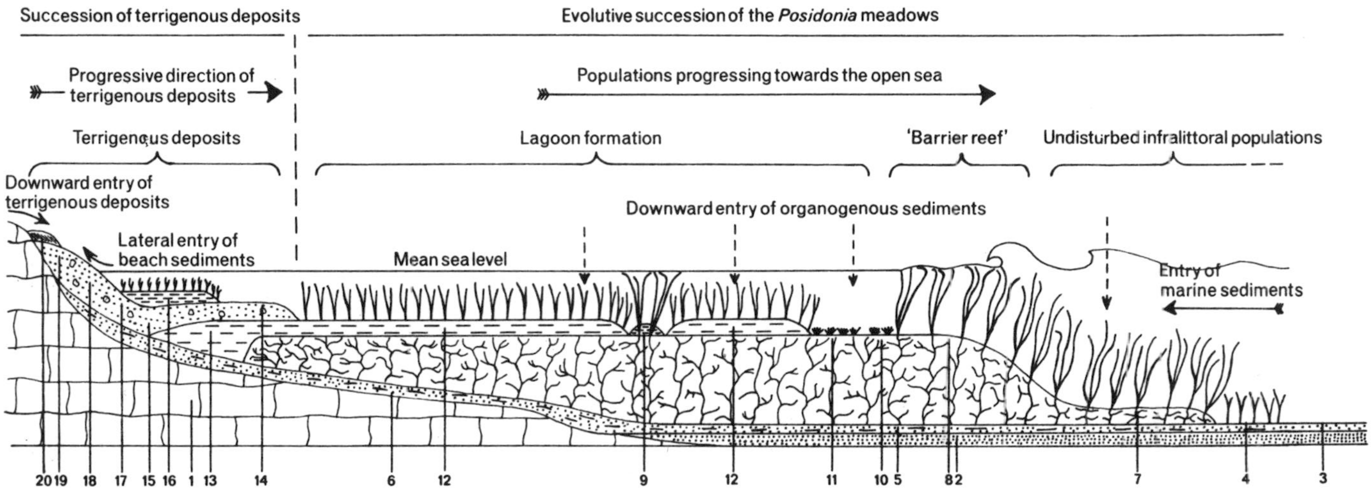

Succession of terrigenous deposits

Evolutive succession of the *Posidonia* meadows

Progressive direction of terrigenous deposits

Populations progressing towards the open sea

Terrigenous deposits

Lagoon formation

'Barrier reef'

Undisturbed infralittoral populations

Downward entry of terrigenous deposits

Downward entry of organogenous sediments

Lateral entry of beach sediments

Mean sea level

Entry of marine sediments

20 19 18 17 15 16 1 13 14 6 12 9 12 11 10 5 8 2 7 4 3

Fig. 87 Scheme of a *Posidonia* barrier reef and its lagoon. (From Pérès 1967.)

1. original rocky substratum
2. original sandy substratum formerly inhabited by SFBC community
3. sediment inhabited by SFBC or SGCF community
4. sediment with humus content increased by retention of organic matter by *Cymodocea* lawn.
5. same, covered by *Posidonia* terraces ('mattes')
6. sediment fastened on rocks by AP community and humus enhanced by *Cymodocea* facies
7. *Posidonia* meadow
8. barrier reef of *Posidonia* due to terrace rising
9. vestigial patch of *Posidonia*
10. *Padina pavonia* facies of the AP community on dead *Posidonia* rhizomes (11)
12. area of silt sedimentation with *Cymodocea nodosa* lawn on dead *Posidonia* matte
13. silty sediment covered by following one
14. coarse detritic but silty and muddy sediment with *Upogebbia* facies of SVMC community
15. same with *Upogebbia* burrows filled by later sediment
16. silty sediment with *Zostera nana*
17. same covered by terrigenous deposits
18. mediolittoral beach
19. supralittoral beach
20. mass of dead *Posidonia* leaves (Banquettes) with supralittoral community

SFBC : fine well sorted sand
SGCF : coarse sand and fine gravel under bottom currents
AP : photophilic algae
SVMC : superficial muddy sands in sheltered areas

build-up may continue until the tips of the leaves reach the surface and there form a 'barrier reef' which slowly migrates out to sea as sedimentation continues. Behind this, in the calmer waters of the 'lagoon', fine sedimentation and considerable temperature fluctuation cause degeneration of the *Posidonia* meadow and the development of photophilic plants, including a smaller marine angiosperm, *Cymodocea nodosa*. Stages in the development of *Posidonia* beds are shown in Fig. 87.

The meadows house a considerable flora and fauna which has been compared and contrasted for deep and shallow sites by Molinier (1960), using phytosociological methods. The epiphytic and associated plants and animals must be divided into those living on and amongst the leaves (photophilic) and those on the rhizomes and substrate (sciaphilic). Molinier noted a change in the composition of the associated communities, with animals becoming of increased importance at greater depths. Whilst investigating the *Posidonia* meadows in Malta, Drew (unpublished) noted a complete absence of algae, other than encrusting Melobesias on the leaves, in the shallow meadows (6 m. depth) whereas at 27 m. the leaves were clothed with other algae besides Melobesias and the rhizomes supported a sparse but diverse flora of the larger algae. These are listed in Table 7 together with an indication of their abundance on the basis of biomass present.

It has been suggested by Odum (1956) that marine angiosperm communities are highly productive. His data for the diurnal changes in oxygen concentration in water flowing over meadows of *Thalassia testudinum* in shallow bays suggest a rate of productivity of about 27 metric tons dry matter/hectare/year. Burkholder *et al.* (1959) determined the biomass in communities of this plant and found an average of 4,685 gm. dry weight per m.^{2}, although they estimated that 80–90 per cent of this was underground rhizomes and roots two to three years old. Thus, with a standing crop of about 725 gm. dry weight of leaves per m.^{2} and taking into account the contribution of the year's production to secondary growth of the older rhizomes, annual production would be of the order of 20 metric tons/hectare/year.

Recently Drew (unpublished) has investigated the biomass of *Posidonia oceanica* by analysis of samples collected by divers from two depths in Malta. It was found that, taking into account only the leaves, there were 954 $\pm$ 118 gm. dry weight/m.^{2} at 6 m. as compared with 742 $\pm$ 199 at 27 m. These values were not significantly different and a possible reason for such similarity, despite much lower ambient light energy at the deeper site, is heavy grazing by echinoderms in shallow

TABLE 7. *Biomass of various algae associated with* Posidonia *meadows at 27 metres depth in Malta*
(Data quoted as mg./m.^{2} of meadow)

Rhizomes

Species	*Dry weight/metre*^{2} ($n = 5$)
Udotea petiolata	707 ± 550
Phylophora nervosa	447 ± 469
Peyssonnelia rubra	99 ± 118
Rhodymenia corallicola?	92 ± 84
Halopteris filicina	57 ± 42
Botrycladia botryoides	25 ± 32
Vidalia volubilis	12 ± 15
TOTAL	1,439 ± 1,015

Leaves (samples particularly heavily covered with filamentous brown epiphytes encrusting Melobesias not counted)

Leaf area (*cm.*^{2})	*Weight Leaf* (*mg.*)	*Weight Algae* (*mg.*)	$\frac{mg.\ Algae}{g.\ Leaf}$
16·5	167·7	80·6	480
17·6	130·5	34·9	280

Summary

Average *Posidonia* leaf biomass (g./m.^{2})	742 ± 199
Average algal biomass on rhizomes (g./m.^{2})	1·4 ± 1·0
Approximate epiphyte load on leaves (g./m.^{2}) (assuming 25 per cent of leaves carry half measured load)	35

water. These were absent from the deeper meadows, and this, together with certain other features of the *Posidonia* crop, is discussed in the next section on Grazing. Taking into account the contribution of the year's production to the development of older rhizomes, *Posidonia* may well produce 20 metric tons/hectare/year.

Grazing

Certain animals, particularly echinoids and some crustacea, gastropods, and fishes, graze extensively on submarine plants. Certain marine angiosperms form essential parts of the diets of larger marine animals; thus *Thalassia testudinum* is the food plant of the green marine turtle and is commonly known as turtle grass.

The effect of grazing on the giant kelp beds off the Californian coast has been thoroughly investigated and Leighton *et al.* (1967) have summarized the relationship between these plants and the dominant echinoids *Strongylocentrus franciscanus* and *S. purpuratus*. These animals are especially important in the vicinity of sewage outfalls on the coast where the combined effect of turbidity, reducing light penetration, and grazing can result in complete destruction of the kelp community. *Macrocystis* plants transplanted by divers into such denuded areas survive well if enclosed in cages to keep out the grazing animals. North (1958) concluded that destruction was not due to toxic compounds in the effluent but that the urchins may have been partially nourished and hence 'encouraged' by organic matter in the effluent. They were certainly able to survive in large numbers after the complete disappearance of macroscopic edible plants. These conclusions are supported by calculations of plant productivity and echinoid requirements in another polluted area where the two organisms co-existed, at least temporarily. Leighton *et al.* (1967) reported that echinoids were present at a density indicating an organic matter requirement of 126 to 375 gm./m.^{2}/year but that local plants could only produce 0·078 to 36 gm. of this. However, some 380 gm. organic matter/m.^{2}/year were available from the sewage and could therefore support the echinoid population.

The relationship between kelp growth, water temperature and echinoid feeding is indicated by the following data:

Q_{10} of kelp photosynthesis	2·0
Q_{10} of frond elongation	1·7
Q_{10} of urchin consumption	3·8
(5–15°C: declines above 17°C)	

This clearly explains the great increase in damage to kelp beds in periods of warm water; other grazing animals probably respond similarly to increased water temperature.

Another important factor is the site of primary grazing attack. Thus, the crustaceans *Idothea resecata* and *Ampithoe humeralis* are canopy feeders and may occur in sufficient densities to consume the upper 3 m. of the plants completely in a few weeks (Jones 1965). However, the echinoids attack the base of the stipe and the greater part of the plant drifts away; their effect is thus much greater than actual consumption would indicate.

Control measures to protect the economically important kelp beds may be either natural or chemical. The sea otter *Enhydra lutris*, the fish *Pimelometopon pulchrum* (sheepshead), and two asteroids, *Pycnopodia helianthoides* and *Asterometris sertulifera*, are all natural predators on the echinoids. The otter is particularly effective but at present this animal is found only in small numbers. The most practical method of control of the echinoids is the use of quicklime 'bombs' (Leighton 1963). Applying this material in lumps into the wake of a motor boat at about 0·5 kg./m.^{2} results in the death of 60 to 100 per cent of the echinoids on the bottom. Actual contact of lumps with the animals is essential. Algal regeneration then occurs if echinoid concentrations are reduced to less than 1 per m.^{2} for *S. franciscanus* and 10 per m.^{2} for *S. purpuratus*. Manual control by divers using small picks, screwdrivers, etc., is only feasible if echinoid concentrations are fairly high; then 1–2,000 animals per hour can be punctured and will subsequently die. However, the method is impracticable if concentrations are less than 4 per m.^{2} or if most of the animals are less than 2 cm. in diameter.

Kain and Jones (1967) reported that the echinoid *Echinus esculentus* can completely prevent the growth of Laminaria forests in the Isle of Man if present at densities of 3·6 per m.^{2}, although at shallower depths forests exist with an average echinoid density of 1·1 per m.^{2} Clearing all the echinoids from the deeper region (8 to 11 m.) resulted in the establishment of considerable algal growth and two years later second year sporelings of *Laminaria hyperborea* were present at 4·4 plants per m.^{2} The area had been cleared of echinoids monthly during the entire period.

In the Mediterranean two echinoids, *Paracentrotus lividus* and *Arbacia lixula*, are very common in the upper 15 m. of the sublittoral. In Malta they occur at densities as great as 13 per m.^{2} (average 5 per m.^{2}) and their maximum occurrence coincides with a marked reduction in algal biomass. The gut contents of both species indicate considerable algal grazing; a study by Neill and Larkum (1966) showed that *Paracentrotus* grazes on adult algal plants, whilst *Arbacia* gut contents consisted of a white calcareous paste containing distinct traces of algal pigments. The latter apparently scrapes algal sporelings off the rock surface, an hypothesis substantiated by the development of a thick 'fur' of algal sporelings on bare rock within an area kept clear of echinoids for one month.

The potential consumption of algae by *Paracentrotus* has been investigated by controlled feeding of individuals in cages kept under water (Riley and Drew, unpublished). Over an eight-day period adult individuals consumed about 25 mg. of algal organic matter per day of the calcareous phaeophyte *Padina pavonia*, which represented most of the alga actually provided. This agal species is common in the region where *Paracentrotus* lives; two calcareous green algae which grow somewhat deeper (*Udotea* and *Halimeda*) were apparently unpalatable. However, the gut contents of the echinoid *Sphaerechinus* contained considerable amounts of *Halimeda*. Since this animal occurs somewhat deeper than *Paracentrotus* it is possible that preference for different algae may influence the vertical distribution of these echinoids.

Algal productivity in the upper 15 m. in Malta is approximately 1·2 g. organic matter per m.^{2} per day (estimates for *Padina*). Thus, echinoid grazing at the intensity indicated above could have a considerable effect on the biomass present, five *Paracentrotus* per m.^{2} consuming 125 mg. of organic matter or 10 per cent of the total algal production. Site of attack is probably again very important, since this rate of consumption of new sporelings (by *Arbacia*) would have much more devastating effects.

In *Posidonia* meadows off the coast of Malta grazing patterns appear to differ considerably with depth. Comparing the animal populations of meadows at 6 and 27 m. Drew (unpublished) found a considerable number of adult *Paracentrotus* in the shallow meadow (3·5 per m.^{2}) whilst there were none in the deep site. Analysis of the gut contents indicated that the echinoids present at 6 m. were grazing on *Posidonia* leaves. It was noted above that the common echinoids in this region were concentrated above 15 m. depth, and although Gamble (1966) suggested that they can penetrate deeper, he found few below 20 m.; this would explain their absence from the deeper meadows. The abundance of encrusting animals (bryozoans, hydroids, and sponges) and algae in these meadows compared with their complete absence in shallower waters may be due to the absence of large herbivores. This feature may also contribute to the surprisingly similar standing crop of *Posidonia* leaves at the two depths, for grazing pressure could possibly reduce that in shallow, well-lit habitats to the same level as

TABLE 8. *Comparison of standing crop (leaves only) of* Posidonia *at two depths in Malta*

Depth (m.)	*Plants/m.²*	*Leaves/plant*	*Leaf area/ plant (cm.²)*	*Leaf area/ metre² (cm.² × 10⁻³)*	*Dry weight leaves/m.² (g.)*
6	365 ± 63	3·6 ± 0·3	227 ± 65	80·7 ± 19·5	954 ± 118
27	278 ± 63	4·4 ± 0·3	237 ± 33	68·6 ± 22·2	742 ± 199
t-test difference $p =$	~0·03	< 0·01	> 0·1	> 0·1	> 0·1

that of the darker, deeper beds. Further evidence for this hypothesis is found in the comparison of other features of the two *Posidonia* crops set out in Table 8. Two features which would not be directly affected by grazing but would be indicative of different growth rates—number of shoots per m.² and number of leaves per shoot—are both significantly different at the two sites. However, features such as leaf area per plant, leaf area per m.², and the total mass of leaves would all be directly altered by grazing and indeed it is these features which show no statistical difference at the two depths investigated.

Echinoids must thus be regarded as important herbivorous grazers; although they appear to be very slow-moving animals, investigations have shown that they are capable of moving 1–2 m. per day under natural conditions (Neill and Larkum 1966), and that they are more active by night than day (Gamble 1966). Furthermore, those associated with the Californian kelp forests can detect kelp plants, transplanted into barren areas, from as far distant as 30 m., presumably by means of chemoreceptors (Leighton 1966, North 1965).

Underwater Physiological Experiments

It is usually possible to draw only tentative conclusions about the relative success of various algae at various depths from simple determinations of the biomass present. Photosynthesis and primary productivity frequently bear little relationship to the standing crop at any one time (Blinks 1955, Ryther 1956). It is therefore important to obtain data about the metabolism of algae under various environmental conditions, and in particular measurements of their photosynthetic rates. Using diving techniques it is now possible to carry out detailed *in situ* experiments to considerable depths, and two series of investigations involving the use of radioactive tracers in short-term growth experiments under water will now be described.

Coral reef studies

Goreau (1963) has investigated the relationship between photosynthetic carbon fixation and calcification in a number of important algae and corals on Jamaican reefs. Material collected by divers was placed in sealed jars containing sea-water labelled with both radioactive $Ca^{45}Cl_2$ and $NaHC^{14}CO_3$. These were positioned 'in shallow parts of the reef' by divers and later collected after 1·5 to 2·5 hours' exposure to light. Some of the jars were painted black (these were the dark exposures) and such jars had to be placed in the shade of coral formations to prevent overheating. Data were expressed both in μg. carbon or calcium fixed/mg. N in the tissue/hour and in terms of per cent accretion:

$$\frac{\text{weight of C or Ca deposited per hour}}{\text{TOTAL C or Ca content of sample}} \times 100.$$

Most of the algae tested were able to fix calcium at a similar rate in the dark as in the light—only in a few did the dark rate fall below 50 per cent of the light rate—and in certain species dark calcium fixation considerably exceeded that in the light.

The actual rates of carbon fixation quoted by Goreau almost certainly err on the low side, since he did not take into account the refixation of already fixed C^{14} subsequently respired as $C^{14}O_2$. However, they serve as a guide to fixation rates and some are tabulated in Table 9 after conversion to μg. C/mg. dry weight/hour for direct comparison with data for Mediterranean species (Drew 1969) discussed below and also set out in Table 9. Fixation rates are certainly of the same order in both places in the four genera compared.

Reef-building or hermatypic corals are also photosynthetic organisms, for the coral polyps contain numerous unicellular algae as symbionts. The photosynthetic rates of such corals, determined by Goreau (1963) and set out in Table 10, are similar to those of the calcareous algae and indeed higher than in the major reef-building algae. Calcification proceeds more slowly, however, and shows a complete difference from the pattern seen in the algae, for it is much slower in the dark. This is taken to indicate direct control of the process by light, possibly mediated via the bicarbonate requirements of photosynthesizing algal symbionts (Goreau 1961a).

TABLE 9. *Rates of fixation of carbon by species of the same genera in the Caribbean and the Mediterranean*

(μg C/mg. dry wt./hour)

(Caribbean data from: Goreau 1963; Mediterranean data from Drew 1969)

Caribbean		Mediterranean		
Udotea verticillosa	3·35	*Udotea petiolata*	2·94	(10 metres)
cyathiformis	1·08			
flabellum	0·86			
conglutinata	0·84			
Halimeda tuna	0·62	*Halimeda tuna*	1·22	(10 metres)
opuntia	0·84			
incrassata	0·58			
goreauii	0·47			
monile	0·44			
simulans	0·35			
Padina sanctae-crucis	3·32	*Padina pavonia*	2·00	(20 metres)
Peyssonnelia spp.	0·24	*Peyssonnelia rubra*	0·37	(30 metres)
Lithophyllum incertum	0·11			
Goniolithon spectabile	0·12			

Caribbean data calculated assuming $N = 10$ per cent organic matter and that ash content similar to that for Mediterranean species.

Mediterranean data for photosynthesis at best depth, indicated in parentheses.

The shape of coral colonies at different depths is probably also controlled by light intensity and is a function of algal metabolism. Kawaguti (1937) showed that the progressive flattening of hemispheroid types with depth was actually light controlled since colonies from deep shade in shallow water were similarly flattened, whilst in the same colony shaded portions grew more flattened (Goreau 1961b). The flattening process effectively decreases body weight without affecting surface area, and living polyps are only found at the surface. Calcium carbonate per unit area of polyp tissue can thus be decreased by as much as 70 per cent. Since it would not be possible for the polyp/alga association to maintain in darker depths the high rate of light-dependent calcification found near the surface, such changes are in fact directly caused by the reduced light intensity of depth or shade.

Thus, although corals are supposedly primarily carnivorous animals

TABLE 10. *Comparison of photosynthetic carbon fixation and calcium deposition in coralline algae and hermatypic corals*
(After: Goreau 1963)

Algae

	Carbon (μg/mg. N/hr.)	*Calcium* μg/mg. N/hr.		Calcium
		Light	*Dark*	Dark/Light %
Halimeda tuna	25·3	365	517	142
Udotea verticillosa	43·2	103	132	128
Penicillus lamoureuxii	17·8	166	66	40
Liagora valida	87·6	4,314	2,355	55
Jania adherens	60·0	722	332	46
Galaxaura oblongata	57·9	1,592	959	60
Amphiroa fragilissima	45·1	1,344	1,498	108
Lithophyllum incertum	3·5	31	28	91
Goniolithon spectabile	3·9	1,435	927	65
Padina sanctae-crucis	97·6	191	102	53

Corals

	Carbon (μg/mg. N/hr.)	*Light*	*Dark*	Dark/Light %
Acropora cervocornis				
— *apical corallites*	10·0	106	27	25
— *lateral corallites*	10·5	56	9	17
Millepora complanata	9·7	75	38	50
Porites furcata	13·8	27	6	21

(Yonge 1963), their colony shape, size, and growth rates are greatly dependent on their algal partners. Recent investigations have shown direct transfer of photosynthetic products from plant to animal in considerable amounts (Muscatine 1967). Their consideration in a botanical chapter is therefore well justified.

Deep growing Mediterranean algae

Drew and Larkum (1967) have reported investigations of the variation in photosynthesis with depth in *Udotea petiolata* in Malta. Their work was carried out using the C^{14} method in *in situ* experiments down the vertical submarine cliffs which have already been described. *Udotea* was the dominant alga at all the depth stations except the shallowest at 10 m. One of the underwater platforms which were erected at each station is illustrated in Fig. 88. In this work allowance was made for refixation of respiratory $C^{14}O_2$ according to measurements of the actual respiratory rates of the algae in the laboratory.

This work has been continued using other species besides *Udotea* in order to investigate the photosynthetic potential of red (*Peyssonnelia rubra*) and brown (*Padina pavonia*) algae. Photosynthesis at depths in excess of those actually accessible with the aqualung was determined by carrying out experiments within the submarine cave described previously. The results shown in Fig. 89 (Drew 1969) are for carbon fixed per day in excess of that respired away. The point of zero net assimilation represents the actual compensation depth; allowance was made for the fact that photosynthesis was restricted to daylight hours whilst respiration was continuous. Clearly the compensation depths of *Udotea* and *Peyssonnelia* are controlled by light intensity since in the cave net photosynthesis of the dominant species is zero at the inner limit of each community. However, *Padina*, which is not found below about 15 m.

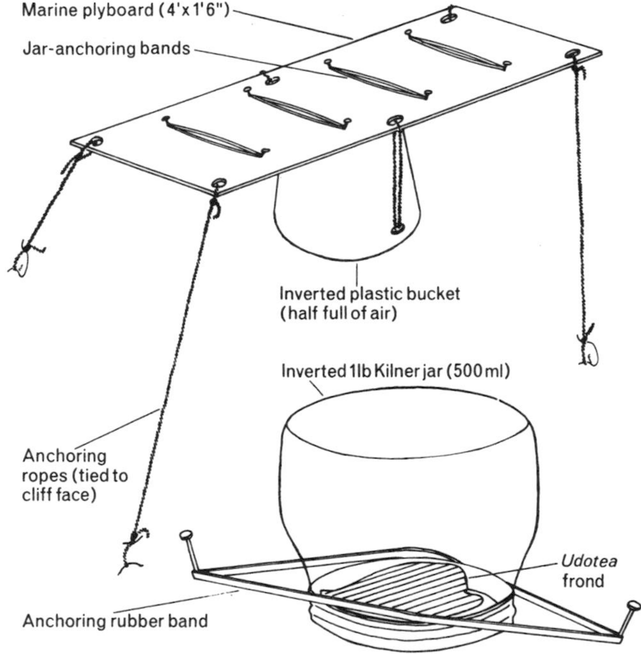

Fig. 88

Apparatus for exposure of algae in ^{14}C-labelled seawater to submarine light at various depths on vertical cliff face. (From Drew and Larkum 1967.)

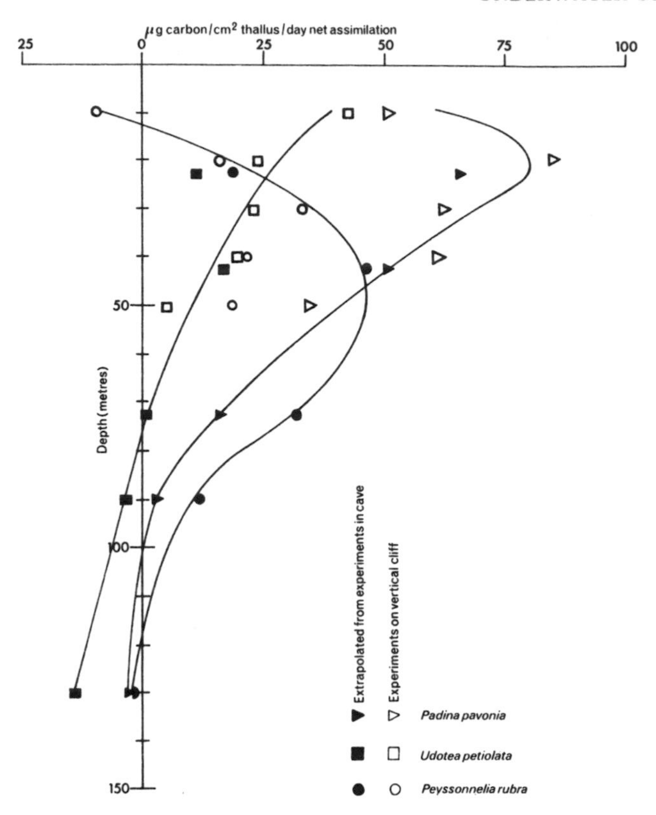

Fig. 89
Net daily photosynthesis in three Mediterranean algae. (From Drew 1969.)

and certainly never inside the cave, should show net assimilation to a depth actually in excess of that for *Udotea*; the distribution of this species must be controlled by factors other than light intensity. Extrapolation of this rate of photosynthesis to other brown algae in Malta explains their ability to grow at the great depths from which they have been recorded both in the Mediterranean and elsewhere. The validity of such extrapolation is indicated by preliminary results for *Dictyopteris membranacea* which photosynthesizes at similar rates to *Padina* to at least 50 m. depth on the open cliffs.

Calculation of net productivity of these algae at the depths of maximal photosynthesis indicate values of 6 (*Padina*), 4 (*Peyssonnelia*), and 2 (*Udotea*) metric tons of organic matter/hectare/year, assuming carbon to represent 50 per cent of organic matter. Johnston (1969) reports similar rates of photosynthesis in comparable species in the Canary Islands, also measured using an *in situ* C^{14} method.

Vertical Distribution of Attached Marine Algae

The postulation of complementary chromatic adaptation

In this chapter much emphasis has been placed on the occurrence of either green, brown, or red algae at different depths in various amounts. The reason for this emphasis will now be explained.

Algae contain certain accessory photosynthetic pigments in addition to chlorophyll *a*. These may be other chlorophylls, carotenoids such as fucoxanthin which is responsible for the colour of the brown algae, or the phycobilins, phycoerythrin, and phycocyanin, in the red algae. Absorption spectra for these are shown in Fig. 90 and it has been clearly demonstrated in laboratory experiments that these different pigments allow the plants to use light of different colours for photosynthesis (Klugh 1931, Levring 1947, Haxo and Blinks 1950). Light energy absorbed by the accessory pigments has been shown to be transferred to the chlorophyll *a* system which is present in all photosynthetic plants and which alone is capable of mediating the conversion of such energy

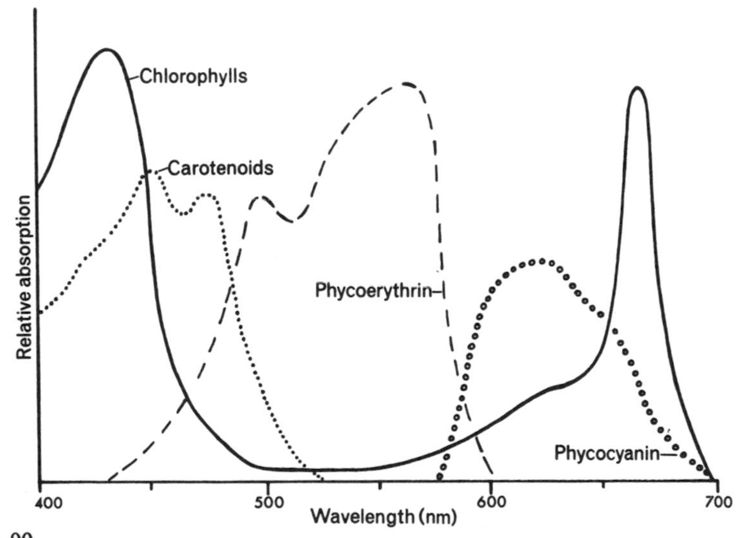

Fig. 90
Absorption spectra of algal photosynthetic pigments. (Generalized from various sources.)

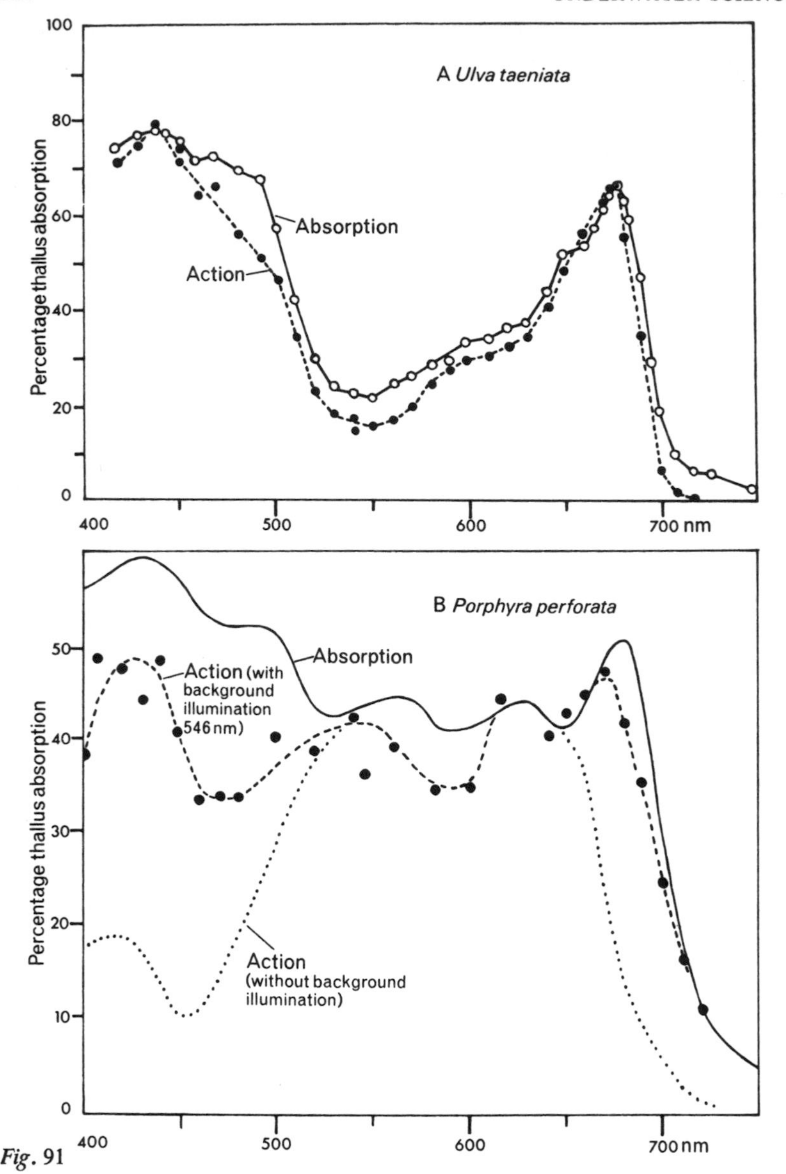

Fig. 91

Light absorption and action spectra of green and red marine algae.
(A. from Haxo and Blinks 1950; B. from Fork 1963.)
(Background illumination provides very little additional light energy in B. but results in considerable enhancement of chlorophyll-mediated photosynthesis. For further discussion of the enhancement phenomenon see Emerson 1957.)

into chemical energy. Close agreement between absorption spectra and photosynthetic action spectra can be seen in the examples in Fig. 91.

Such observations are in agreement with the hypothesis of complementary chromatic adaptation, first suggested by Englemann (1883) and later set out formally by Gaudokov (1903). It was proposed that the occurrence of red algae at lower depths than other algae in coastal waters represented adaptation to the green light predominant in such waters and absorbed by their red pigments. *In situ* experiments have been carried out on the photosynthesis of algae in temperate coastal waters by several workers who have lowered jars containing algae to various depths and later measured changes in the oxygen content of the water therein. Printz (1939) and Levring (1947) have both found the maximum depth for detectable photosynthesis in these waters to be between 30 and 40 m., and also that the red algae were best suited to conditions near this depth. However, the differences between the three algal types were not great and Levring concluded that several other factors were involved in addition to the spectral composition of the ambient submarine light. One of the most important of such factors was the effect of reduced over-all light intensity, originally postulated by Bertholt (1882) and Oltmanns (1892) as an alternative hypothesis to that of chromatic adaptation.

Only recently has direct observation of the sublittoral vegetation been really feasible and extension of the kinds of investigations described in this chapter should allow clear solution of these controversies. It seems that in temperate coastal waters the massive brown Laminariales are usually dominant down to at least 20 m., exceptionally nearly 40 m., with an undergrowth of red species. These may give way to smaller brown or coralline red species extending to about 45 m. under favourable conditions. Green algae are not quantitatively important except in the littoral zone. Other factors being favourable, light is the controlling factor in this vertical distribution.

Effect of reduced light intensity without spectral modification

In clear waters found around the shores in more tropical latitudes, where light penetrates to a much greater depth and is blue rather than green, a different situation is found. Green algae, often species of *Halimeda*, are dominant from near the surface to about 80 m., with a cover of coralline red algae below this depth. However, brown algae may also penetrate very deeply here. The maximum depth of penetration

TABLE 11. *Efficiency of utilization of submarine daylight by a green and a red alga at increasing depths in oceanic water*
(From: Larkum *et al.* 1967)

Depth (m.)	*Relative number of incident quanta*	*% of incident quanta absorbed in photosynthesis*	
		Porphyra perforata (Red)	*Ulva taeniata (Green)*
15	158·0	64·5	54·0
30	53·4	68·0	60·0
60	6·7	68·5	64·0
150	0·02	68·5	68·5

Daylight calculated on quantum intensity basis for colour temperature of 7500°K. Absorption of light by sea-water from Jerlov (1951) data for type II oceanic water. Action spectra for photosynthesis from Haxo and Blinks (1950).

of algae in the clearest waters is not yet agreed upon; 150 to 200 m. is probably a reasonable estimate.

It can be calculated from published data that the proportion of ambient light energy available for photosynthesis by the different types of algae in clear oceanic waters is virtually the same down to at least 150 m. Data for red and green species are set out in Table 11. Thus chromatic adaptation would not be expected to play an important role in such waters. Nevertheless, despite very deep penetration by green and brown species, red algae seem to have an advantage even here at extreme depths. Below 40 m. there is little change in the spectral composition of submarine light in such waters so that the ability of red algae to utilize light at intensities below those limiting for other types seems to be their important ecological advantage.

Reduced attrition

Both Goreau (1963) and Drew (1969) have shown that calcareous red algae have in fact lower rates of primary production than do brown or green species, suggesting that they owe their success at low light intensities to a slower rate of attrition rather than to exceptionally efficient utilization of the available light. Respiration rates in *Peyssonnelia* (red) were indeed found by Drew to be very much lower than those in *Udotea* (see Table 12) whilst the calcareous nature and prostrate habit of the deeper growing red algae may be deterrent to both herbivorous grazing

TABLE 12. *Oxygen uptake in respiration by green and red Mediterranean algae at various temperatures*

		μl. oxygen/cm.² thallus/hour					
		15°C		20°C		25°C	
Udotea petiolata	a	2·40	2·50	3·31	3·18	4·25	4·00
	b	2·60		3·05		3·76	
Peyssonnelia rubra	a	0·83	0·88	1·14	1·20	2·10	2·18
	b	0·92		1·27		2·25	

and physical erosion. The effect of reduced respiratory rate at low temperatures has already been suggested to explain the growth of massive Laminariales at considerable depths in the Mediterranean as well as the deep penetration of algae in polar seas even under almost permanent sea ice.

Photosynthetic efficiency

It can be calculated, from the data on photosynthesis in Mediterranean algae presented earlier in this chapter, that the efficiency with which they convert available light energy into organic matter (photosynthetic efficiency) increases markedly with depth. Near their compensation depths these algae may be using light energy at an efficiency approaching the maximum theoretically possible.

Similarly, the data of Bellamy and Whittick (in press), which are presented in Table 2, indicate a considerable increase in photosynthetic efficiency with depth in *Laminaria hyperborea*.

Such increases are obviously of great advantage to the algae, and elucidation of the mechanism bringing them about may be of considerable scientific and practical importance.

References

Aleem, A. A. (1956), 'A quantitative study of the benthic communities inhabiting the kelp beds off the California coast, with a self-contained breathing apparatus.' *Proc. IInd Int. Seaweed Symp.*, 149–52.

Anderson, E. K. and W. J. North (1967), '*In situ* studies of spore production and dispersal in the giant kelp, *Macrocystis*.' *Proc. Vth Int. Seaweed Symp.*, 73–86.

Baardseth, E. (1954), 'Kvantitative tare-undersokelser i Lofoten og Salten sommeren 1952.' *Report N. 6, Norwegian Institute of Seaweed Research.*

Bailey, J. H., A. Nelson-Smith, and E. W. Knight-Jones (1967), 'Some methods for transects across steep rocks and channels.' *Underwater Assn. Rep., 1966–67*, **2,** 107–11.

Beebe, W. (1926), *The Arcturus Adventure*. New York.

Beebe, W. (1928), *Beneath Tropic Seas*. New York.

Bellamy, D. and A. Whittick (1967), 'Operation Kelp.' *Triton*, April 1967.

Bellamy, D. and A. Whittick (1968a), 'Operation Kelp.' *Triton*, February 1968.

Bellamy, D. and A. Whittick (1968b), 'Kelp forest ecosystems as a "phytometer" in marine pollution.' *Underwater Assn. Rep.*, **3,** 79–82

Bellan-Santini, D. (1966), *2nd Int. Oceanogr. Congr., Moscow*, Abstracts 29–30.

Bertholt, G. (1882), 'Uber die Verteilung der Algen im Golf von Neapel.' *Mitt. zool. Stn. Neapel*, **3,** 393–536.

Blinks, L. R. (1955), 'Photosynthesis and productivity of littoral marine algae.' *J. Mar. Res.*, **14,** 363–73.

Boden, B. P., E. M. Kampa, and J. M. Snodgrass (1960), 'Underwater daylight measurements in the Bay of Biscay.' *J. Mar. Biol. Ass., U.K.*, **39,** 227–38.

Braun-Blanquet, J. and J. Pavillard (1922), *Vocabulaire de sociologie végétale.* Montpellier.

Burkholder, P. R., Gillian M. Burkholder, and J. A. Rivero (1959). 'Some chemical constituents of Turtle Grass, *Thalassia testudinium.*' *Bull. Torrey. Botan. Club*, **86,** 88–93.

Clendenning, K. A. (1964), 'Photosynthesis and growth in *Macrocystis pyrifera.*' *Proc. IVth Int. Seaweed Symp.*, 55–65.

Crossett, R. N. and A. W. D. Larkum (1966), 'The ecology of benthic marine algae on submarine cliff faces in Malta.' *Proc. Symp. Underwater Ass. Malta '65*, 57–61.

Delephine, R., I. MacKenzie Lamb, and M. H. Zimmermann (1967), 'Preliminary report on the marine vegetation of the Antarctic Peninsula.' *Proc. Vth Int. Seaweed Symp.*, 107–16.

Drew, E. A. (1969), 'Photosynthesis and growth of attached marine algae down to 130 metres in the Mediterranean.' *Proc. VIth Int. Seaweed Symposium*, 151–9.

Drew, E. A. and A. W. D. Larkum (1967), 'Photosynthesis and growth of *Udotea*, a green alga from deep water.' *Underwater Assn. Rep., 1966–67.*

Emerson, R. (1957), 'Dependence of yield of photosynthesis in long-wave red light on wavelength and intensity of supplementary light.' *Science*, **125,** 746.

Englemann, T. W. (1883), 'Farbe und Assimilation.' *Bot. Ztg.*, **41,** 1–29.

Fane, F. D. (1959), 'Skin-diving in polar waters.' *Polar Record*, **9,** 433–5.

Fork, D. C. (1963), 'Observations on the function of chlorophyll *a* and accessory pigments in photosynthesis.' In *Photosynthetic Mechanisms in Green Plants*, Symp. Nat. U.S. Acad. Sci., Publ. 1145. Washington, 352–61.

Forster, G. R. (1954), 'Preliminary note on a survey of Stoke Point rocks with self-contained diving apparatus.' *J. Mar. Biol. Ass. U.K.*, **33**, 341–4.

Forster, G. R. (1955), 'Underwater observations on rocks off Stoke Point and Dartmouth.' *J. Mar. Biol. Ass. U.K.*, **34**, 197.

Gamble, J. C. (1966), 'Some observations on the behaviour of two regular echinoids.' *Proc. Symp. Underwater Ass. Malta '65*, 47–50.

Gaudokov, N. (1903), 'Die farbenveranderung bei den Prozessen der komplementaren chromatischen Adaptation.' *Ber. dt. bot. Ges.*, **21**, 517–39.

Gautier, Y. and J. Picard (1957), 'Bionomie du banc du Magaud (Est des Iles d'Hyères).' *Rec. Trav. Stat. Mar. Endoume, Bull.* 12, Fasc. 21.

Gilmartin, M. (1960), 'The ecological distribution of the deep water algae of Eniwetok Atoll.' *Ecology*, **41**, 210–21.

Goreau, T. F. (1961a), 'On the relation of calcification to primary productivity in reef building organisms.' In *The Biology of Hydra*, 269–85. Edited by H. M. Lenhoff and W. F. Loomis. University of Miami Press.

Goreau, T. F. (1961b), 'Problems of growth and calcium deposition in reef corals.' *Endeavour*, **20**, 32–9.

Goreau, T. F. (1963), 'Calcium carbonate deposition by coralline algae and corals in relation to their roles as reef builders.' *Ann. N.Y. Acad. Sci.*, **109**, 127–67.

Greig-Smith, P. (1964), *Quantitative Plant Ecology*, 2nd Edition. London: Butterworth.

Haxo, F. T. and L. R. Blinks (1950), 'Photosynthetic action spectra of marine algae.' *J. Gen. Physiol.*, **3**, 389–422.

Jerlov, N. G. (1951), 'Optical studies of ocean water.' *Rep. Swed. deep sea Exped.* (Physics & Chem.), **3**, 1–59.

Johnston, C. S. (1969), 'Studies on the ecology and primary production of Canary Islands marine algae.' *Proc. VIth Int. Seaweed Symposium*, 213–22.

Johnston, C. S., I. A. Morrison, and K. Maclachlan (1969), 'A photographic method of recording the underwater distribution of marine benthic organisms.' *J. Ecol.*, **57**, 453–9.

Jones, G. (1965), *Ann. Rpt. Kelp Habitat Improvement Project. Calif. Inst. Tech.*, 62.

Kain, J. M. (1960), 'Direct observations on some Manx sublittoral algae.' *J. Mar. Biol. Ass. U.K.*, **39**, 609–30.

Kain, J. M. (1962), 'Aspects of the biology of *Laminaria hyperborea*. I. Vertical distribution.' *J. Mar. Biol. Ass. U.K.*, **42**, 377–85.

Kain, J. M. (1963), 'Aspects of the biology of *Laminaria hyperborea*. II. Age, weight, and length.' *J. Mar. Biol. Ass. U.K.*, **43**, 129–51.

Kain, J. M. (1964), 'A study on the ecology of *Laminaria hyperborea* (Gunn.) Fosl.' *Proc. IVth Int. Seaweed Symp.*, 207–14.

Kain, J. M. and N. S. Jones (1967), 'Subtidal algal colonisation following the removal of *Echinus*.' *Wiss. Helgoländer Meeresunters*, **95**, 460–6.

Kawaguti, S. (1937), 'On the physiology of reef corals. II. The effect of light on colour and form of reef corals.' *Palao Trop. Sta. Stud.*, **2**, 199–208.

Kershaw, K. A. (1964), *Quantitative and Dynamic Ecology*. London: Arnold.

Kitching, J. A., T. T. Macan, and H. C. Gilson (1934), 'Studies in sublittoral ecology. I. A submarine gully in Wembury Bay, S. Devon.' *J. Mar. Biol. Ass. U.K.*, **19**, 677.

Kitching, J. A. (1941), 'Studies in sublittoral ecology. III. Laminaria forest on the west coast of Scotland. A study of zonation in relation to wave action and illumination.' *Biol. Bull.*, **80,** 324.

Kjellman, F. R. (1877), *Nova Acta Soc. Sci. Upsal.*, Vol. Extraord.

Klugh, A. B. (1931), 'Studies on the photosynthesis of marine algae. I. Photosynthetic rates of *Enteromorpha*, *Porphyra umbilicalis* and *Delesseria sinuosa* in red, green and blue light.' *Contr. Can. Biol. Fish.*, **6,** 41–63.

Laborel, J. (1960), *Rec. Trav. Stn. Mar. Endoume. Bull.* 20, Fasc. 33, 117–73.

Larkum, A. W. D., E. A. Drew, and R. N. Crossett (1967), 'The vertical distribution of attached marine algae in Malta.' *J. Ecol.*, **55,** 361–71.

Laubier, I. (1966), *Annls. Inst. Oceanogr., Monaco*, **44** (7), 1–406.

Leighton, D. L. (1963), *Kelp Habitat Improvement Project Final Report, 1962–63. Univ. Calif. Inst. Mar. Res.*, IMR Ref. 63–13.

Leighton, D. L. (1966), University of California Press.

Leighton, D. L., L. G. Jones, and W. J. North (1967), 'Ecological relationships between giant kelp and sea urchins in southern California.' *Proc. Vth Int. Seaweed Symp.*, 141–53.

Levring, T. (1947), 'Submarine daylight and the photosynthesis of marine algae.' *Gotenborgs K. Vetensk. Vitterh. Samh. Handl. ser B.*, **5** (6), 2–89.

MacFarlane, C. I. (1967), 'Sublittoral surveying for commercial seaweeds in Northumberland Strait.' *Proc. Vth Int. Seaweed Symp.*, 169–76.

Molinier, R. (1960), 'Etude des biocenoses marines du Cap Corse, I and II.' *Vegetatio*, **9,** 121–92 and 217–312.

Molinier, R. and J. Picard (1952), 'Recherches sur les herbiers de Phanerogames marines du littoral mediterraneen français.' *Ann. Inst. Ocean.*, **27,** Fasc. 3.

Morrison, I. A., (1970), 'An inexpensive photogrammetric approach to the reduction of survey diving time.' *Underwater Assn. Rep.*, **4,** 22–8.

Muscatine, L. (1967), 'Glycerol excretion by symbiotic algae from corals and *Tridacna*, and its control by the host.' *Science*, **156,** 516–19.

Neill, S. R. St. J. and H. Larkum (1966), 'Ecology of some echinoderms in Maltese waters.' *Proc. Symp. Underwater Ass. Malta '65*, 51–6.

Neuschel, M. (1959), *Studies on the Growth and Reproduction of the Giant Kelp Macrocystis pyrifera.* Ph.D. Thesis, Univ. Calif., Los Angeles.

Neuschel, M. (1961), 'Diving in Antarctic waters.' *Polar Record*, **10,** 353–8.

Neuschel, M. (1963a), 'Reproductive morphology of Antarctic kelps.' *Bot. Mar.*, **5,** 19–24.

Neuschel, M. (1963b), *Brit. Phycol. Bull.*, **2.**

Neuschel, M. (1965), 'Diving observations of subtidal Antarctic marine vegetation.' *Bot. Mar.*, **8,** 234–43.

North, W. J. (1958), 'Quantitative measurements of importance for ecological evaluations in beds of the giant kelp, *Macrocystis pyrifera*.' *Proc. IIIrd Int. Seaweed Symp.*, 37.

North, W. J. (1961), 'Life span of the fronds of the giant kelp, *Macrocystis pyrifera*.' *Nature*, **190,** 1214–15.

North, W. J. (1964), 'Experimental transplantation of the giant kelp, *Macrocystis pyrifera*.' *Proc. IVth Int. Seaweed Symp.*, 248–55.

North, W. J. (1965), *Kelp Habitats Improvement Project*. Am. Rep. Calif. Inst. Tech. p. 33.
Odum, H. (1956), 'Primary production in flowing waters.' *Limnol. & Oceanogr.*, **1,** 102–17.
Oltmanns, F. (1892), 'Uber die Kultur und Lebensbedingungen der Meeres-algen.' *Jb. wiss. Bot.*, **23,** 349–440.
Oosting, H. J. (1956), *The Study of Plant Communities*. 2nd Edition. San Francisco: Freeman.
Peckham, V. (1964), 'Year-round SCUBA diving in the Antarctic.' *Polar Record*, **12,** 143–6.
Pérès, J. M. (1967), 'The Mediterranean benthos.' *Oceanogr. Mar. Biol. Ann. Rev.*, **5,** 449–533.
Pérès, J. M. and J. Picard (1964), 'Nouveau manuel de bionomie benthique de la Mer Mediterranee.' *Rec. Trav. Stn. Mar. Endoume. Bull.*, **31,** fasc. 47, 5–137.
Printz, H. (1939), 'Uber die Kohlensaureassimilation der Meeresalgen in verschunden Tiefen.' *Skr. norske Vidensk-Akad. Kl. I*, **1,** 1–101.
Ryther, J. H. (1956), 'The measurement of primary photosynthesis.' *Limnol. & Oceanogr.*, **1,** 72–84.
Sculthorpe, C. D. (1967), *The Biology of Aquatic Vascular Plants*. London: Arnold.
Strickland, J. D. H. (1958), 'Solar radiation penetrating the ocean. A review of requirements, data, and methods of measurement, with particular reference to photosynthetic productivity.' *J. Fish. Res. Bd. Canada*, **15,** 453–93.
Wilce, R. T. (1967), 'Heterotrophy in Arctic sublittoral seaweeds: an hypothesis.' *Bot. Mar.*, **9,** 185–97.
Wilkinson, C., J. Bevan, and C. Balaam (1967), 'Distributional studies of marine molluscs in Malta.' *Underwater Assn. Rep., 1966–67*, **2,** 89–99.
Yentsch, C. S. (1962), 'Marine phytoplankton.' In *Physiology and Biochemistry of the Algae*. Edited by Lewin. New York: Academic Press.
Yonge, C. M. (1963), 'The biology of coral reefs.' *Adv. Mar. Biol.*, **1,** 209–60.
Zaneveld, J. S. (1967), 'The occurrence of benthic marine algae under shore fast-ice in the Western Ross Sea, Antarctica.' *Proc. Vth Int. Seaweed Symp.*, 217–31.
Zinova, A. D. (1958), 'Composition and floral character of the algae along the shores of Antarctica and the islands of Kerguelen and Macquartia.' *Soviet Antarctic Exped., Information Bull.*, **3,** 47–9.

7 Archaeology

J. du Plat Taylor

Underwater exploration has added a new region to the realm of archaeology. During the pioneer years of the 1950s divers discovered ancient wrecks and other antiquities which had formerly been outside the reach of the archaeologist.

Early attempts to undertake underwater excavation and to explore the sea bed before the archaeologist learnt to dive, have been described in my *Marine Archaeology* (1964) and George Bass's *Archaeology Underwater* (1966). Now that diving is accepted as a regular method of exploring underwater archaeological sites, I propose in this chapter to consider only selected projects.

The principal fields in which exploration has been carried out are: coastal research closely allied to geology and geomorphology; the study of ancient harbours and roadsteads; exploration of inland waters and rivers, and, by far the most extensive, the study of ancient wrecks. Experience has shown that recording standards under water can be just as high as on land; but surveying and photographic methods have had to be adapted both to the different element and also to the fact that things do not remain static under water, so that speed becomes of importance both in diver-time and in immediate recording.

Coastal Research

Exploration of the continental shelf is being undertaken by many of the sciences; but from the archaeological approach the rise and fall of the sea in the last million years, caused by the growth and decay of the ice caps, is of considerable importance. Raised beaches have been observed all over the world, but with diving and modern underwater techniques, it is now possible to study the sea levels when the sea had retreated. These changes have not been so extensive in the last 10,000 years; nevertheless, there are inaccessible cave sites and islands off the coast of Brittany and Wales, for instance, which contain human and animal remains belonging to the Stone Age, indicating that there must have been large areas of land around them suitable for hunting and fishing.

In many examples of geomorphological coastal research, archaeological and historical factors play a considerable part in the evidence, for instance, of the *Making of the Broads* (Lambert *et al.* 1960). Miss Akeroyd's unpublished thesis (Akeroyd 1966) deals with the question of changes of sea level round the coast of Britain using the data from archaeological sites.

Ancient harbours

Submerged as well as silted-up sites can contribute much evidence for the chronology of sea-level changes. So far, underwater exploration has not been used by archaeologists but Dr. Flemming has shown that submerged sites can be used to furnish data for sea-level research (p. 276). The discovery of submerged sites, apart from indicating changes in coastline, adds information to the distribution maps of any one period, and in coastal regions may well indicate small ports or fishing villages. For these shallow sites, the system of surveying is usually an adaptation of land methods combined with underwater planning of the settlement.

During the excavations of the neolithic settlement at Saliagos island off Antiparos in the Aegean, underwater exploration showed that the site had been joined to the main island by a neck of land now submerged. A survey carried out by Mr. Morrison's photographic methods revealed buildings of the hellenistic period on this neck and around the island, indicating that the submergence had been comparatively recent (Evans and Renfrew 1968).

At Halaeis in Greece, Mr. Julian Whittlesey (1968) has used balloons and kites to carry cameras for air survey; in clear water this can be a very useful preliminary over-all survey but needs to be followed up by detailed underwater checking and exploration.

Submerged sites lead directly to the exploration of *ancient harbours*. The positions of many of these are known; some are silted up and their exploration would require a land excavation. Of others only the landward part is visible and the remainder is submerged. When Lehmann-Hartleben (1923) prepared his work on ancient Mediterranean harbours, he was hampered by these factors; and it was not until Père Poidebard (1939) applied modern techniques, using air photographs and helmeted divers for the exploration of Tyre and Sidon before World War II, that the first accurate survey and study of the construction of a Phoenician port was made. His work remains outstanding and modern exploration is but an expansion of his methods.

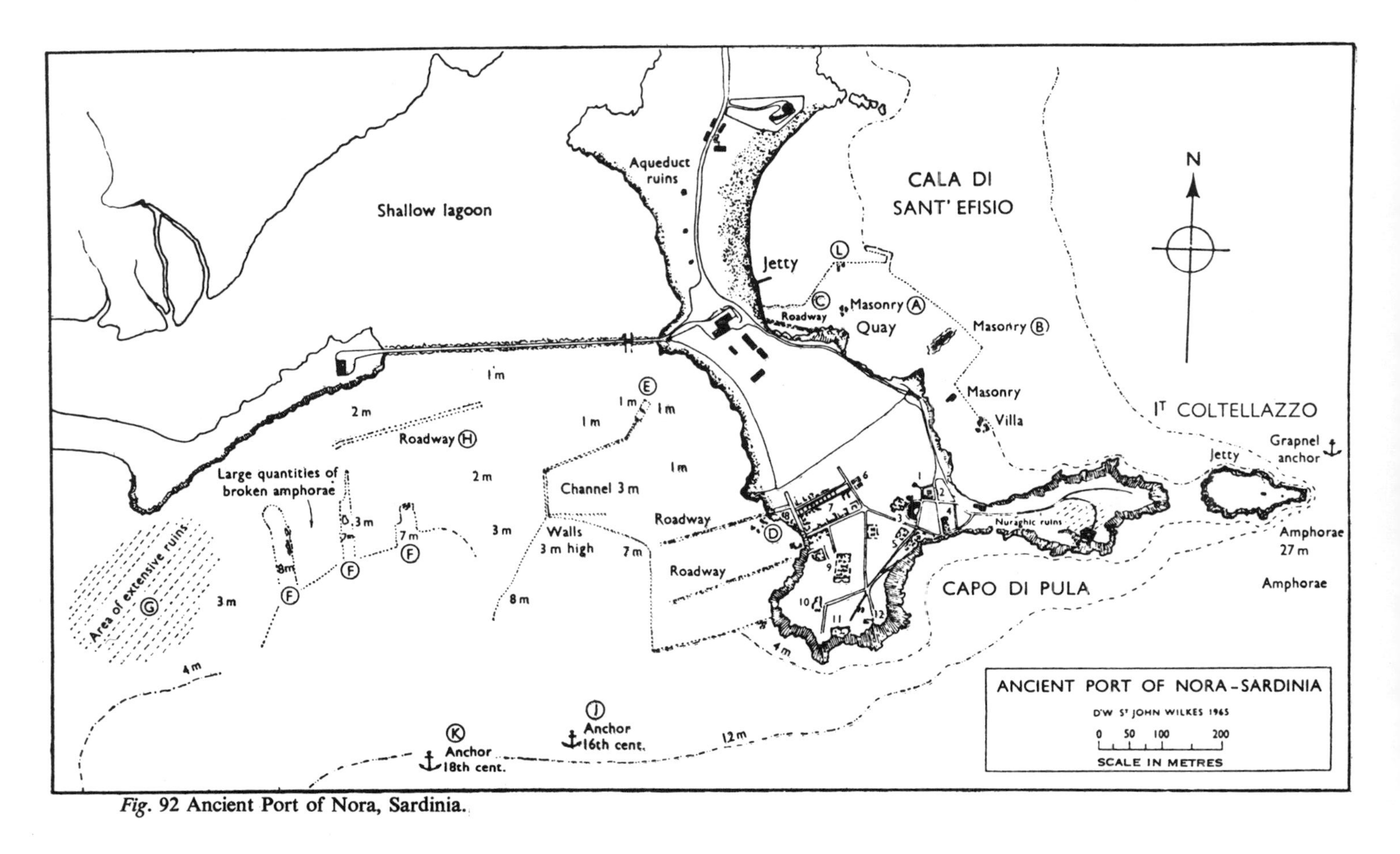

Fig. 92 Ancient Port of Nora, Sardinia.

Harbour surveys have been the subject of a number of university and service expeditions, but to obtain really useful results a considerable study of the history and documents of the area chosen is required beforehand; and the results should be published. Air photography can play an important part in the selection of sites for exploration. General Schmiedt's paper (Schmiedt 1964) on the ancient ports of Italy is an excellent example of the type of information that forms the background for the selection of an underwater site; but attempts to interpret from air photographs should be made with caution, for mudbanks and weed can often give a false impression of structure.

The Mensura Diving Team's expedition to Nora, a Phoenician port in southern Sardinia (Macnamara and Wilkes 1967) is an example of the information that a careful underwater survey can give. The town of Nora now stands on a promontory with bays to east and west; that some of the town was submerged was apparent from the fact that a few of the Roman streets ran out into the sea; but the new survey of the somewhat weed covered sea bed, outlined quay walls and small inlets, perhaps docks surrounding a much shallower bay (Fig. 92). Few of these could be seen on the air photo; and in one case what had been taken for a quay wall was in fact the line of an ancient street (Schmiedt 1964, Figs. 3–4). The submergence of the site was shown to have taken place since Roman times and probably caused its abandonment.

A group of divers from Imperial College, assisted the British expedition excavating on the Island of Motya in western Sicily in 1962 (Taylor 1964). This site was also a Phoenician port set in the middle of a lagoon; the aim of the underwater exploration was to map and study a submerged causeway joining the island to the mainland and to search for other structures. Drawings were prepared of the construction of the causeway and from indications first noted on an air photo, the plan of a small boat harbour was discovered and surveyed (Fig. 93). The

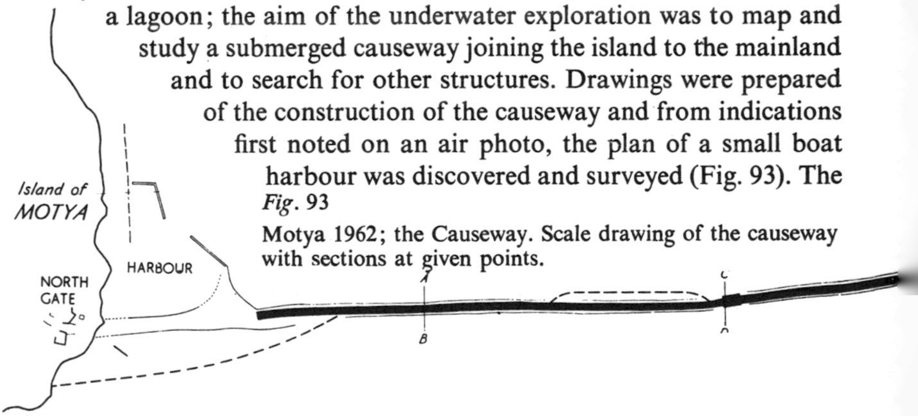

Fig. 93

Motya 1962; the Causeway. Scale drawing of the causeway with sections at given points.

lagoon was very shallow, but an investigation of the mud by means of core samples, as well as a study of the biological material, has helped to confirm that the silting has taken place in the lagoon since Phoenician and Roman times.

Miss Frost's studies of the offshore island of Arwad in Syria (Frost 1964, 1966) were much more intensive and extended over two seasons. The island and its accompanying outlying reefs extended for several miles, so that a topographical survey from the ground and under water only would have been impossible; but with the aid of low-level air photos taken from a light aeroplane covering fixed points measured on the ground, she was able, using graphic photogrammetric methods, to make a detailed plan of structures both above and below water. This was followed by intensive underwater exploration and study of the causes of submergence.

Her survey showed that the huge ashlar walls surrounding the island were built on an artificially levelled rock platform as a protection against the sea. Their destruction was due, not to human agency as had been formerly believed, but to tectonic damage when the island was torn by geological rifts, and parts were submerged. Underwater examination of the spread and direction of the stone from the fallen walls

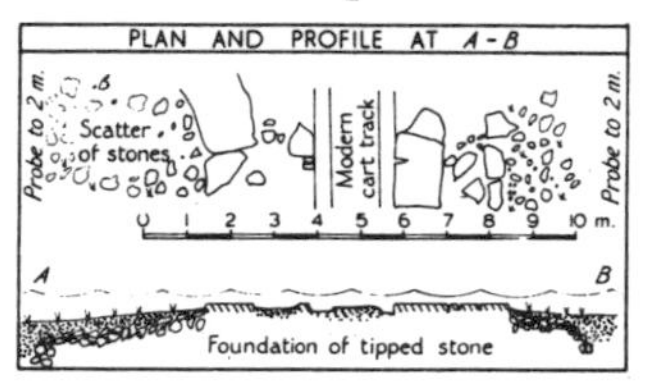

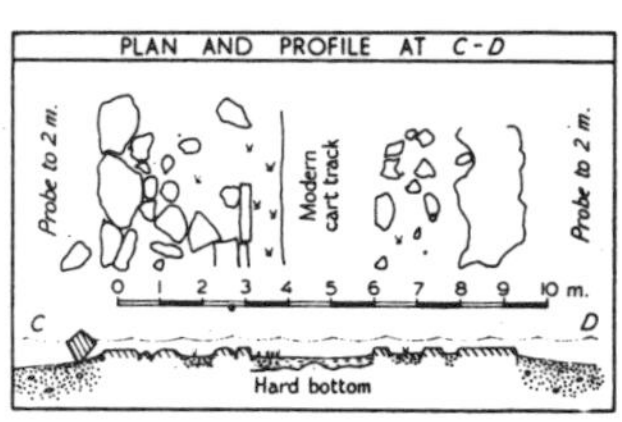

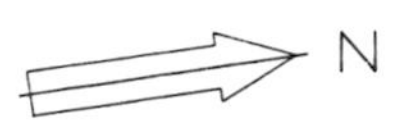

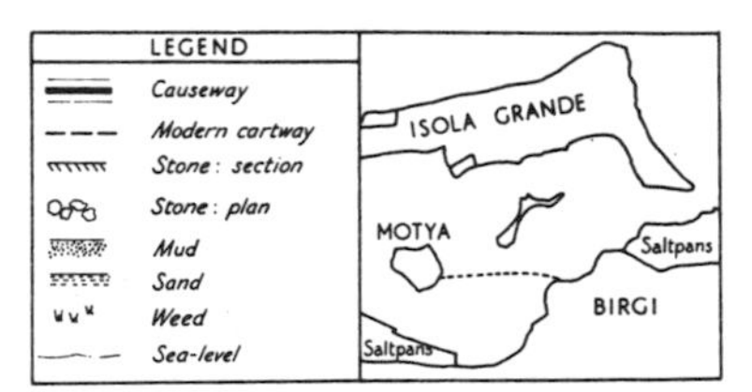

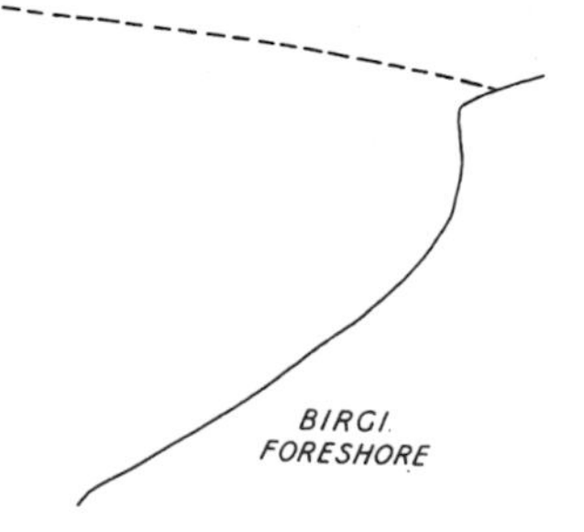

indicated that the island had tilted towards the mainland, and that a small section to the north had split off, carrying part of the wall with it.

The outer harbour or roadstead of Arwad was protected by a dunary reef several miles long, of which only a few small islands indicate the line. Miss Frost made a special study of one, Machroud, which is now half submerged. A central area was partly hewn from the rock, partly enclosed by a line of huge blocks. Four openings now lead under water to a long rock-cut channel descending to at least 9 m. These channels could not have been cut under water and indicate that they must have been made at a time of much lower sea level. The tops are now overgrown with *vermetus* trottoir, and there is lower trottoir which may indicate an intervening sea level. Concretion almost obscures the paving within the enclosure; fissures caused by earthquake have split the island and some of the stone blocks.

The sea is known, from geological sources, to have risen during the first millennium B.C. to a higher level than the present day, and that there has been a subsequent retreat. This fact, in conjunction with the underwater study of the Arwad structures, has enabled a date to be ascribed to them which could not be ascertained in any other way. The walls of Arwad must therefore have been built in the Bronze Age at the end of the second millennium, when the town was already historically famous. A date for the submergence of the reef is indicated by a number of sunken cargoes of amphorae along it; none of these is earlier than the fifth century B.C., from which it can be assumed that the passages through the reef had been forgotten as the sea had risen over it. Machroud now stands at 6 m. above the sea, at an intermediate level.

Miss Frost stresses the need to study the formation of rock pools, and the *vermetus* trottoirs as indications of sea level, as well as the geological phenomena, as guides to archaeological chronology, whereby the geologist, marine biologist, and archaeologist can assist one another in working out a chronological sequence.

Roadsteads and Anchorages

The study of roadsteads and other places where ships have anchored can furnish information on methods of navigation and ancient sea routes. These offshore moorings have marked the place where sailing ships waited for a favourable change of wind to enable them to proceed, or may indicate fishing grounds. Other types of anchorage are temporary shelter in a bay, or an offshore position awaiting the turn of the tide before entering a river estuary or harbour. All these sites contain numerous lost anchors and rubbish thrown overboard. Dumas pointed out that roadsteads have been used over long periods and that the collection of material from them would yield an important cross-section of shipping customs.

Few surveys of this kind have been done. At Gallinaria island off Albenga in Italy, A. Pederzini, under the direction of Prof. Lamboglia, by careful mapping and recording of the finds round the island, demonstrated the use of such a survey (Taylor 1965). The grouping of lost anchor stocks and amphorae on the west side indicated that it was the most sheltered and frequently used; the pottery collected showed that the anchorage was in use throughout Roman times. A similar anchorage was mapped by a NATO diving group from Naples within the Galli islands off Salerno. Eighteen lead anchor stocks, one stone stock, and other parts of anchors, in four groups were recovered but no other finds. It remains an unsolved problem why Roman ships anchored there. An R.A.F. team some years ago explored the small bay off Lara point in Cyprus. Much debris was scattered on the bottom; so they adopted the system of swimming on compass bearings and counting the potsherds and other objects they found in every 5 m. In this way they were able to define the anchorage; and from the pottery recovered it was possible to say that the site was used from sixth century B.C. to fifth century A.D.; but no later.

Though anchors are part of a ship's equipment, they are frequently found separately and it seems better to consider them in connection with roadsteads. Anchor 'cemeteries' have been mentioned off the British coasts, which may indicate fishing grounds rather than roadsteads. Miss Frost (1963a) has made a study of anchors in the Mediterranean showing that stone anchors were first used in neolithic times and often dedicated as votives in temples; they continued to develop throughout the Roman period and indeed are still used by fishermen in many places. The Romans developed the wooden anchor with leaden stock, but it was the Veneti off Brittany who were the first to use the iron anchor and chain which remained almost unchanged until the present day. Many anchors of all these types have been found, but the sources of few of them have been recorded. A stone anchor has recently been recovered from the south coast of Britain, which has prompted the search for its origin as no other has been reported from Britain. A preliminary examination of the stone by the Institute of Geological Sciences has revealed that it is not a British rock, but possibly from Brittany. Though such anchors may still be in use by fishermen, the fact that the stone is not local indicates a cross-channel trade. A closer study of the distribution of such dated finds is important for trade connections as well as the location of the coastal sites.

Inland Waters

The exploration of lakes, rivers, and lake dwellings has been undertaken in France, Germany, Switzerland, and Poland. Since 1957 a broad survey of underwater archaeological remains was organized for the lakes of Central Poland. Exploration was difficult owing to the layer of liquid mud which covered the bottom of many lakes. Nevertheless, parts of pile dwellings, bridges, and piers were recorded on a number of sites.

Poland

The first specific exploration was carried out in Lake Lednickie, north-east of Poznan in 1959–60 (Bukowski 1965). Surveying methods had to be devised and the initial plotting of structures was done by triangulation from two fixed points on shore, with the additional help of compass bearings for special objects. In the course of this preliminary exploration, the remains of two sections of a wooden bridge or causeway was discovered: the west part 438 m. long, the east 187 m.

In 1961 divers, under the direction of archaeologists, started to work at a depth of 10 m., but conditions were difficult as visibility did not exceed 3 m. and at 8–10 m. it was only 1 to 1·2 m. Much plankton in the water also made observation and photography difficult, and the lake bottom was covered with semi-liquid mud which limited full visibility. To facilitate planning, a rope marked off at 5-m. intervals was laid down from east to west to serve as a base line. Measurements and drawings were made with the help of a 3 × 2 m. grid frame.

As there were no traces of buildings near the shore, investigation started 10 m. out, and followed a line 26 m. long by 3 m. wide towards the centre of the lake. Every square metre was drawn on roughened glass plates at a scale of 1 : 5 and then transferred to mm. paper on the general plan.

The piles of the bridge originally found were no more than a dozen cm. long and seldom showed above the mud, whereas in the centre of the lake they stood up to 3–5 m. above the mud. In the deepest parts of the lake (some 10·70 m.), the piles were at least 13 m. long including the part driven into the lake bottom.

The stratigraphy of the lake mud was also examined: the third layer was recognized, from the rubble and traces of fire, as the destruction of the bridge; the fourth layer was considered to be after the building of the bridge, and the fifth contained many scraps of wood which were believed to come from its construction. The sixth layer contained no objects.

Finds during 1961 included an iron helmet of Czech type, several iron spears, one of them inlaid with silver and still attached to part of the wooden shaft, an iron battle-axe, several whole and many fragmentary vessels dating from the tenth and eleventh centuries, a number of tools and wooden objects, bone, iron, and iron carpenter's axes probably lost during the building of the bridge. The group of artifacts indicate that the bridge was probably destroyed during the Czech invasion under Prince Bretislav in 1038. Several wooden boats were found and also a dug-out canoe 10 m. long, which was brought up for exhibition.

In 1961–2 a team of archaeologists from the Polish Academy of Sciences in Warsaw, assisted by two groups of skilled divers, undertook excavations in Lake Pilakno near Rybno. Pottery had been brought up in the 1930s and other interesting objects were recovered in 1959–60.

Lake Pilakno is a post-glacial lake with strongly corrugated bottom, in places reaching a depth of 55·6 m. The clear surface water is particularly noticeable. The settlement lies in the southern part of the lake, some 20–30 m. from the swampy shore. Preliminary geological borings showed that the settlement had been built on a limestone lake shoal, which archaeological and geological examination indicated was at that time at water level. According to local information, that part of the lake directly over the settlement did not freeze in winter, or was only covered by a very thin crust of ice, possibly because of springs rising from the lake bottom.

Preliminary geological examination established the following stratigraphy:

0–15 cm. sediment humus with lake chalk mixture.
15–25 cm. cultural layer containing timber, vegetable and other remains.
25–350 cm. lake chalk.
350–450 cm. chalk gytia (gyttja).
450–5 cm. humus.
455–65 cm. sand with gytia.
465–500 cm. sand and gravel (natural lake bottom).

Archaeological excavation only penetrated the culture layer and the upper chalk. The site to be examined under water was covered with reeds growing on a thick layer of liquid mud containing vegetable matter.

The excavation party was based on two floating rafts, 3 × 3 m. and able to support a load of over a ton, moored right over the excavation. One of the rafts held a tent for the use of divers and archaeologists, the other carried all the heavy equipment, compressed air bottles and fire engines (Fig. 94). Wooden boats and small rubber rafts and dinghies were also used. The first job was to cut away the reeds and remove all the rubbish covering the bottom. Uprooting the reeds and pulling out the bare roots proved impracticable and harmful, since they had often grown right into the old wooden building, and beams lying underneath could easily be destroyed.

Squares of 5 × 5 m. were marked out on the bottom with contrasting nylon line, and related to permanent survey points. To clear the layers of sediment from the construction and reveal the artifacts a strong jet of water was used. This was supplied by the fire engine on one of the rafts. At first, pumping out the liquid mud was attempted, but this was found to be slow and uneconomical as the pump was constantly clogged by plant remains.

Fig. 94

Diagram of underwater explorations in Lake Pilakno, showing fire engine on raft, water jets with multiple nozzles, and narghile air supply to diver. The archaeologist watches from a rubber dinghy.

A new method was therefore tried. This consisted in altering the direction of flow to give a strong jet of water at 3–4 m. This stream of water was made to flow 20–30 cm. above the layer to be examined. A diver swimming immediately above the bottom, stirred up the water flowing just over the bottom with his hands, and lifted the mud and vegetable remains. This allowed the diver and the supervising archaeologist to see through clear water and make their observations. The size and direction of the stream of water was governed by a series of three to five water-jets of various sizes fixed to a heavy iron frame lowered to the bottom. It was absolutely necessary to fix this frame permanently, as the jet stream caused it to move about. The water jets were set at various angles, both upwards and to the sides so that the mud did not settle nearby. The underwater work was carried out by divers wearing bottles, which enabled them to stay an hour and a half on the bottom. Later the 'narghile' system was introduced served by large bottles on the raft. Suits were indispensable for this kind of work in Poland, as in summer the water temperature was seldom higher than 15–20°C. With a rubber suit the diver could work for two hours under water.

The diver's work was continually supervised by an archaeologist overhead, paddling along in a rubber dinghy and wearing mask and snorkel. He could see the diver at work all the time and give instructions when necessary. His range of vision was wider but no worse than that of the diver and from a different angle.

The artificial stream of water carried away all rubbish and waste, pushing the heavier pieces along the bottom, sometimes far out of the excavated area; a strong vertical screen of close mesh had to be set up at the limit of the excavation, from which much valuable botanical material was recovered. When the excavated area had been cleaned, it was covered by a 1 m. grid of nylon line, suitable for the planning and photographic record. Every square metre was photographed vertically and the more interesting parts also from an oblique angle. All artifacts were recorded in three dimensions.

For more precise drawing, 1 × 1 m. wooden or metal grid frames were used, painted in contrasting colours to make them visible at a depth of 5–6 m.; they were in turn gridded at 10 cm. intervals with nylon line. The 'Plexiglass' drawing boards were covered with a grid of 10 × 20 cm. divided into 2 cm. squares to facilitate drawing at a scale of 1:5.

Sections of the plan were joined to the over-all plan. Every layer was recorded separately under water, or floating on the surface. In the latter case, the draughtsman worked in mask and snorkel from a dinghy with his head and drawing board under water; in this way he had a better view of the area, provided the depth was not more than 2–3 m. and the water clear. The diver's help was required in moving and setting up the frames and also in recording the deeper finds.

When the occupation layer had been cleared of timber, it was thoroughly searched for artifacts which had accumulated beneath; many had sunk into the ground well below the construction and included pottery and grindstones as well as bones and sherds. Organic samples were also collected; parts of the wooden structures, charcoal, seeds, and other vegetable remains, some of which were contained in pots.

In addition to investigating a particular section of the site, the clear water enabled the team to establish its limits by observation with snorkel and mask. Many more wooden structures were noted, as well as threshing floors and hearths. The excavated area was about 110 m. by 40 m. usually at a depth of about 1½ m. extending to 2 m. No structures were found near the present shore which confirms the idea that the settlement was built on a shoal. The water level at that time could be determined by the remains of half-burnt stakes.

Summarizing the results, besides testing new methods, the excavation showed that the settlement was built on a system of superimposed logs laid on the shoal. The logs were fixed by stakes driven in, either vertically or at an angle. The logs, planks, and stones formed no systematic shape, but had fallen in a confused heap. In the western section some beech and alder beams were lying parallel and in the eastern part, a flat paved area about 9 sq. m. bore traces of a hearth. The complex was thought to belong to a farm building with a road adjoining. The cultural layer, 15–30 cm. thick, was found beneath the collapsed building and on the frame of the structure. The occupation of the site was not a long one and it was burnt before it was deserted.

The scanty remains of pottery, and a bronze necklace, date the site to about 350–200 B.C., a date confirmed by the palynological and geological evidence. A C^{14} date obtained for the timber gives 230 $\pm$120 B.C. A final test by core sampling was made for the geological and palynological evidence which also confirmed the findings.

North America

In north America at the height of the fur trade era, trade goods were brought to Grand Portage on Lake Superior from eastern Canada and the northern parts of the United States (Kenyon *et al.* 1963–7). These materials were then packaged into bales for distribution into remote trading posts in present-day Minnesota, North Dakota, Ontario, Manitoba, and Saskatchewan. Canoes of birch bark were used to transport the goods along the waterway of lakes and rivers which form the modern boundary between Minnesota and Ontario.

Numerous hazards were present along the route, and a number of accounts relating canoe upsets have been preserved in fur trade literature. The Minnesota Historical Society jointly with the Royal Ontario Museum of Toronto is now engaged in the underwater recovery of highly significant fur trade artifacts from locations where such accidents took place. The project is directed by Mr. Robert Wheeler of the Minnesota Historical Society and Mr. Walter Kenyon of the Royal Ontario Museum.

The project began in the summer of 1960 (Kenyon *et al.* 1963–7) when E. W. Davis took three divers on a search of likely spots for canoe accidents. They found a nest of seventeen graduated brass trade kettles at the foot of the rapids, which were donated to Minnesota Historical Museum. Up to 1962 other materials recovered included two flint-lock trade muskets, thirty-six iron trade axes, and twenty-four chisels and spears. Items from other locations included more kettles, axes, musket balls, lead shot, buttons, thimbles, cloth fragments, knives, beads, whetstones, gun-flints, and a large assortment of iron objects.

Search in the turbulent waters of the rapids was difficult, and meant crawling from rock to rock and crevice to crevice, but it was soon found that turbulence was only a problem in relatively shallow water; on the bottom from thirty to thirty-five feet below the surface there was no turbulence, only a steady current.

On the Canadian side, divers in the French River were most successful, bringing to the surface ancient muskets, copper kettles, and ice chisels. A cluster of more than a hundred axes was found to have been packed in wooden boxes, each full box weighing ninety to a hundred pounds. The voyageurs were so pressed for space that they filled even the eyes of the axes with gun-flints. The collection included several hundred musket balls, over 200 gun-flints and an assortment of strike-a-lights, hinges, double-pointed awls, wooden handled iron knives, and a variety of files. Most of the items are remarkably well preserved, and though they cannot yet be precisely dated, they would appear to be largely of French origin between 200 and 250 years old.

From 1963, with financial support, the project was set on a firm footing, for which six inter-related phases of research were planned:

1. A search for fur trade artifacts in the rivers and lakes which form the Minnesota–Ontario boundary, correlated with pertinent fur trade literature.
2. The study of recovered artifacts by an archaeologist experienced in fur trade materials, with co-operation from metallurgists and other specialists.
3. The establishment of a type collection of dated fur trade artifacts, and a photographic file of other fur trade materials. Both would be made available for scholars for research purposes.
4. A search for documentary materials relating to manufactured goods and trade goods (i.e. fur trade inventories, list of goods distributed to Indians at treaties, etc.) in both published and manuscript accounts. This information would be collated and cross-related in the hope of establishing trends in trade goods used in North America from about 1690 to 1850.
5. The development of improved methods for underwater recovery of artifacts as the project matures.
6. The establishment of an advisory committee of a dozen or more prominent students of the fur trade and its artifacts to assist in aspects of research. The publication and dissemination of information under phases 2 and 4.

The Quetico-Superior Underwater Research Project, as it was called, has continued its researches on sites in Canada associated with the eighteenth- and nineteenth-century British forts, where land excavation is also taking place. At St. Joseph's island they raised a sunken batteau which had been covered with stones, and was thought to have been sunk as a form of cribbing for the wharf. On the Richelieu river a quantity of pottery of some 2,000 years ago was found in a confined area and probably represents an early boating accident. Most of the material belongs to the seventeenth, eighteenth, and nineteenth centuries from trading voyages with the Indians; but in one instance military equipment included two broad axes stamped with a broad arrow, and badges of the troops stationed near the river in 1782 and 1814.

In 1966, the first attempt was made by the Minnesota Historical Society to search for a particular canoe accident which had been graphically reported by Alexander Henry of the Northwest Company in 1800. It was lost in the north channel of Boundary Falls in the Winnipeg River on 9 August, and a list of the trade goods was known.

Within twenty minutes of entering the water, a diver had reported a whole bundle of iron axes on the bottom, and others rapidly located knives, musket balls, copper kettles, and beads. The discoveries, marked by plastic buoys, were found to lie in two groups indicating two canoe accidents. One group corresponded closely with the list of objects given by Henry, while the other contained tin dishes among other objects, made by Townsend and Compton, not before 1801. This clearly indicated another accident a few years later than that which happened to Henry's canoe. In this later canoe, the wooden handles of a quantity of jack-knives were also preserved, as well as a twist of tobacco which the voyageur had smoked as a cigar.

This 'White Water' archaeology in the American and Canadian rivers is now a regular part of the annual archaeological programme and is furnishing a large number of artifacts to illustrate the history of the period.

Wrecks

The study of ships not only includes their construction and equipment, but also their cargoes. When a ship sinks she represents a 'capsule in time' wherein she and her contents are all of contemporary use and her equipment and cargo as well as the possessions of the crew and passengers form one of the most valuable pegs for archaeological chronology and the study of ancient trade.

Wrecks already discovered cover a wide range of time. The amphora wrecks of the Roman period in the Mediterranean were the first to be observed; but soon other types of pottery of different periods were found together with cargoes of stone, sculpture, sarcophagi, and bronzes.

Outside the Mediterranean bronze and iron guns are the most readily identified objects and many have been reported. These do not necessarily indicate warships, for merchant ships continued to be armed till the end of the eighteenth century. These are also associated with the so-called 'treasure ships' which have furnished a number of wrecks of Spanish ships, Dutch East Indiamen and other trading vessels around the coast of America, Australia, and Africa, as well as around our own shores. Lastly there has been the search for named vessels such as the *Vasa*, sunk on her maiden voyage in Stockholm harbour, in Sweden, the *Liefde* and the *Association* and some Armada wrecks in U.K. and Eire.

The study of wrecks begins with the initial discovery and reporting of an ancient wreck, or the discovery of a named wreck as the result of the study of ancient records. On the careful survey and recording at this stage will depend the decision whether the wreck is worth further study, and also the assessment of its date. Wrecks which are worthy of further work can be divided into two groups: broken-up wrecks, and those well-preserved by sediment.

Experience has shown that broken-up wrecks of which little but the armament or cargo remains are well worth survey and excavation for even though the hull may have been destroyed and the contents scattered, the disposition of the cargo, anchors, the crew's possessions, and the armament form, nevertheless, a very valuable archaeological group.

Very few wrecks reported are in as good condition as the *Vasa*; if they have become well covered in sand or silt there is a good chance that they will be worth total excavation, but in this case, very little will be showing on the sea bed, and the clues to the position of the wreck may be few.

When not discovered by chance, wrecks have usually been found by diving groups working regularly along an area of coast, possibly where there are known hazards. The discovery of an anchor, a gun, or some fragments of pottery may indicate the presence of a wreck.

But more organized searches have borne the most fruit. One may cite Mr. Throckmorton's (1965a) original exploration of the Turkish coast, in which he rightly assumed that the sponge divers would have seen many ancient wrecks on the sea bed and that useful information might be gained from them. After several months of diving in their company he was able to record some forty ancient wrecks; and as a result of the photographs, sketch plans and drawings of amphorae and other finds made by Throckmorton and by Frost (1963b, Chapter 10) two of the most important were selected for the first total underwater excavation. Mr. Throckmorton has continued this method successfully in Greece and south Italy, basing his exploration on local information gathered from fishermen and others.

In Britain, an area in North Wales was selected by the North Wales Archaeological Exploration Group as a likely position for wrecks. A prolonged search over several months accompanied by a traverse survey led to the discovery of a nineteenth-century cargo of lead ingots with which were stowed ivory tusks.

Land techniques for planning and recording have had to be modified for use under water and the diver's time in the sea being limited, photographic and mechanical means have been devised to economize this to the utmost. Primary exploration was visual, with the diver towed behind a boat and dropping markers to indicate items to be investigated. This was uneconomic in manpower, and scientific sonar and other devices have been adopted.

Dr. Edgerton's mud-penetrator or 'pinger' has been used in harbour sites to define the ancient sea bed and to search for structures; more recently it was used to search for the buried hull of the seventeenth-century warship *Mary Rose* in the Solent. Though a strong anomaly was registered, it remains to be seen whether this is really the hull of the ship which lies beneath 20 feet of mud.

Dr. E. T. Hall of Oxford (Hall 1966a) has adapted the proton-magnetometer for the detection of wrecks and other underwater anomalies. This instrument being dependent on the presence of iron, it is necessary to estimate the approximate detection distance required when undertaking a survey, since this knowledge will affect the planning of the search. Iron and steel ships of ten tons and upwards can be detected from 45 m. and more; but the wooden wreck with a limited amount of iron, cannon balls, fittings, and anchors, or amphora wrecks will need much closer survey at perhaps 3 m. or less from the site. Trial runs in the Mediterranean in 1966 were very promising, but in the open sea a more precise position fixing of the anomalies was required, and Dr. Hall has at the same time developed a sea-going sonar position-fixing system (Hall 1966b).

For search in deeper water, Dr. George Bass (1965) of Pennsylvania University designed and used a two-man submarine *Asherah* for archaeological exploration. She is fitted with viewing ports, motors to move her in all directions, and lights and cameras to make stereo surveys; she can reach a depth of 600 feet and remain under water for up to ten hours. She has been used in conjunction with Dr. Hall's proton-magnetometer to investigate anomalies at depths beyond the easy reach of divers and can with her cameras make a preliminary survey of a site in one or two runs, which can subsequently be converted to a scale plan on land.

These methods are still in the experimental stage, and have to prove their real value against empirical search; but they may well have a useful place among the tools of archaeological exploration.

The initial survey of a wreck when discovered may be made by the methods described by Ian Morrison, p. 23, and should include photographs of the more important aspects. From these it is possible for the archaeologist to assess their value for further study. (Cf. Frost 1963b, Chapter 10.)

Broken-up wrecks are most frequently found close in-shore. Mr. Throckmorton has planned a number of stone carriers with columns and sarcophagi using simple tape triangulation methods. One of the more interesting was at San Pietro di Bavagna in south Italy (Throckmorton 1965a), containing a cargo of marble sarcophagi; at least a dozen of these roughed out marble blocks were shipped, some packed one inside the other, others with slabs of stone for the lid still attached. The general outlines of the Roman marble trade, and that stone was quarried and stored in bulk both at the quarries and in the marble yards of the principal ports of the Roman Empire, are known from finds on land; but until now, the method of shipment of that material and the tonnage of the bulk cargo carriers was unknown. If the stone can be identified, this will be an indication of the port of origin.

Another example is that of a wreck broken up on a reef, where the cargo and contents have become scattered. The heavy objects of metal such as guns, coins, and equipment, though they may no longer lie exactly in their original positions on the ship, as groups still have some significance and as a collection of artifacts of one period are valuable.

On the reef beside the Great Basses lighthouse in Ceylon, an ex-pearl diver discovered the remains of a wreck carrying a hoard of silver coins and some guns. Enlisting the help of Mr. Throckmorton and Mr. A. C. Clarke in 1962–3, he planned and raised the finds and from the resulting investigations, these deductions were made (Clark 1964, Throckmorton 1964). The ship was about 120 feet long, she carried a British bronze cannon weighing 333 lb. and 21 other guns, one of 10 feet and the others 8 feet long. There were also two small swivel guns and a wooden pistol stock. The silver coin was packed in bags round the cannon, and more than three hundredweight was brought up in lumps retaining the shape of the original bags. The coins were Surat rupees in excellent condition, bearing the Arabic date equivalent to 1701–2. The final study may well identify the ship.

Excavation

The selection of a wreck for total excavation must be based on the evidence of the preliminary survey. As the work will be both costly and lengthy, the choice must depend on:

(a) the condition, enclosed in sand or mud, which may indicate the likely state of preservation;

(b) whether the period of the wreck will contribute to archaeological knowledge and not, at present, be a repetition of a type already examined;
(c) its position and convenience for the conduct of the excavation;
(d) the funds and team available.

So far total underwater excavation under archaeological direction has only taken place on four occasions: on the Bronze Age wreck and the Byzantine wreck in Turkey, on the Viking ships at Roskilde in Denmark (in which the work was completed as a land excavation within a coffer dam), and the *Vasa* in Stockholm harbour.

The Bronze Age ship was selected because she was the earliest ship known (and still is). This important find caught the imagination of the archaeologists and a joint American–British expedition was formed (Bass 1967). With George Bass of Pennsylvania in charge, Peter Throckmorton, the discoverer, Honor Frost, who had already worked in Turkey, and the experienced advice of Frédéric Dumas, the party worked with two sponge-diving boats from a nearby shore-based camp. The cargo of ingots lay on a rocky bottom free of sand but masked by a heavy growth of marine concretion which made it difficult to free and hid many of the finds. The site did not lend itself to the use of a grid and the planning was carried out by triangulation with tapes. Nor could the concretion be removed under water, so that it was necessary to cut large sections of the cargo up into lumps, before bringing them to the surface to join them together on the beach where the final careful dissection could take place.

The heavy weight of the copper ingots, over 1,000 kg., had held the cargo in position on the sea bed together with any objects beneath them; but of the hull, only a small piece under some ingots remained. Reconstruction of the size and shape of the ship is tentative, but from the cargo and finds the conclusion could be drawn that she was carrying bronze craftsmen and traders from the Levant towards Greece. Large numbers of broken tools, as well as tin oxide showed that with the copper, probably from Cyprus, the ship carried all the ingredients for trading new tools for old. The cabin end of the ship was established and from it came pottery and mortars for the crew's use, weights for use in trade, scarabs, and a cylinder seal. The scarabs gave the closest evidence for date, which was subsequently supported by the pottery study and C^{14} samples. She is believed to have sunk about 1200 B.C.

In the following year Bass commenced work on the Byzantine

amphora wreck off Yassi Ada near Bodrum (Bass 1966, Chapter 8). This was more conventionally situated on a reef, and buried in sand. After some experiments a fixed iron grid was erected over the whole site which remained in position throughout the excavation. With it, the most accurate plan and photographic record was made throughout the three years of exhaustive excavation. Every item from the ship was tagged and raised, and ultimately a large part of the hull was uncovered and brought to the surface in many sections. From these detailed records a reliable reconstruction of the ship, its cabin, the position of the masts and anchors has been made possible.

The main hold contained about 900 globular wine amphorae of a type known in the Aegean. The cabin aft had a tiled roof and floor, contained the galley, a drinking-water jar, and metal and pottery vessels for cooking. About forty coins, of which four were of gold, from the reign of Heraclius II, gave a date for the sinking in the mid-seventh century A.D. Scales and weights were again among the finds, also a steelyard in fine condition bearing the name of 'George senior sea-captain'. Though the excavation has been completed, the final report has not yet been published.

Simultaneously in 1962, Danish archaeologists began the excavation of the Viking ships in Roskilde fjord (Olsen and Crumlin Pedersen 1958, 1967). These sunken blockships filled with stone had been discovered by divers in 1957. Though in shallow water, visibility through the mud stirred up was very poor and planning, though successful, was very difficult. Five ships were identified and their excavation was considered sufficiently important to warrant the construction of a coffer dam and complete the work in the open. The five ships when cleaned were found to be three merchantmen, a converted warship, and a passenger ferry. Thousands of pieces of timber were recorded in position, for all the iron fastenings had corroded, and each piece of wood had to be raised and placed in an airtight plastic bag for removal to the conservation laboratory. Conservation will take a number of years but it is hoped ultimately to reconstruct the ships. They were sunk about A.D. 1100 to block the channel against invasion.

The magnificent salvage of the *Vasa*, sunk on her maiden voyage in 1628, remains unique. The type of research carried out by Anders Franzen (1960) to identify the position in which she sank was rewarded after two years of exploration in the dark and turbid waters of Stockholm harbour. In 1961, after the position of the ship was confirmed, it was thought her full recovery might be possible. A salvage committee was set up, and with the help of naval divers and a salvage company, the complete ship was raised and brought to dock in 1961. Of necessity this was a professional job and archaeologists could not have undertaken it without their help: nevertheless the recording, and now the conservation of the finds, fell to their share, and the preparation of the special exhibition dock and her final conservation are still in progress. She remains a unique find, well preserved owing to the conditions obtaining in the Baltic which is free of the teredo; and, in spite of the damage by earlier salvage operations, she furnishes a complete picture of conditions on a vessel of the seventeenth century.

The excavations described illustrate the very full information that can be obtained, if at the cost of a considerable amount of time and funds. For a different reason, Throckmorton (1965b) undertook in 1963 the excavation of some recent, dated ships in Porto Longo harbour near Methone in the Peloponnese. His view was that this would yield:

(1) the chemical analysis of dated material, giving a good idea of corrosion processes over a time span impossible to duplicate by experimental means;
(2) a first-hand study of the destruction of a wooden ship by marine organisms;
(3) reconstructions made from archaeological sectioning which could be checked against existing data and serve as a check on excavation techniques;
(4) calibration of experimental devices on sections of modern wrecks, increasing our knowledge of how to find wrecks invisible under the mud;
(5) reliable data on actual construction methods and materials of eighteenth- and nineteenth-century ships, of which there is a scarcity though many plans exist.

The ships explored were the *Colombine* and the *Asia* sunk in 1824, an Austrian brig sunk in 1860 (Fig. 95) which he discovered by chance, and the *Heraclea*, a schooner sunk in 1940. All were covered in deep mud; by his system of sections, Throckmorton was able to show not only that underwater trenches will remain stable, but also the manner and time in which a wooden vessel will break up under given conditions. In Porto Longo harbour, from the evidence of the *Heraclea*, it appeared she flattened out on the bottom and became covered by mud within twenty years.

F. Dumas (1962, p. 18) suggested some years ago that in order to understand a buried wreck, to be able to dig in the right place and to

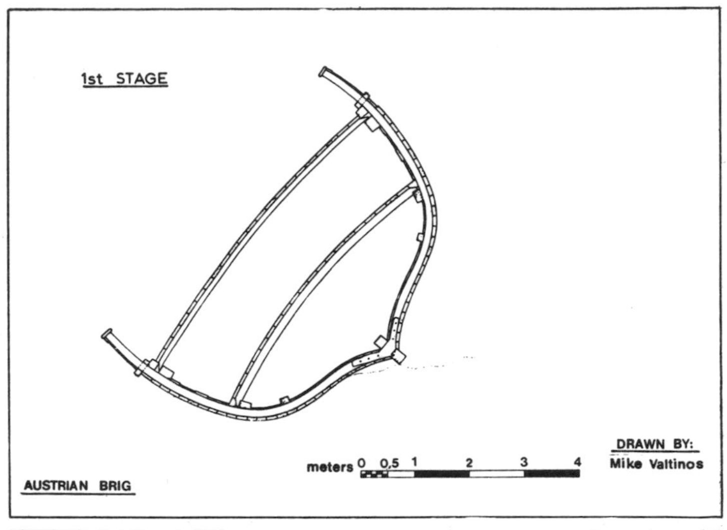

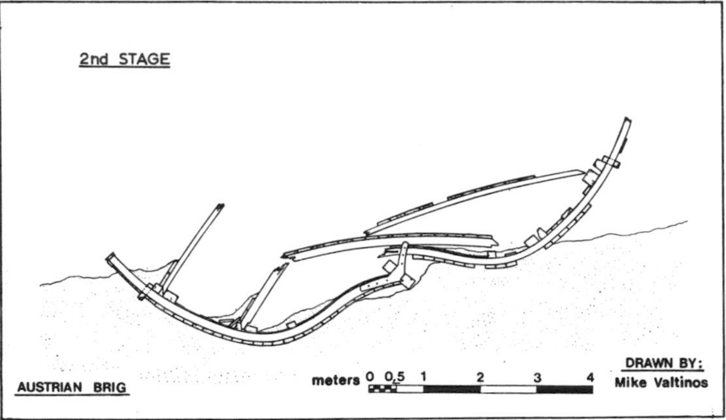

Fig. 95

Reconstructed stages of the destruction of the Austrian brig sunk in Porto Longo harbour.
(After Throckmorton.)

The first stage, in which the wreck lay intact on the bottom, cannot have lasted more than ten or fifteen years, until the hull, weakened by teredos on the starboard side, was pushed outwards by one of the south-east storms which sank the ship.

The second stage, during which the wreck's disintegration must have been accelerated until the fore-and-aft members collapsed of their own weight, cannot have lasted more than five to ten years.

Thus the third stage, during which the wreck gradually fades into the bottom, will have been reached about twenty years after sinking.

The stabilized section shows the wreck as she was when excavated.

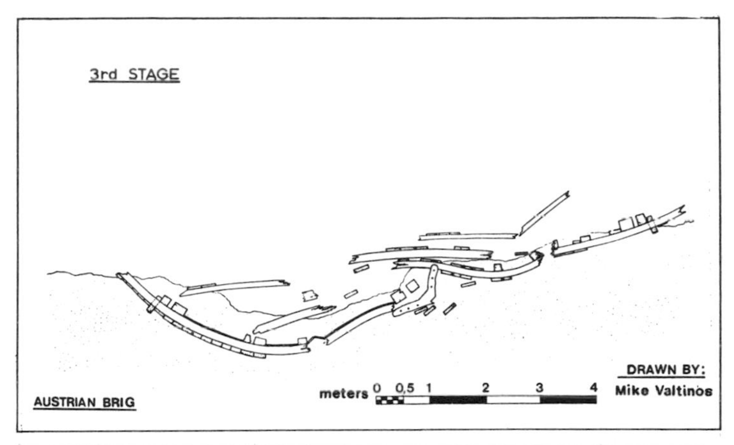

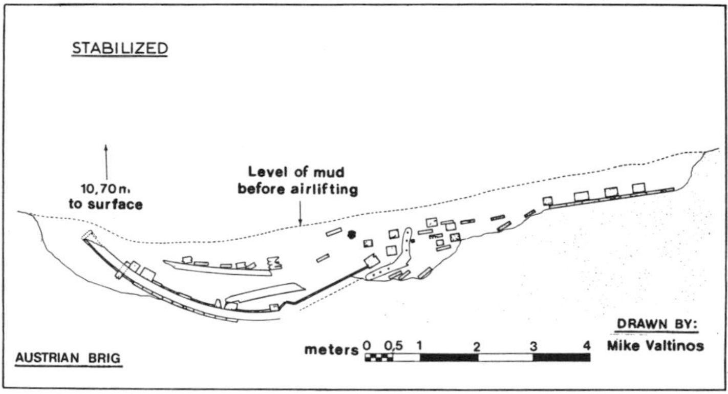

have a basis for the future reconstruction of the finds, the following points have to be established at the outset:

the axis of the visible wreck mound (usually its crest);
the keel axis, which should be found before digging starts;
the area of spilt cargo below the surface;
lastly, the list of the boat before collapse, or an axial plane taken at right angles to the keel. If the boat came to rest upright, this plane must correspond to the axis of the keel; but if she listed, the plane will be more difficult to determine, though it may emerge from the excavation and reconstructions.

He suggested that the outline might be defined with a core sampler, as shown on Fig. 96. When the area had been defined, it could then be

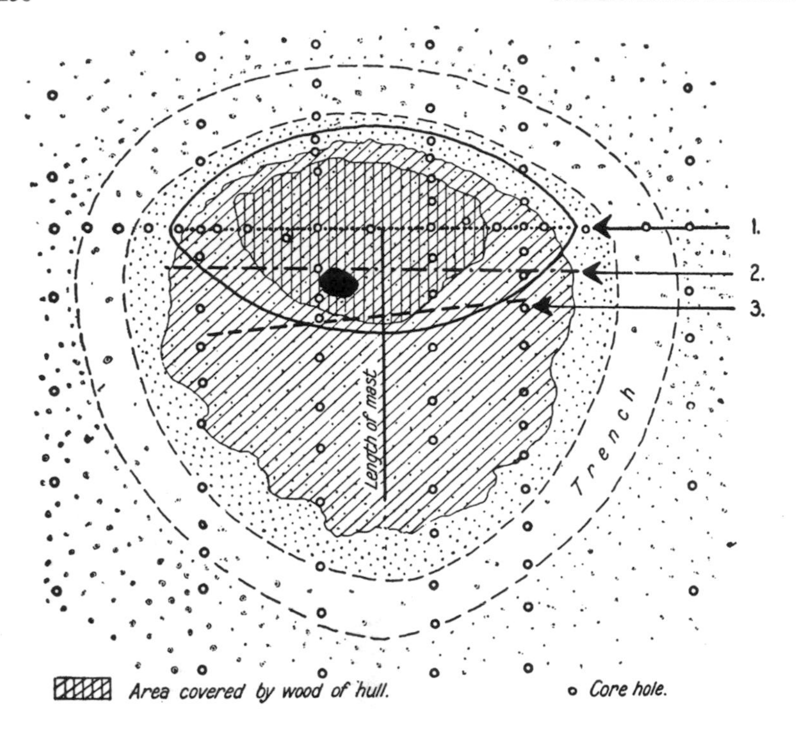

Fig. 96

Schematic plan of a buried wreck-area.

After a preliminary analysis which has been tested by coring, a trench is dug with the air-lift, around the significant area. Arrows: 1. axis of buried keel; 2. surface intersection of plane through mast and keel; 3. visible crest of tumulus.
(From *Deep-Water Archaeology* by Frederic Dumas, reproduced by permission of Messrs. Routledge & Kegan Paul.)

surrounded by a peripheral trench dug with the airlift, into which the spoil could then be drawn.

This method of defining a wreck before excavation has now been used, substituting magnetometer, metal detector and probe for the core sampler, which might be rather destructive if used on the cargo.

Following the suggested method (Green *et al.* 1967), Pennsylvania University Expedition led by Michael Katzef, explored a fourth-century B.C. amphora wreck off the north coast of Cyprus in preparation for its excavation. Only a small number of amphorae were seen emerging from a sandy sea bed in 30 m. of water (Katzef 1969, 1970).

In order to outline as far as possible the extent of the wreck without disturbing the site, a survey was made by magnetometer, metal detector and probe on a grid pattern of one metre squares laid out with nylon rope. Of the three, the magnetometer was not very satisfactory through the lack of ferrous metal, but the probe yielded the outline of the amphora cargo and the metal detector the areas of metal. From the accompanying plans (Fig. 97) the axis and limits of the ship could be defined preparatory to excavation, and it is apparent that the bulk of the cargo had spilled to one side. How far this type of survey has proved to be accurate will be seen after the excavation.

The Future

It is evident that underwater exploration without excavation is still the first and most important archaeological requirement. Looting in all waters has already destroyed many sites, but nations have been rather slow in providing legislation to protect underwater discoveries.

Another approach which has had some success in Britain and France, is to train interested divers to become amateur archaeologists, and co-operate with their land counterparts in recording discoveries. In Britain a *Committee for Nautical Archaeology* has been established to bring divers and archaeologists together. It organizes lectures and weekend courses given by archaeologists, related scientists, and divers. As a result, diving groups are working with local archaeological organizations, and finds are being recorded to acceptable standards.

In other countries, too, divers are being called in to assist archaeologists; but the principal safeguard for underwater sites is for archaeologists to organize surveys round their coasts. In this way the most valuable sites will become known and can be scheduled for future study and, perhaps, excavation. Minor sites which are identified and individual finds on land, furnish valuable contributions to the geographical distribution of archaeological and historical material.

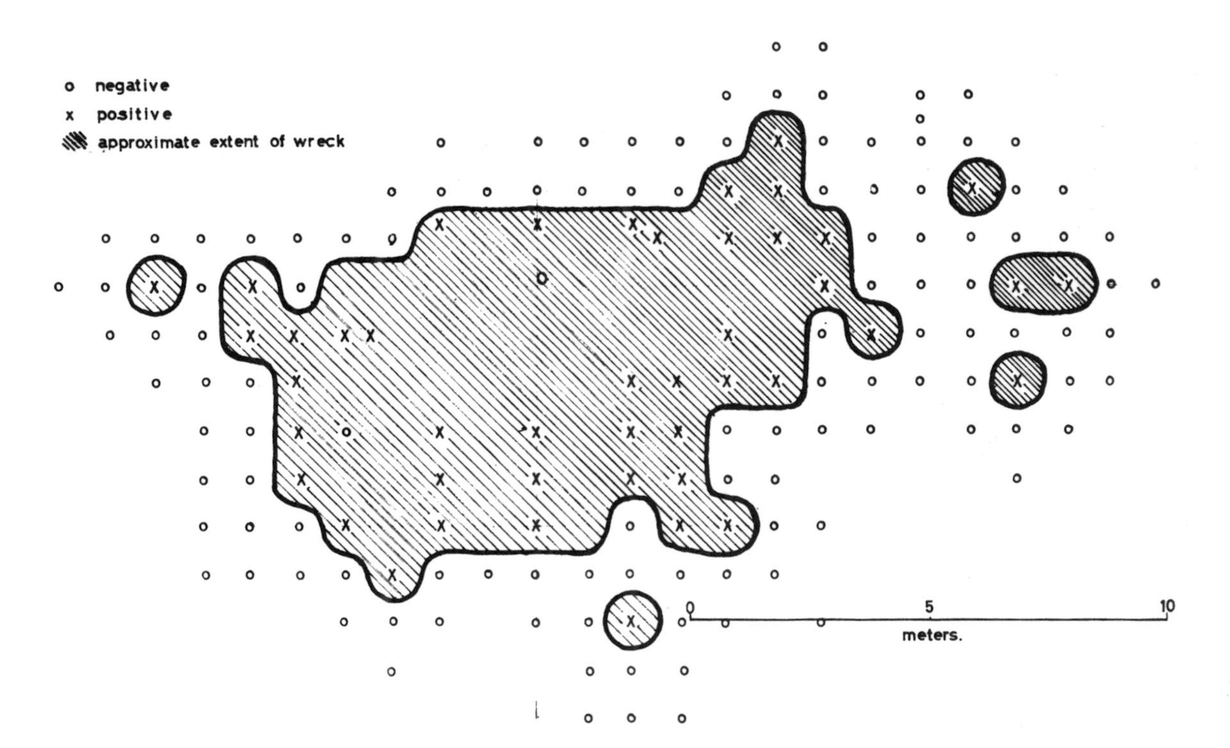

Fig. 97 *Above* Probe Survey. *Below* Metal Detector Survey. (Drawn by Jeremy Green.)

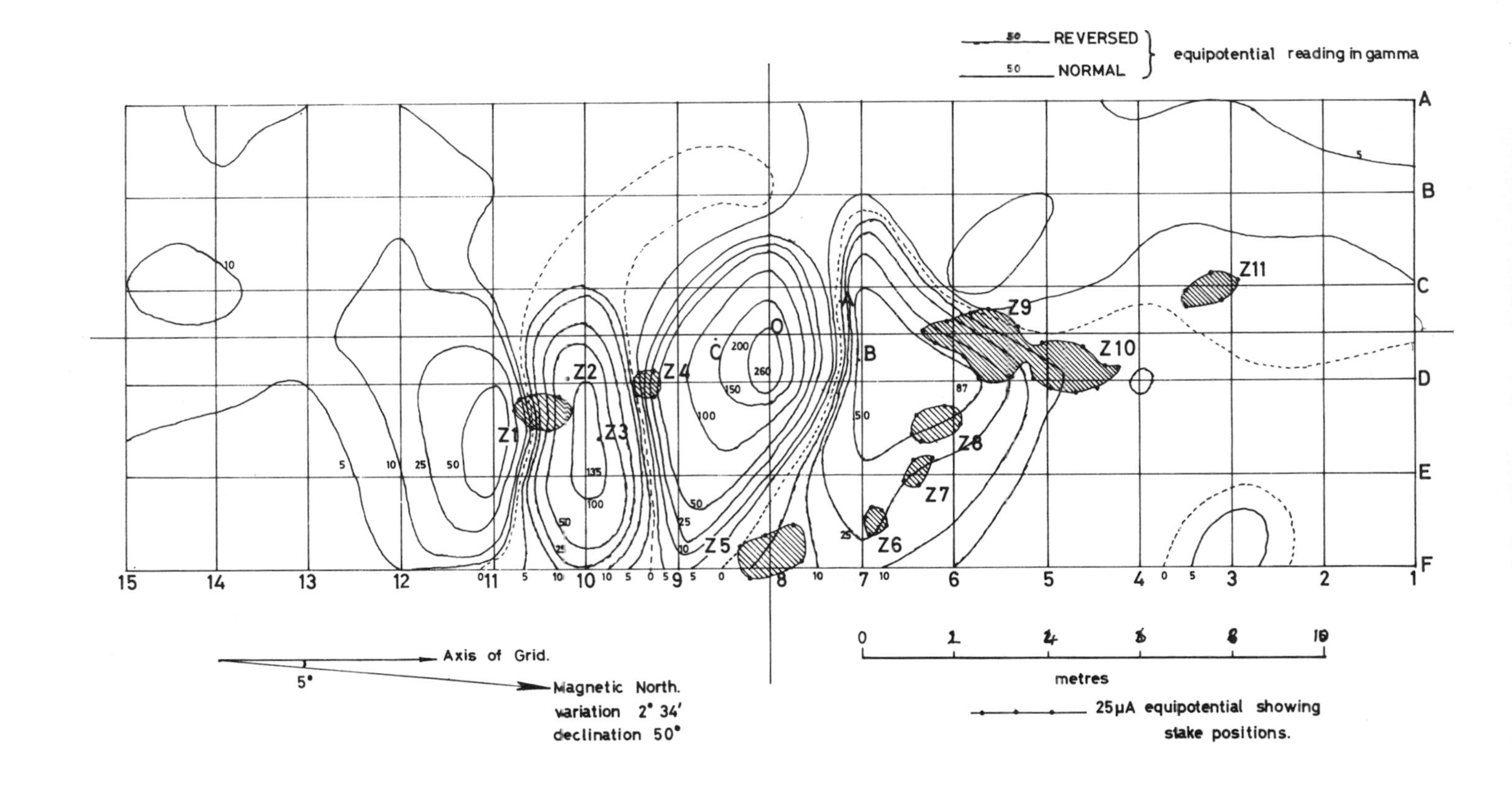
REVERSED
NORMAL
equipotential reading in gamma
Z1
Z2
Z3
Z4
Z5
Z6
Z7
Z8
Z9
Z10
Z11
metres
25μA equipotential showing stake positions.
Axis of Grid.
5°
Magnetic North.
variation 2° 34′
declination 50°

References

Akeroyd, A. (1966), 'Changes in relative land- and sea-levels during the post glacial in southern Britain with particular reference to the post-mesolithic period.' Thesis, M.A. London University.

Barkman, L. (1967), *On Resurrecting a Wreck*. Stockholm: Wasa Dockyard.

Bass, G. (1965), 'The Asherah: a submarine for archaeology.' *Archaeology*, **18,** 7.

Bass, G. (1966), *Archaeology Underwater*. London: Thames & Hudson.

Bass, G. (1967), 'Cape Gelidonya: a Bronze Age shipwreck.' *Trans. Amer. phil. Soc.*, **57,** 8.

Bukowski, Z. (1965), 'Remarks on underwater archaeological research based on the example of a settlement within Lake Pilakano, Mragowo District (north-eastern Poland).' *Arch. Polona*, **8,** 105–23.

Clarke, A. (1964), *The Treasure of the Great Reef*. London: Barker.

Dumas, F. (1962), *Deep-water Archaeology*. London: Routledge & Kegan Paul.

Evans, J. D. and C. Renfrew (1968), *Excavations at Galiagos near Antiparos*. Appendices I and XI. London: Thames & Hudson.

Franzen, A. (1960), *The Warship Wasa*. Stockholm: Norstedt & Bonnier.

Frost, H. (1963a), 'From rope to chain.' *Mariner's Mirror*, **49,** 1.

Frost, H. (1963b), *Under the Mediterranean*. London: Routledge & Kegan Paul.

Frost, H. (1964), 'Rouad, ses récifs et mouillages; prospections sous-marine.' *Annales arch. Syrie*, **14,** 67–74.

Frost, H. (1966), 'The Arwad plans 1964: a photogrammetric survey of marine installations.' *Annales arch. Syrie*, **16,** 13–28.

Green, N. J., E. T. Hall, and M. L. Katzev (1967), 'Survey of a Greek shipwreck off Kyrenia, Cyprus.' *Archaeometry*, **10,** 47.

Hall, E. T. (1966a), 'The use of the proton-magnetometer in underwater archaeology.' *Archaeometry*, **9,** 32–44.

Hall, E. T. (1966b), 'The sea-going sonar position-fixing system.' *Archaeometry*, **9,** 45–50.

Katzev, M. L. (1969), 'The Kyrenia shipwreck.' *Expedition*, **11,** 55–9.

Katzev, M. L. (1970), 'Resurrecting the oldest known Greek ship.' *National Geographic*, **137,** 841–57.

Kenyon, W. A., J. D. Holmquist, and A. H. Wheeler (1963–7), *Diving into the Past*. Section 7. Minnesota Historical Society, 1964; and personal communication.

Lambert, J. M. *et al.* (1960), *The Making of the Broads: a Reconsideration of their Origin in the Light of New Evidence*. London: Royal Geographical Society; Murray. (R.G.S. Research Series: no. 3.)

Lehmann-Hartleben, K. (1923), 'Die antike Hafenanlagen des Mittlmeeres.' *Klio*, Beiheft XIV. Leipzig, 1923.

Macnamara, E. and W. St. J. Wilkes (1967), *Nora*. Papers of the British School at Rome, **35,** 4.

Olsen, O. and O. Crumlin Pederson (1963), *The Viking Ships in Roskilde Fjord.* Copenhagen: National Museum.

Olsen, O. and O. Crumlin Pedersen (1958), 'The Skuldelev ships I.' *Acta Arch.*, **29**, 164.

Olsen, O. and O. Crumlin Pedersen (1967), 'The Skuldelev ships II.' *Acta Arch.*, **38**, 734.

Poidebard, A. (1939), 'Un grand post disparu: Tyr. Recherches aériennes et sous-marines. *Bibliothèque archaeologique et historique* No. **29** (Paris).

Schmiedt, G. (1964), 'Contribution of photo-interpretation to the reconstruction of the geographic-topographic situation of the ancient ports of Italy.' *10th Congress of the International Society of Photogrammetry*, Commission VII.

Taylor, J. du Plat (1964), 'Motya: a Phoenician trading settlement in Sicily.' *Archaeology*, **17**, 91.

Taylor, J. du Plat (1965), *Marine Archaeology.* London: Hutchinson.

Throckmorton, P. (1964), 'The Great Basses wreck.' *Expedition*, **6**, 3, 21.

Throckmorton, P. (1965a), *The Lost Ships.* London: Cape.

Throckmorton, P. (1965b), 'Wrecks at Methone.' *Mariner's Mirror*, **51**, 305.

Throckmorton, P. and J. B. Ward Perkins (1965), 'The San Pietro wreck.' *Archaeology*, **18**, 201.

Whittlesey, J. (1968), 'Balloon over Halieis.' *Archaeology*, **21**, 667.

8 Geomorphology

N. C. Flemming

Introduction

The form of the earth's surface is the result of the interaction of geological structure and properties, rock type, bedding, hardness, folding, fractures, faults, etc., with the physical processes of erosion, deposition, and deformation. The geometrical forms at present beneath the sea may have been caused by purely submarine processes, purely subaerial processes, or one modified by the other. The subaerial processes at work are wave action at the shoreline, chemical solution, the action of frost, ice, rivers, and wind, and biological action. The submarine processes are wave action—though greatly reduced—current and tidal action, chemical solution, and biological action—again, greatly reduced at depth. The geological processes of faulting, folding, and igneous activity occur both on land and under the sea.

Geophysical studies of the earth's crust beneath the ocean tend to suggest that continental and oceanic crust areas can interchange by either lateral or vertical movements, and therefore all submarine areas, except those created by upwelling and never subsequently lifted above sea level, are potentially affected by both subaerial and submarine forces. Although submarine forces, apart from vulcanism, are generally slower in action than fluvial and subaerial forces, the rate of interchange of oceanic and continental conditions is probably an order of magnitude slower, so that vast areas of the deep ocean basins can be regarded as possessing geomorphic form resulting solely from submarine processes. Near the continental margins, however, the situation is quite different.

In some areas the margin seems to have been downwarped in the last few million years, e.g. California (Emery 1960, pp. 94–6; Roberts and Stride 1968, p. 49), so that regions now at depths of several thousand feet still bear traces of subaerial landforms. In other areas, e.g. Scandinavia (Rossiter 1967, p. 284), the rate of uplift is so high that the subaerial landscape is in part of marine origin.

There is growing evidence that glaciation of Greenland and Antarctica started in the Miocene and was already producing a world-wide lowering of sea level (Tanner 1968). This implies that a slight eustatic drop in sea level may have started 10 million years ago. On a shorter time scale, about 2 million years, the relatively rapid glaciation and deglaciation of the continents has produced fluctuations of world-wide sea level with an amplitude of the order of 100 m. and a period of the order of 100,000 years (Flint 1957, Fairbridge 1961). Thus, during the Pleistocene, subaerial and submarine erosion alternated over a broad band of the continental margin, and the zone of maximum erosion or deposition, the surf zone, migrated backwards and forwards leaving terraces or beaches wherever the sea level halted or reversed its direction of change. At the lower limit of this Pleistocene zone of alternating processes the landform is dominantly of submarine origin with traces of subaerial influence; at the upper limit the situation is reversed. In the intervening area, near present sea level, forms created by both processes have been superimposed several times.

Although it is clear that erosion and deposition do take place on the deep ocean floor to depths of thousands of metres (Laughton 1968), such processes are slow compared with those on the continental shelf and slope, and are at present largely beyond the reach of submersibles and divers. This chapter will therefore be restricted to the continental shelf and slope, where direct underwater observation has produced many interesting results.

The following figures are taken from Shepard (1963, p. 257). The average width of the continental shelf is 60 km., though this varies from zero to 1,500 km.; the average depth at which the slope changes most rapidly is 150 m., though this varies from 50 to 400 m.; the average slope of the shelf is 0° 07′. The gradient of the continental slope is usually several degrees (Shepard 1963, pp. 279–310).

The origin of the continental shelf is composite. Its total form results from warping and faulting of the continental margin, glaciation, and the migrating shoreline erosion. While present wave action does move sediments all over the continental shelf, the power of erosion is several orders of magnitude lower than in the surf zone. Most erosion in water deeper than 10 m. is caused by currents.

Little geological observation by divers or from submersibles has been carried out in polar waters, and submerged glaciated terrain will not be considered further in this chapter. The ideal areas for such study would be Norway, Scotland, or Canada. Observations in the Caribbean by Goreau and Link (personal communications) show that traces of submerged Pleistocene coastal forms can be detected there, but most observations have been made in temperate climatic regions, and consequently this chapter will be devoted principally to the interaction of submarine processes and fluvial or frost landforms.

In order to anticipate the types of form which may be found under water it is necessary to consider briefly the general nature of shoreline processes (King 1959, Zenkovich 1967). The two extreme types are the erosional and the accretional. The former is typified by a terrace cut into the rock, a cliff, and a narrow beach which is rapidly comminuted or transported laterally or seaward. The latter is typified by broad, flat, sandy beaches backed by dunes, or by cuspate forelands or spits of sand and gravel. Coastal peat bogs are useful indicators of previous shorelines as they can be dated by C^{14} (Jelgersma 1961, Coleman and Smith 1964) as can mangrove swamps (Scholl and Stuiver 1967), but such deposits are liable to be altered by a rapidly rising sea level, or successive changes of sea level. Studies are usually carried out by coring, and divers have not been employed. Coastal bars of sand and gravel have been found submerged (Curray 1960), and calcified submerged sand dunes (Flemming 1968a, Fig. 8). Deltas tend to build up great thicknesses of sediment so that earlier levels can only be detected by deep coring or drilling.

Low-level erosion terraces have been studied from charts and echo-soundings by Yabe and Tayama (1934), Shepard and Wrath (1937), Parker and Curray (1956), Emery (1958), Heezen (1959), Flemming (1965, 1968a), and Martineau (1966, 1968). Since sea level has varied so much in the last million years it is important to establish how terraces form at a fixed sea level, and how long features take to develop, and then to consider how the rate of change of level affects this. The form of the final profile is greatly affected by the initial average gradient. Assuming subaerial weathering to be minimal, a cliff will in general be higher on a steep coast than on a gentle one, and the terrace correspondingly narrow. Terraces will tend to amalgamate or eliminate one another. Conversely, the same wave action and sea-level-time-curve on a gentle coast will tend to produce low cliffs and broad terraces, possibly with bars and dunes. Most features will remain on the record and small terraces may be superimposed on large ones without eliminating them.

The rate of change of sea level, up or down, during the Pleistocene has characteristically been about 1 m. per century, and this is such that in general the sea does not remain sufficiently long at one level to significantly alter the coastline, except at turning points in the sea-level-time-curve. Even assuming that the turning point of the curve is not a plateau or step, but a smooth sinusoidal function, the increased time which the surf-zone spends near the turning point is sufficient to erode a distinct terrace (Flemming 1968a). The width of terrace depends on the rate of erosion in the particular case. If one turning point is just below a previously eroded terrace, then the second terrace may remove the first.

If a turning point brings the sea level against a cliff face some way from the bottom the wave energy will be totally reflected. Field evidence suggests that limestone cliffs can, in the absence of frost action, survive uneroded for 5,000 years in this situation. If the rock is soluble a solution notch will form (Kaye 1957). Where such notches clearly cut across the bedding or structural planes in the rock they are excellent sea-level indicators. Caves, wave-cut notches at the foot of cliffs, and stacks, are also valuable indicators.

The interaction of subaerial and coastal processes has determined the forms which occupy the surface of the continental shelf, both in general and in detail, and upon which modern sediments move or are deposited. Waves and currents transport sediment over this surface in different ways, creating different forms and patterns, and waves and currents interact to produce still further variations. Wave transport is most active in shallow water and tends to generate ripple patterns parallel to the crests of the water waves, which are often more or less parallel to the shore. Current action generates ripples (Potter and Pettijohn 1963, Allen 1968), or sand waves (Stride 1958), with their crests normal to the direction of the current. Where currents are close to the shore current-generated ripples will therefore tend to be perpendicular to the shore, and current- and wave-generated patterns will be intersecting perpendicular systems. Similar complex patterns are found where two wave trains intersect, as in the lee of islands, or where an oblique wave train is totally or partially reflected by a cliff or submarine reef. In deeper water the currents tend to form the sediments into elongated banks or flat ribbons of material (Stride 1958, 1963). Again, wave-generated features may lie perpendicular, or at any angle, to the boundaries of the current features (Flemming 1965, Flemming and Stride 1967). Large-scale physiographic surveys of the United States continental shelf areas have been published by Emery *et al.* (1965), and Uchupi (1968).

In general geomorphological features are difficult to date unless they are associated with shells or peat which can be fixed by C^{14}. However, during much of the Pleistocene, parts of the continental shelf were occupied by stone-age man, and the caves which he inhabited and the tools and debris left in them serve as sea-level indicators. Thus, on a time scale of the last 5,000 years, towns and coastal settlements can provide accurate indicators of eustatic sea-level change and local earth movements (Günther 1903, Haffemann 1965, Flemming 1968b).

The Application of Underwater Science

An authority on eustatic sea-level changes stated recently, 'The record of the glacial low stands (of sea level) is submerged and largely inaccessible.' (Alt 1968, p. 91.) This underlines how little we know about the history of the continental shelf in spite of many years of work. Basic understanding of the topography has come from normal sounding and charting, while precision echo-sounding revealed further details (Emery 1958). In the last two decades the use of seismic penetration methods has elucidated the sub-bottom structure, and commercial petroleum surveys have contributed much to this.

Interpretation of underwater geomorphology depends on really detailed studies of small areas in three dimensions, including the relation of geometrical form to rock type, sedimentary material, and process. Discrete point observations, or even linear observations such as echo-sounding, entail that the interpretation is always based on interpolation and guesswork. Are sediment patches linked or separate? Is a terrace continuous and horizontal? Does a terrace slope, or are there two terraces? Three important technical developments have increased the accuracy of answers to such questions: scanning sonar, diving, and manned submersibles.

Visual observations and experiments carried out on the sea floor by divers or submersibles are rather expensive, and in an age which sets great store by quantitative data it is important not to set an observer to tackle a job which is better done by a machine. In geomorphology it is not difficult to separate the two phases. The initial stages of a project always depend on good mapping and survey, often of quite a large area. Only from these surveys can one select the areas to be checked by observation. Whether the research area has been charted or not, in principle the study starts with a bathymetric survey, and in practice it is often necessary to run a number of echo-sounding profiles to reveal details of local relief.

Since mapping depends on accurate navigation, the survey boat should be equipped either with a Decca Navigator, or, in the absence of shore stations, use can be made of horizontal sextant angle fixes. If there is no Decca chain and the shore is not visible, and it is not possible to utilize one of the portable radio navigation systems such as Hi-Fix, and if exceptional accuracy is required both on the surface and while submerged, it is now becoming possible to consider use of acoustic beacons and positioning systems (*Hydrospace* 1968).

Geomorphological interpretation depends also on geological mapping, and examples of such work are: Curry *et al.* (1962), Worzel (1968), Donovan (1968), Emery (1960), Emery *et al.* (1965), using coring, grab sampling, and gravimetric and magnetic surveys. In some cases divers have been used to confirm observations and take measurements (Stride 1963, Dill and Shumway 1954, McKeown *et al.* 1968). Interpretation of geomorphic forms, erosional and depositional, depends on the geology and the local wave and current conditions. Although basic information on these can be obtained from published charts and papers, use may have to be made of current meters and wave measuring equipment for more detail. Usually one is interested in conditions very close to the sea floor, or in areas where rocks are being vigorously eroded, and suitable equipment is not available. Current meter studies with submersibles (Busby and Costin 1968) have shown that bottom currents vary significantly over lateral distances of only one metre, and this suggests that interesting experiments could be carried out with an array of meters implanted by divers or a submersible.

Erosional Forms

Submarine Canyons

The origin of submarine canyons is discussed at length by Shepard (1963, pp. 311–48). Canyons typically have an upper section cut into rock with steep or even overhanging walls. The head of the canyon may lead back towards the coast and into an estuary or valley, or it may start from a straight coast, or part way across the shelf. The outer part of the canyon is often of shallower and broader cross-section cut into a sediment fan. In the case of the La Jolla canyon the transition is at 600 m. depth, and the outer end of the canyon is at 3,000 m. The origin of the rock-cut section of a canyon is difficult to explain, but it is probable that they are subaerial valleys strongly modified by submarine erosion processes. In the shallower ranges these processes have been observed by divers, and in deeper water by submersibles. Dill (1964) and Limbaugh and Shepard (1957) observed the topography of the Scripps Canyon, the movement of sediment in it, and the erosion produced by the sediment, using divers. Dill has also made a number of observations from the Diving Saucer. The movement of sand in canyons appears to be considerable, though intermittent, and in the cases where the walls are markedly overhanging and bevelled at the foot it is clear that sand transport is currently modifying the canyon profile.

Fjords and submerged valleys

The entrance to a fjord or loch is often marked by a shallow sill which is presumed to be the result of glacial overdeepening of the valley. Acoustic and underwater photographic and TV surveys would fill in the details, while work by divers or submersibles would aid interpretation of microstructures. Submerged subaerial river valleys have been detected bathymetrically on many shelves, especially the Sunda Shelf (Daly 1934, p. 185). The detailed history of submerged valleys, river terraces, etc., could be analysed first by acoustic penetration techniques, followed by coring and possibly detailed surveys by divers or submersibles.

Terraces and related features

The typical marine erosional terrace includes some or all of the following features: wave-cut notch, terrace surface, terrace outer edge break of slope, caves, stacks and arches, scour notch, and solution notch. These will be treated separately.

Cliffs

The steepness of the cliff is determined by lithology and the degree of subaerial weathering and the rate of wave erosion. To obtain rock samples from a cliff face it is necessary to employ divers. The height of the cliff is determined by the cliff slope, distance of erosion inland, and the initial mean land slope. The submerged cliff terrace junction is the most likely place for sediment to collect, and thus sonar profiles are unlikely to give an accurate indication of the depth of the cliff foot. Owing to multiple reflections near steep surfaces acoustic penetration methods are equally inaccurate in such locations. Shepard (1968) employed divers to clear the sediment to reveal the true cliff-terrace junction, but this is in general too expensive and time consuming. However, the sediment lens is not usually of constant thickness, and by swimming, or cruising in a submersible, it is often possible to find zones where the sediment is absent.

Since cliffs can be caused by other factors than beach erosion, it is important to establish absolute horizontality of the cliff foot, and, if possible, to identify corroboratory features such as caves and stacks.

Link (personal communication) has observed extensive submerged terraces in the Caribbean from a submersible, and Dill (personal communication) has similarly observed them off South California. Flemming (1965) has shown how the terrace-cliff geometry is affected by a continuously varying sea level, and the errors which may be introduced by ignoring this factor.

Wave-cut notch

Zeuner (1952) states that the wave-cut bevelled notch at the foot of a cliff is the most accurate indicator of mean sea level, but, owing to sediment cover, this is often not visible. I have observed such notches near the entrance to submerged caves at Gibraltar and Marseille, but not elsewhere.

Terrace surface and outer edge

A rising sea level can cut the cliff back as it rises, thereby destroying the original cliff-terrace junction which indicated the sea-level stillstand. In these circumstances the sea level must be deduced from the terrace form alone. The profile of a terrace is limited to a minimum mean slope of 1/100, and the depth of erosion at the outer edge to 10 m., and thus the maximum width is 1,000 m. Since the profile tends to be convex upwards, much of the outer section is almost flat. This flat section remains almost unaltered by subsequent changes of sea level, apart from the deposition of marine or subaerial sediments on it. The outer edge of the terrace may be removed by erosion at a lower sea level, leaving only the mid-section. Sonar survey, combined with diving observation to determine the nature of sediment cover, outcrops, continuity, etc., is the best way to identify such relict terraces. Once identified it is possible to say that the originating sea level was within 10 m. vertical range above the flat. Projection of the terrace landwards to the most likely position of the cliff, may give more precise determination.

Caves

The existence of caves at a cliff-terrace junction is an indicator of marine origin, unless the caves are shown to be karstic. Caves can only be detected by diving or submersible observation, and the internal survey to distinguish between marine or solution origin can only be done by divers. The form of a cave is partly determined by lithology, bedding, and intrusions, and these can be detected by sampling by divers if necessary. Sometimes karstic origin is indicated by the outflow of a submarine flow of fresh water, which can be detected either by its effect on the refractive index, or by its temperature. Surveys of the insides of caves for biological or geological purposes have been carried out by Laborel and Vacelet (1958), Crosset and Larkum (1966), Zibrowius (1968), Martineau (1966, 1968). Inside the cave the roof may rise up in the form of a chimney which opens either on to the cliff face or a higher terrace. Chimneys are formed by waves closing the mouth of a cave and impacting a lens of air against the roof, and are thus most common in regions of high tidal range, or in caves formed at a low sea-level stillstand. In the latter case the chimney is formed during the rise from the stillstand. I have swum into such chimneys in submerged caves off Corfu, and there is one in the Grotte de l'Ile Plane off Marseille (Flemming 1968a).

In karstic limestone a cave formed by freshwater solution may be modified by subsequent marine action. Along the Jugoslav coast there are many underground rivers which emerge offshore, and there are occasional features of this kind in the South of France, Greece, and Turkey. The Grotte de la Triperie between Marseille and Cassis is a solution cave containing stalactites submerged by 10–15 m. Karstic caves are interesting in their own right, but in the cases where the mouth of the cave has been modified by marine action, possibly during a fairly rapid passing of the sea level, they can be confused with true marine caves formed during a long stillstand.

The accumulated sediments and breccia in caves may contain bones and implements from periods of occupation by Stone-Age man. Water-level caves at Gibraltar and at Palinuro in South Italy contain bone breccia below the sea level on the walls, and similar deposits might be found in completely submerged caves. Air-lifts, water hoses, and rock-drills would probably be needed for a proper investigation.

Stacks and arches

Arches are most common in horizontally bedded rock when two neighbouring caves coalesce. Alternatively, in vertical bedded rock a hard bed may be isolated as a standing ridge, and then punctured by the waves to form an arch. If the top of the arch collapses, or if a sharp promontory of the cliff is cut away from the mass of the cliff, then a residual tower of rock is left, known as a stack. Submerged stacks which reach near the surface from deep water are known as 'secs' in the Mediterranean, and are frequently the cause of shipwrecks. They can be seen from the air or from a cliff by the local change of colour of the water. Examples are Le Veyron, off Marseille; Secca di Cano, off Syracuse; the pinnacles off Cap du Vent, Palinuro; the reefs between Corsica and Sardinia; and Eddystone in the English Channel. Sometimes a large stack has secondary caves around its base.

Although stacks and arches are of shoreline origin, and indicate a period of vigorous erosion, they are not perfect sea-level indicators, as the outer edge may have been some way out on the terrace, and below mean sea level. The use of such features to aid sea-level determination is described by Flemming (1968a).

Scour notch

Wave action below the surf zone can move sand and gravel, and in the mouth of a cave or round a stack this movement can be sufficiently vigorous to erode and undercut a notch down to depths of 20 m., and possibly more. It follows that such a notch is not a certain indicator of a sea-level stillstand.

Solution notches

The form of solution notches in calcareous rocks has been discussed by Günther (1903), Kaye (1957), and Martineau (1968). Submerged solution notches must be distinguished from differential weathering due to bedding, and this can only be done where the notch is shown to run horizontally across bedding, structure, and faults. A solution notch only forms on a vertical or near vertical surface, and on a high cliff several notches may be preserved at different levels, indicating many stillstands (Martineau 1966, 1968; Flemming 1968a). Solution notches can only be detected and followed by divers or submersibles.

Coral reefs

I have no personal experience of diving on coral reefs, or of using fossil reefs as sea-level indicators. Guilcher (1953, 1958) has described some aspects of the geomorphology of reefs, and Goreau (1961) has used diving to study the rate of growth and calcium deposition. Shinn (1963) describes the use of diving to place dynamite charges for breaking off large chunks of reef to analyse the internal structure to depths of 25 ft. off the Florida Keys. Werth (1953), Emery (1958), and MacIntyre (1967) have used fossil submerged reefs as sea-level indicators.

Depositional Forms

Depositional forms will be considered in two categories; ancient forms which are no longer in their original relationship to water level, and then those forms which are the direct result of present water movements. There is the intermediate case of, say, a submarine bank which was formed under water when the sea level was 8 m. lower than now, and which is being modified by the present sea conditions. However, in general, it is possible to distinguish the two categories.

Ancient land features

Submerged depositional land features include calcified dunes and near-shore ridges, peat-bogs, mangrove swamps, and other organic materials, deposits of animal bones and other archaeological remains. Jelgersma (1961) gives a detailed analysis of the use of peat deposits as sea-level indicators, while Kaye and Barghoorn (1964) have studied the effects of auto-compaction of peat. All the features mentioned in this section are usually studied by coring or grabbing, but it is possible that more detailed surveys by divers or submersibles would reveal microstructures and boundaries between types of deposit which would enable more accurate sampling and interpretation.

The occurrence of submerged forests has been described by North (1957), and by Steers (1946), while Scholl and Stuiver (1967) have made a detailed study of recent changes of sea level on the coast of Florida from the rate of submergence of mangrove swamps.

Bones of land animals, including those of men, and human artifacts, are all indicators of dry land. We have already considered the possible occurrence of bone and shell breccia in caves. Emery (1966) describes the finding of mammoth bones, oyster middens, and other signs of land occupation during survey of the surface geology of the East Coast continental shelf of the United States. Moriarty and Marshall (1964) also found Indian middens, shell deposits, and corn-grinding stones at many sites to a depth of 30 m. off the California coast. Emery's data were obtained by coring, dredging, and photography, while Moriarty used diving.

At Er Lanic in Brittany (Crawford 1927) there is a Neolithic circle standing half submerged, and submerged walls and structures have been found in the Isles of Scilly. Prehistoric walls were found off Antiparos in the Aegean, and an entire Helladic town near Elaphonisos in the southern Peloponnese at a depth of 3 m. (Flemming 1968b, Jones and Gubbins 1969, in press). On a shorter time scale, changes of land or sea level in the last 2,000 years are indicated by submerged ruins in many parts of Europe, but outside the Mediterranean few of these have been surveyed by divers.

Ancient coastal depositional features

Coastal depositional features include beaches, deltas, bars and ridges, beach rock, and certain archaeological features. Apart from the last-mentioned, the initial discovery of such features is not likely to be by diving or from submersibles. Identification will probably be from detailed bathymetry, sonar profiles, and coring. In all cases, details might be revealed by diving, especially in the case of beach rock. Off the coast of Cyrenaica and various parts of the Peloponnese I have frequently found slabs of beach rock *in situ*, and sometimes extensive strips of it, many metres below sea level. In some cases, such as off Petalidhi, southern Greece, the beach rock rests on ribs of bedrock, with voids large enough to swim through. This unusual structure would never be suspected or measured except by diving.

Certain coastal archaeological deposits are directly related to sea level, and can be used as indicators. In particular the classical harbours of the Mediterranean have been so used (Negris 1904, Haffemann 1960, and Flemming 1968a, 1969).

Modern beach and near-shore features

Beach structures and associated phenomena are described by Johnson (1919), King (1959), and Zenkovich (1967). Most research on beaches and the water movements under breaking waves has been done in wave tanks, though a number of large-scale field experiments have been carried out. Films taken by David Owen of Woods Hole Oceanographic Institute show the violent movement of sand over ripples under a breaking wave, and similar films could be used for quantitative experiments, especially if tracers or neutral density floats were included in the picture. Details of the grain-by-grain mechanisms by which beaches adjust to changes in wave conditions and the tidal cycle are not understood, and at least some clues to the process might be gained by close visual observation under water, and emplacement of instruments at critical locations.

Vast quantities of material move onshore and offshore, and along the coast on the sea floor out to depths of about 100 ft. This depth is quite arbitrary, but it is convenient to separate the phenomena associated with coastal processes from those which are more typical of the continental shelf as a whole, and work below this depth also requires more advanced diving techniques. Structures associated with sediment movement in the near-shore zone include ripple marks, sand waves, and elongated banks. Although all these features also exist in deeper water, the generalized type of ripple occurrence with wave and current ripples combined is largely confined to coastal waters, and so will be dealt with in this section.

Ripple marks and internal cross-bedding of sediments are studied in detail by Potter and Pettijohn (1963) and by Allen (1968). *In situ* studies by divers have been carried out by Inman (1957), and Kidson, Steers, and Flemming (1962). Ward (personal communication) used diving to measure the movement of large current generated ripples off Dungeness, while wave generated ripples in gravel were observed at a depth of 180 ft. off Plymouth (Flemming and Stride 1967). There is great scope for more detailed field observation and *in situ* experimentation on this subject, and utilization of divers is almost essential.

Generalized studies of sand and gravel movement in the offshore zone have been facilitated by the use of marked tracers, observed and recovered by divers (Joliffe 1962a, b; and Inman and Chamberlain 1959).

Modern continental shelf sediments

Large-scale surveys of the sedimentary environment of the continental shelf have been made by Emery (1960, 1966), Stride (1963), and Uchupi (1968). Features which could be studied by underwater observation and experiment include sand waves, sand ribbons, patchiness of sediments, and dunes and banks.

Stride (1963) has shown that while the southern North Sea is covered by sand waves, parts of the English Channel and southern Irish Sea are patterned with long ribbons of sand aligned with the tidal streams. Other areas show sheets of sediment alternating in size, with varying degrees of patchiness and elongation (Fig. 98). Problems which require further study are the relation between the migration of sand wave crests and velocity of tidal currents; the precise internal structure of sand ribbons; the exact nature and stability of the boundaries of ribbons and sediment patches; the rates of transport of sediments on the various parts of waves and ribbons; and possible size and shape sorting factors of mineral grains. The importance of shape and density sorting is increased by the presence of shell fragments, sea urchin spines, lithothamnion fragments, and other material having shapes which do not occur in normal clastic sediments, and hence most unusual hydraulic properties. For the relation between shape and transport characteristics, see Zingg (1935), Sneed and Folk (1958), Flemming (1965b), and Humbert (1968).

Fig. 98
Sediment waves showing separation into coarse and fine material. Photograph taken by the author during a survey in the English Channel.

Sand banks and elongated dunes have been studied by Cloet (1954) using aerial photographs, current meters, tracers, and ancient maps. The currents around such features are usually strong, over 2 knots, and so diving is not an attractive method.

Sand waves and ribbons occur out to the edge of the continental shelf, and although they can be studied to a depth of 200 ft. by diving (Flemming and Stride 1967), more extensive work will require submersibles.

Underwater weathering and biological action

Although some study has been made of the effect of biological agents on the weathering of rock in the intertidal zone, relatively little has been done in deeper water. The coralline algae of the Mediterranean can completely cover a rock substrate at depths below 100 ft. (Vacelet 1967), and boring mulluscs, and possibly the attachments of algae, may weaken rocks, while chemical action of the water leaches out soluble minerals. I have observed that submerged outcrops of slates and phyllites off Cornwall appear to be fragile and weathered at depths of 40 m., and may be surrounded by fragments which seem to have recently fallen.

TECHNIQUES FOR THE UNDERWATER GEOLOGIST

Vision

Underwater geomorphological features have been discussed at some length largely because one of the diver's most effective tools is his vision, combined with a clear vocabulary of ideas about the forms and processes which he is observing. Although the most obvious characteristic of a diver as compared with unmanned systems is vision, vision itself is useless, unless the diver knows where he is, can identify what he sees, record or memorize what he sees, and is in a fit state to do all these things efficiently. To be fit to think and work he must be relaxed, and thus the choice of correct equipment and breathing system is an essential factor (see Chapter 1).

Visibility (Hemmings and Lythgoe 1964, 1966; Ross 1966) is usually limited to 2–10 m. in British waters, and 10–30 m. in the Mediterranean. In America the visibility near the coast of the northern states is comparable to that in Britain, while that of the southern states, especially Florida, is of Mediterranean type. The limitation of visibility places severe restraints on the planning of geological and geomorphological work. Most species of animal or algae, and a wreck or an amphora, are all sufficiently small for their form and orientation to be readily identifiable in good visibility, and usually identifiable in 2–10 m. visibility. However, a cliff or large stack, a sand ribbon or a sand wave, cannot be seen all at once in good visibility. The diver on first contacting the feature may not realize what he has found, and will have to swim some way before he can be sure that the crest of a sand wave is continuous, or the foot of a cliff is horizontal. In this sense the diver's vision, combined with his ability to recognize and integrate the parts of large features with his knowledge of position, becomes a real technique, even before we embark on anything which could be called surveying. Vision, backed up by compass, depth-gauge, and writing tablet, are the basic essentials with which a diver can identify geomorphic features. It is amazing how much work can be done with such simple equipment.

To aid the search and identification of features the divers may swim on a fixed course, or alter course in a programmed way. If a feature is suspected of being in a certain locality the diver should be dropped well to one side so that a traverse in one direction is bound to take him across it. If the features are linear the divers should search on a line normal to the trend until they intersect a trend line. In this stage of work an aquaplane may be used (Martineau 1966), or an electric diver

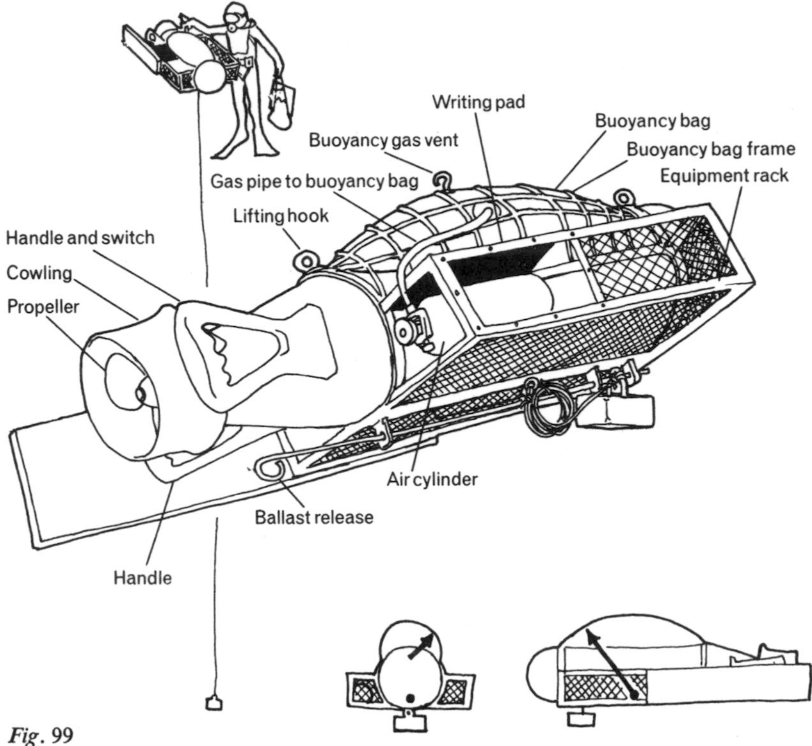

Fig. 99
Electric tug equipped to carry the tools needed by a diving geologist and fitted with an adjustable buoyancy bag to enable the diver to carry heavy specimens.

tug (Flemming 1966, Flemming and Stride 1967) (see Fig. 99). If the water is deep it is safer to lay a bottom line between buoyed shot weights, and if this is marked at intervals it aids position fixing. The buoys can be fixed by sextant or Decca. Again, in deep water, it helps to have a shot line for the diver to descend even if he is only giving a visual report on the bottom material or collecting a single sample, especially if there is a current.

In currents over 1 knot the diver cannot maintain his position by swimming during a search. He can be towed on an aquaplane modified to have a screen which protects him from the water flow, or he can work on a fixed bottom line. If the fixed line leads up to the surface vessel the vertical component of the tension will tend to lift him off the bottom, which is very tiring. The best method is to attach a light line to the anchor before dropping it, keeping the free end on deck.

The diver pulls himself down the combined cable and when he reaches the anchor streams back on the light line, while the free end is held by an attendant. In this manner the diver can swim from side to side downstream of the anchor, searching quite a wide area in complete relaxation. He is also in continuous contact with the boat, and can be pulled back to it when he ascends. This system can be worked in 2–3 knots, after which the face-mask gets pulled off by the current.

Observation can include not only identification, recognition, and delineation, but also the observation of processes and the detection of change. It is immediately apparent to a diver if there is size-sorting between crest and trough of ripples, or if the particles at the crest of ripples are discoloured by biological growth while those in the troughs are not, or if bevelled recesses in a cliff line up to form a continuous notch, or how sand moves over ripples as a wave passes, or whether dead shells on sand are more or less quickly broken up than those on gravel.

There are many useful aids to visualization of processes, such as dumping coloured sand or gravel, or releasing dyes (Woods and Fosberry 1968, Woods 1968) or neutral density floats into the water to reveal the movement. The best way to prepare a marked sample which will have the same hydraulic characteristics as the natural bottom material is to take a bottom sample, wash it, dry it, spray with aerosol paint, dry again, sieve it to break up agglomerates, and replace it on the sea bed at the point where it was removed. There is the inevitable risk that many of the smallest particles do remain stuck together by paint, but this introduces less error than when dumping completely unnatural or artificial materials, and the method is simple to work in the field.

Visual inspection and maintenance of recording equipment is an effective use of divers, though this is common to many disciplines. Divers are also employed during the launching and recovery phases in submersible operations.

Photography

As with vision, photography is severely limited in its application by the large scale of geomorphic features in relation to visibility. Ripples and similar features can be photographed, as can sediment boundaries, though the pictures are not usually of quantitative value in the latter case. Photographs of caves are impressive, but again, not usually quantitative. Notches can be photographed, but one notch looks much like the next, and it is the depth which matters most. Terraces, stacks, and sand waves are too big to be photographed at all. Photogrammetric mapping of large features can be carried out (Rebikoff 1969) but this is too expensive for most research programmes.

Photography therefore has to be applied to the recording of small sedimentary features, or the forms of erosion or weathering, and to active processes such as plumes of sand moving above ripples, or particles inside breaking waves. Probably the most useful application of photography would be in the quantitative recording of the interaction between moving water and the sea floor, either by time lapse pictures, or movies. Although David Owen's film mentioned above shows most dramatically the action of breakers running in over ripples, very few underwater movies have been carried out quantitatively. Inman (1957) used still photography of ripples against graduated marker boards, and submersibles have been used frequently to photograph small-scale sedimentary structures.

Surveying

The simplest form of surveying consists of using a ruler or tape measure and a compass. I have used for many years a fibre-glass 50 m. tape in a bakelite case with brass carrying handle, and have found it excellent. Ranging poles make good underwater rules for rough work, being strong, easily visible, and only slightly buoyant. With a stone or weight tied to one end they float vertically and make good vertical scales in photographs.

The simplest way of fixing location on the bottom is to drop two buoyed shots with marked bottom line between them, and fix the buoys by horizontal sextant angles or Decca. This technique can be used in 60 m. of water, and one buoy swung round the other if it is necessary to extend the traverse. A suitable spacing for the buoys in this depth is 200 m. for swimming, and 300–400 m. with a diver tug. In water depths of 15–30 m. the divers can leave the bottom line and swim on a compass bearing, locating features by judgement of distance, and then return to the bottom line to continue the traverse. The usefulness of this will depend on the visibility. At greater depths they should not get out of visual range of the line, but can swim some way off the line on a separate distance line or marked tape. In 15 m. depth it is possible to lay a survey grid of ropes up to 100 m. wide, in order to map ripple patterns, spreading of marked samples, etc., but in greater depths the difficulty of laying the grid might not justify the effort. Special methods have been developed for surveying in three dimensions within caves (Crossett and Larkum 1966). This involves laying a central reference line and measuring the profile radially from it.

It is extremely desirable for both divers and submersibles to know exactly where they are in three dimensions, throughout an operation, and acoustic systems developed by various navies are now becoming commercially available.

Quantitative physical measurements

The properties of sediment, rock, and water movements can be measured quantitatively *in situ*, and in some cases this can only be done by divers or a submersible. Dip and strike of rock can be determined with a hand-held compass and simple clinometer. The latter are usually made personally for a given project and consist of a semi-circular protractor scale inscribed on a rectangular board of 30–60 cm. base, and fitted with a weighted pendulum to indicate the vertical. Other quantitative measurements might consist of direct measurement of ripple amplitude, orientation, wave length, crest length, sinuosity. Divers are particularly valuable in estimating the 'truthfulness' of a sample, that is to say, noting whether it came from a typical area, or from the immediate neighbourhood of a rock outcrop or sediment boundary.

If a study is being made of rate of movement of material, then a fixed point on the bottom is needed. This may either be a steel bar driven several feet into the bottom, or a concrete block of at least 50 kg. wt. Care should be taken to warn fishermen of the marked point, and the weight should be buoyed. It is wise not to secure the buoy to the sea floor marker, but to a separate anchor. If the buoy is dragged by mistake or by a storm, the bottom marker is not moved, and there is a chance of re-locating it and continuing the survey without loss of accuracy. If this is not done, a slight error half-way through a six-month programme may wreck the whole project.

Secondary points can be marked on the sea floor with painted or numbered stones, plastic sheets or boards, or tapes pegged to the bottom. If sediment movement is to be detected by dumping coloured materials, then a search must be made on a grid about the dumping point, usually with the aid of a torch (Kidson, Steers, and Flemming 1962). If radioactive tracers are used to mark fine sand (Inman and Chamberlain 1959), greased plates can be used to pick up the samples, and these are placed against photographic plates to reveal the number of marked grains per unit area. In 1960 an experiment was carried out by I. Jolliffe using fluorescent marked tracer material and a UV lamp modified for use under water. Owing to extremely bad weather the potential of this technique was not fully tested.

When there is net transport of material into or out of an area the level of the surface must alter accordingly. Stakes driven into the floor can be marked so that such change is detected. The rate of erosion of solid rock under water is virtually unknown at all water depths, although preliminary experiments have been carried out by Jolliffe, using bars fixed in a chalk sea bed. Rates of erosion and sediment transport are most meaningful when correlated with current and wave measurements. Continuous recording meters are available but expensive, and in many cases reliance must be placed on available published tables, backed up by simple field measurements with floats or non-recording meters.

Sampling

Samples of bedrock can be obtained with a heavy hammer, or hammer and cold chisel. A pneumatic hammer is efficient for this work, and special models have been developed which can be used under water with minimum maintenance. Rocks under water are attacked corrosively by sea-water and weathered, as rocks on land are attacked by rainwater and vegetable acids. To study this effect it is necessary to have core samples of various kinds of rock from different depths and locations, so that the composition of the rock at the surface of the material can be compared with the interior. Such a project is in the planning stage. In extremely difficult conditions explosives may be used (Shinn 1963), and this method has also been employed to break up concretions or remove rock-falls from archaeological sites.

The simplest method of sediment sampling is to scoop it up by hand or with a trowel, and to store samples in numbered polythene bags. This provides the equivalent of a grab sample from a surface ship, but with the advantage of highly selective sampling and choice of location, e.g. exactly on a sediment boundary or the crest of a ripple, or the foot of a cliff. Analysis of the samples by sieve or settling column (Pettijohn 1957, Schlee 1964), gives an indication of the hydraulic behaviour of the sediment particles. It is usually important to record stratification of the sediment, and diver-operated corers can be used. For cores up to 45 cm. long in sand or mud a plastic lined corer with retaining device can be driven in with a hammer and removed by hand or with a trowel. But divers often have to work in coarse sediments containing gravel and pebbles, where it is both difficult to drive in the corer, and to retain the sample. In experiments off Plymouth (Flemming and Stride 1967) a 10 cm. diameter 45 cm. long corer was driven into gravel with a 5 kg. hammer in water depths of 50–70 m. The corer was dug out with a trowel and the end blocked with a lid before lifting. There is a wide variety of commercial powered corers, some diver-operated, which can take cores of up to 6 m. length in several hundred metres of water. These are usually only operated by commercial diving companies or contractors, but can be most valuable in research.

A core through stratified sediment may not give a truly representative picture if the stratification varies considerably from place to place. McMullen and Allen (1964) showed that undisturbed bulk samples can be taken with a box corer, and then impregnated with resin for sectioning. I attempted to use a large box corer under water in this way, and although the test was unsuccessful owing to the difficulty of closing the box, the method could be valuable with better equipment.

Removal of material

Where material must be removed prior to study of the underlying beds or strata it is common to use either a water-jet or air-lift. These methods have been used most comprehensively by archaeologists, but may also be applied for geomorphological reasons to clear caves, or reveal the junction of a terrace and cliff.

Submersibles

Submersibles are a recent addition to the range of tools available to the marine scientist, and it is quite impossible to give a fair assessment of their importance in a short section such as this, which is just a tailpiece to a chapter on diving techniques. In general, a submersible is applicable in all the cases where divers are employed in research, but where there is an added requirement for depth capability, endurance, range, power, or weight carrying.

A review of submersibles constructed up to 1965 is given by Abel (1965), and a more detailed inquiry into their design and performance by Busby (1968). Flemming (1968c) and Busby and Costin (1968) investigate the methods of use of submersibles in different kinds of project, and the latter in particular summarizes the experience of American scientists using submersibles on a number of projects.

The decision to use a submersible is usually based on cost, unless the trial is simply an investigation of the performance of the vehicle. In some cases it is clear that, in spite of the high cost of operating and supporting a submersible, the vehicle can achieve in one or two weeks what would otherwise require months or years, or be totally impossible.

Conclusions

Underwater geomorphology depends very much on the recognition of forms, the relation of microstructure to large forms, and on the correlation of samples with observed forms. Although conventional surveying and sampling techniques are an essential prerequisite to efficient use of divers, there are many types of observation and experimental work which can only be carried out by divers.

References

Abel, R. (1965), *Undersea Vehicles for Oceanography.* Interagency Committee for Oceanography Report, 18 Oct. 1965.

Allen, J. R. L. (1968), *Current Ripples: their relation to patterns of water movement and sediment motion.* Amsterdam: North Holland Publishing Co.

Alt, D. (1968), 'Pattern of post-Miocene eustatic fluctuation of sea level.' *Palaeo-Geography/Climatology/Ecology*, **5**, 1, 87–94.

Busby, R. F. (1968), *Design and Operational Performance of Manned Submersibles.* U.S. Naval Oceanographic Office, Informal Report, IR, No. 68–62.

Busby, R. F. and J. M. Costin (1968), *Ocean Surveying from Manned Submersibles.* U.S. Naval Oceanographic Office, Informal Report, IR. No. 68–82.

Cloet, R. L. (1954), 'Hydrographic analysis of the Goodwin Sands and the Brake Bank.' *Geog. J.*, **120**, 203–15.

Coleman, J. M. and W. G. Smith (1964), 'Late recent rise of sea level.' *Bul. Geol. Soc. Am.*, **75**, 833–40.

Crawford, O. G. S. (1927), 'Lyoness.' *Antiquity*, **1**, 5–14.

Crossett, R. N. and A. W. D. Larkum (1966), 'The ecology of benthic marine algae on submarine cliff faces in Malta.' (*Malta '65*) *Underwater Assn. Rep.*, **1**, 57–61.

Curray, J. R. (1960), 'Sediments and history of Holocene transgression, continental shelf, Northwest Gulf of Mexico: Recent sediments N.W. Gulf of Mexico, 1951–1958.' *Am. Assoc. Petr. Geol.*, 221–56.

Curry, D., E. Martini, S. J. Smith, and W. F. Whittard (1962), 'The geology of the western approaches of the English Channel. Chalky rocks from the upper reaches of the Continental Slope.' *Phil. Tran. Roy. Soc.* (B), **245**, 267–90.

Daly, R. A. (1934), *The Changing World of the Ice Age.* Newhaven: Yale U.P. Reprinted, New York: Hafner, 1963.

Dill, R. F. (1964), 'Underwater mapping and observation of the sea floor by geologists.' In *Submarine Geology*, by F. P. Shepard (q.v.), p. 36.

Dill, R. F. and G. A. Shumway (1954), 'Geologic use of SCUBA.' *Bull. Am. Assoc. Petr. Geol.*, **38,** 148–57.

Donovan, D. T. (1968), 'Geology of the continental shelf around Britain: a survey of progress. Geology of shelf seas.' *Proceedings of 14th Inter-university Geological Congress*. Edinburgh: Oliver & Boyd, pp. 1–14.

Eden, R. A. and Anne V. F. Carter (1969), 'Detailed examination of sub-marine rock sections.' *Underwater Assn. Rep.*, **3,** 8.

Emery, K. O. (1958), 'Shallow water submerged marine terraces of South California.' *Bul. Geol. Soc. Am.*, **69,** 39–60.

Emery, K. O. (1960), *The Sea off Southern California*. New York: John Wiley & Son.

Emery, K. O. (1966), 'Early man may have roamed the continental shelf.' *Oceanus*, **12,** 3–4.

Emery, K. O., A. S. Merrill, and J. V. A. Trumbull (1965), 'Geology and Biology of the sea floor as deduced from simultaneous photographs and samples.' *Limnology and Oceanography*, **10,** 1–21.

Fairbridge, R. W. (1961), *Eustatic Changes of Sea Level: Physics and Chemistry of the Earth*, **4,** pp. 99–185. London: Pergamon.

Flemming, N. C. (1965), 'Form and relation to present sea level on Pleistocene marine erosion features.' *J. Geol.*, **73,** 799–811.

Flemming, N. C. (1966), 'Operational diving with oxy-helium self-contained diving apparatus.' (*Malta '65*) *Underwater Assn. Rep.*, **1,** 3–12.

Flemming, N. C. (1968a), 'Derivation of Pleistocene marine chronology from morphometry of erosion profiles.' *J. Geol.*, **76,** 280–96.

Flemming, N. C. (1968b), 'Mediterranean sea level changes.' *Science Journal*, April 1968, 51–5.

Flemming, N. C. (1968c), 'Functional requirement for research/work sub-mersibles.' *Aeronautical J. of R. Ae. Soc.*, **72,** 123–31.

Flemming, N. C. (1969), 'Archaeological evidence for eustatic change of sea level in the Western Mediterranean.' *Bull. Geol. Soc. Am.* special paper 109.

Flemming, N. C. and A. H. Stride (1967), 'Basal sand and gravel patches with separate indications of tidal current and storm-wave paths, near Plymouth.' *J. Mar. Biol. Assoc. U.K.*, **47,** 433–44.

Flint, R. F. (1957), *Glacial and Pleistocene Geology*. New York.

Goreau, T. (1961), 'Problems of growth and calcium deposition in reef corals.' *Endeavour*, **20,** 77, 32–9.

Guilcher, A. (1953), 'Problèmes et méthodes de l'étude géomorphologique des récifs Coralliens.' *Bul. de la Societé des Sciences de Nancy*, 1–10.

Guilcher, A. (1958), *Coastal and Submarine Morphology*. London: Methuen.

Günther, R. T. (1903), 'Earth movements in the Bay of Naples.' *Geog. J.*, **22,** 121, 269.

Haffemann, D. (1960), 'Ansteig des Meeresspiegels in gesichtlicher Zeit.' *Die Umschau*, **60,** 7, 193–6.

Heezen, B. C. (1959), 'Submerged ancient beaches of the Atlantic.' Abstract, *1st Internat. Oceanographic Congress*, 622–3.

Hemmings, C. C. and J. N. Lythgoe (1964), 'Better visibility for divers in dark waters.' *Triton*, **9,** 28–31.

Hemmings, C. C. and J. N. Lythgoe (1966), 'The visibility of underwater objects.' (*Malta '65*) *Underwater Assn. Rep.*, **1,** 23–9.

Humbert, F. L. (1968), *Selection and Wear of Pebbles on Gravel Beaches.* Rijksuniveritait Groningen.

Hydrospace (1968), 'SURV trials, valuable lessons learned.' Editorial, **1,** 3, 8–11.

Inman, D. L. (1957), *Wave Generated Ripples in Nearshore Sands.* Beach Erosion Board Technical Memo., 100.

Inman, D. L. and T. K. Chamberlain (1959), 'Tracing beach sand movement with irradiated quartz.' *J. of Geophysical Research*, **64,** 41–7.

Jelgersma, S. (1961), *Holocene Sea Level Changes in the Netherlands.* Maastricht: Meded. van de Geologische Stichting, Ser. C.

Johnson, D. W. (1919), *Shore Processes and Shoreline Development.* New York: John Wiley & Son.

Joliffe, I. P. (1962a), 'Free diving and coastal research.' *Triton*, **8,** 27–31.

Joliffe, I. P. (1962b), 'The use of tracers to study beach movement.' *Revue de Geomorphologie Dynamique*, **12.**

Jones, R. and D. Gubbins (1969), 'Survey of Elaphonisos.' Unpublished expedition report, Cambridge University Underwater Exploration Group.

Kaye, C. A. (1957), 'The effect of water movement on limestone.' *J. Geol.*, **65,** 35–46.

Kaye, C. A. and E. S. Barghoorn (1964), 'Late Quaternary sea level change and crustal rise at Boston, Mass., with notes on the autocompaction of peat.' *Geol. Soc. Am. Bul.*, **75,** 68–80.

Kidson, C., J. A. Steers, and N. C. Flemming (1962), 'A trial of the potential value of aqualung diving to coastal physiography.' *Geog. J.*, **128,** 49–53.

King, C. A. M. (1959), *Beaches and Coasts.* London, Arnold.

Laborel, J. and J. Vacelet (1958), 'Etudes des peuplements d'une grottee sous-marine du Golfe de Marseille.' *Bul. Inst. Oceanog. Monaco*, **1114.**

Laughton, A. S. (1968), 'New evidence of erosion on the deep ocean floor.' *Deep Sea Research*, **15,** 21–9.

Limbaugh, C. and F. P. Shepard (1957), 'Submarine canyons.' *Geol. Soc. Am. Mem.*, **67,** 1, 633–9.

MacIntyre, I. G. (1967), 'Submerged coral reefs, west coast of Barbados.' *Canadian Journal of Earth Sciences*, **4,** 461–74.

McKeown, M. C., R. A. Eden, and Anne V. F. Carter (1968), 'Detailed examination of submarine rock sections by the Institute of Geological Sciences.' *Underwater Assn. Rep.*, **3,** 8.

McMullen, R. M. and J. R. L. Allen (1964), 'Preservation of sedimentary structures in wet unconsolidated sands using polyester resins.' *Marine Geology*, **1,** 88–97.

Martineau, M. P. (1966), 'Marine terraces in Malta.' (*Malta '65*) *Underwater Assn. Rep.*, **1,** 69–71.

Martineau, M. P. (1968), 'The formation and significance of submarine terraces off the coast of Malta.' *Underwater Assn. Rep.*, **2,** 19–23.

Moriarty, J. R. and N. F. Marshall (1964), 'Principles of Underwater Archaeology.' *Pacific Discovery*, Sept.–Oct. 1964.

Negris, P. (1904), 'Vestiges antiques submergées.' *Athenischer Mitteilungen*, **29,** 340–63.

North, F. J. (1957), *Sunken Cities*. University of Wales Press.

Parker, R. H. and J. R. Curray (1956), 'Fauna and bathymetry of banks on the continental shelf, N.W. Gulf of Mexico.' *Am. Assoc. Petr. Geologists. Bul.*, **40,** 2428–39.

Pettijohn, F. J. (1957), *Sedimentary Rocks*. New York: Harper Brothers.

Potter, P. E. and F. J. Pettijohn (1963), *Paleocurrents and Basin Analysis*. Berlin: Springer-Verlag.

Rebikoff, D. (1969), 'Underwater mapping telemetry for deep offshore installations.' *Oceanology International 1969*, **4** (unpaginated).

Roberts, D. and A. H. Stride (1968), 'Late Tertiary slumping in the continental slope of southern Portugal.' *Nature*, **217** (5123), 48–50.

Ross, H. E. (1966), 'Size and distance judgements under water and on land.' (*Malta '65*) *Underwater Assn. Rep.*, **1,** 19–22.

Rossiter, J. R. (1967), 'An analysis of annual sea level variations in European waters.' *Geophysical Journal. Royal Astronomical Society*, **12,** 259–99.

Schlee, J. (1964), 'New Jersey offshore gravel deposit.' *Pit and Quarry*, Dec. 1964.

Scholl, D. W. and M. Stuiver (1967), 'Recent submergence of Southern Florida: a comparison of adjacent coasts and other eustatic data.' *Geol. Soc. Am. Bulletin*, **78,** 437–54.

Shepard, F. P. (1963), *Submarine Geology*. New York: Harper & Row.

Shepard, F. P. (1968), *The Earth Beneath the Sea*. London: O.U.P.

Shepard, F. P. and W. F. Wrath (1937), 'Marine sediments around Catalina Island.' *J. Sed. Petr.*, **7,** 41–50.

Shinn, E. (1963), 'Spur and groove formation on the Florida reef flat.' *J. Sed. Pet.*, **33,** 291–303.

Sneed, E. D. and R. L. Folk (1958), 'Pebbles in the Lower Colorado River Texas: a study in particle morphogensis.' *J. Geol.*, **66,** 114–60.

Steers, J. A. (1946), *The Coastline of England and Wales*. Cambridge University Press.

Stride, A. H. (1958), 'Large sand waves near the edge of the Continental Shelf.' *Nature*, **181** (4601), 41.

Stride, A. H. (1963), 'Current-swept floors near the southern half of Great Britain.' *Quart. J. Geol. Soc. London*, **119,** 175–99.

Tanner, W. F. (1968), 'Tertiary sea level symposium—Introduction.' *Palaeo-Geography/Climatology/Ecology*, special issue 'Tertiary Sea-level Fluctuations,' **5,** 1, 7–14.

Uchupi, E. (1968), *Altantic Continental Shelf and Slope of the United States—Physiography*. U.S. Geological Survey Professional Paper. 529–C.

Vacelet, J. (1967), 'The direct study of the populations of underwater cliffs and caves.' *Underwater Assn. Rep.*, **2,** 73–6.

Werth, E. (1953), 'Die eustatische Bewegungen des Meerespiegels während der Eiszeit und die bildung der Korallenriffe.' *Akad. Wiss. u. Litt. Mainz. Abhandlung. Math. Nat.*, **8.**

Woods, J. D. (1968), 'Wave-induced shear instability in the summer thermocline.' *J. Fluid Mech.*, **32,** 791–800.

Woods, J. D. and G. G. Fosberry (1968), 'The structure of the thermocline.' *Underwater Assn. Rep.*, **2,** 5–18.

Worzel, J. L. (1968), 'Advances in marine geophysical research of continental margins.' *Canadian Journal of Earth Sciences*, **5**, 963–83.

Yabe, H. and R. Tayama (1934), 'Bottom relief of the sea bordering the Japanese islands and Korean peninsula.' Tokyo Imperial University: *Earthquake Research Inst. Bul.*, **5**, 12, 539–65.

Zenkovich, V. P. (1967), *Processes of Coastal Development*. London: Oliver & Boyd.

Zeuner, F. E. (1952), 'Pleistocene shorelines.' *Geol. Rundschau.*, **40**, 39–50.

Zibrowius, H. (1968), 'Etude morphologique, systematique et écologique des Serpulidae (Annelida, Poychaeta) de la région de Marseille.' *Receuil des Traveaux de la Station Marine d'Endoume*. Fasc. 59, Bul. 43, 81–252.

Zingg, T. (1935), 'Beitrag zur Schotteranalyse.' *Scheiz. Mineral. Petrog. Mitt.*, **15**, 39–140.

9 Micro-oceanography

J. D. Woods

Since the principal object of going under water is to see what is there, the underwater oceanographer will be concerned almost exclusively with looking at motions whose dimensions are smaller than his visual range (and with time scales shorter than his endurance in the sea). The classical oceanographer, on the other hand, is mainly concerned with currents on a global scale and has only recently begun to consider oceanic variability on a scale of a few kilometres and over periods of a few days. Drawing on the classification used by meteorologists we may divide physical oceanography into three overlapping scales, namely 'climatic', 'synoptic', and 'micro' and we expect to find parallels between atmospheric and oceanic motions within the same division. Micro-meteorology (see the classic text by Sutton 1953) offers well-established and powerful techniques for studying the atmosphere by concentrating upon the effects of turbulence; its greatest success has been in describing the transport of heat, momentum, and water vapour through the atmospheric boundary layer, but micro-meteorological methods are equally valid at higher levels wherever the air is made turbulent by the release of buoyancy (in clouds) or by strong wind-shear (clear air turbulence—generally referred to as CAT).

The largest (vertical) eddies in atmospheric turbulence may reach 100 or even 1,000 m. across, but the corresponding motions in the sea will be reduced by a factor equal to the square root of the ratio of the densities of air and sea-water. Thus oceanic eddies comparable with the largest seen at the tropopause will be about 15 m. across. The size of the smallest eddies (λ_v) depends upon the viscosity of sea-water and rather weakly upon the energy dissipation rate (ε); in a fully developed inertial subrange

$$\lambda_v = \left(\frac{\nu^3}{\varepsilon}\right)^{\frac{1}{4}},$$

so for $\varepsilon = 10^{-2}$ erg/gs., $\lambda_v \simeq 1$ mm. We conclude that oceanic turbulence is likely to comprise eddies which are smaller than the limit of visibility in clear ocean water and yet not so small that they cannot be resolved by a photograph taken at reasonable range. Furthermore, the time scale of the largest motions is likely to be about 10 minutes or so, which may comfortably be followed by a diver, who can usually stay down for 30 minutes, still more by the observer in a research submersible with an endurance of several hours.

Although the essential ideas of micro-oceanography are presented in most textbooks on physical oceanography (notably Sverdrup, Johnson, and Fleming 1942; Defant 1960), recent reviews by Bowden (1962 and 1965) and Phillips (1966) have shown how little field evidence is available to support the application of these theoretical ideas. In particular there have been virtually no experimental field tests of the validity (and limitations) of such universally used parameters as 'eddy viscosity' and 'eddy conductivity', yet the micro-meteorologist has shown how limited are the applications of these parameters in the atmosphere (Pasquill 1962).

References to some recent investigations into oceanic turbulence are given in Table 1.

TABLE 1. *A classification of oceanic turbulence into sites and causes*

Site	*Cause*	*Author*	*Location of study*	*Method*
1. Sea surface boundary layer	a. Surface waves b. Wind drag c. Free convection	Stewart and Grant (1962) Schmitz (1962a, b) Kraus and Rooth (1961)	Vancouver Island — —	Hot film flowmeter Theoretical Theoretical
2. Sea bed boundary layer	a. Bottom drag on tidal current (i) shallow sea (ii) deep ocean b. Turbidity currents	Grant, Stewart, and Moilliet (1962) Bowden (1962b) Munk, W. (unpublished) Thorpe, S. (unpublished) Brooke-Benjamin (1968)	Vancouver Island Irish Sea California Bay of Biscay —	Hot film flowmeter Mechanical current meters Direction from vanes Theoretical
3. Water mass boundaries	Current shear	Defant (1948)	Straits of Gibraltar	Temperature soundings
4. Thermocline	Internal waves	Grant, Moilliet, and Vogel (1968) Phillips (1966) Woods (1968a)	Vancouver Island — Malta	Hot film flowmeter on submarine Theoretical Dye tracers

Until very recently, the diver's only contribution to micro-oceanography lay in his use as a labourer responsible for setting up current meters and thermometers arrayed on buoy cables near the surface or on frames standing on the sea bed. However, there has existed for many years an extensive diving folklore concerning the temperature and current microstructure of the sea and occasionally reports have reached the scientific literature (e.g. Limbaugh and Rechnitzer 1955, Banner 1955). The first serious attempts to use dye tracers to reveal flow patterns for divers to photograph (e.g. Lafond 1962; Zhukov, Mayer, and Rekhtzamer 1964) took place in shallow, coastal waters that are, by deep ocean standards, extremely turbulent so that the tracers were rapidly diluted and lost to view. This rapid dispersion of the trace material was further aggravated in these experiments by the use of unduly turbulent sources, which virtually eliminated any hope of measuring the far weaker ambient turbulence by classical diffusion methods. This has led Zhukov *et al.* (1964), amongst others, incorrectly to ascribe differences in the dispersion rate of tracers released at selected levels in the thermocline to differences in the local intensity of the turbulence, whereas they were in fact due to the different rates at which the artificial turbulence from the source lost energy by working against the varying temperature gradient through the thermocline. A more satisfactory experimental technique will be described on page 295.

These early experiments stimulated a micro-oceanographical investigation of the summer thermocline in which divers used carefully designed dye tracers to measure the detailed structure of the steady and turbulent components of the internal flow. Since it is, as yet, the only extensive use of divers in micro-oceanography, this investigation will form the subject for the remainder of this chapter.

The Temperature Microstructure of the Summer Thermocline

The first stage in a study of turbulent transport processes in the summer thermocline is to measure the temperature and velocity structure to a sufficiently fine resolution. At the start of the investigation in 1965 it was not at all clear just how fine a scale would have to be plotted, although it was suspected that the resolution of the Spilhaus bathythermograph and the Sippican expendable bathythermograph (the principal instruments for routine operational study of the thermocline), would be too coarse: it subsequently turned out that even the best research tool, the Bisset-Berman Hytech STD probe, does not faithfully record the finest microstructure. By the end of the first season it had been established that the major temperature features are about 10 cm. thick, so a vertical resolution of 1 cm. was specified for the new microstructure sounding system.

The principal design limitation for a microstructure probe is the response time of the sensors. For example, the fastest oceanographic thermometer (of the platinum resistance type) gives a 63 per cent response to a step function of temperature in about 0·1 s., so the probe's rate of descent should be rather less than 10 cm./s. if it is to achieve the required 1 cm. vertical resolution. This design criterion leads to two important considerations. First, a complete sounding from the surface down to a maximum depth of 100 m. will take nearly 20 minutes. Secondly, it will be necessary to isolate the probe from vertical ship motions exceeding 10 cm./s.; the solution is to allow the probe to sink freely through the sea, rather than to lower it in the conventional manner at the end of a taut cable. A new microstructure sounding system, which achieves this objective (Woods 1969a), has been designed almost exclusively on the evidence of divers.

The apparatus used to obtain the results quoted below differed from the new probe by depending on slow lowering by hand: this is quite adequate in calm weather, but becomes unreliable beyond about force 4. In addition to the sensitive platinum resistance thermometer for absolute temperature measurement, both the old and the new probes include a pair of platinum resistance thermometers set 10 cm. apart, on a vertical frame to give a direct measurement of temperature-gradient averaged over 10 cm. Profiles of temperature and temperature-gradient are drawn continuously against depth on an XY_1Y_2 plotter. A typical plot is shown in Fig. 100, where we see the series of sharp steps which is the most important feature of the thermocline's temperature microstructure. The regions of relatively weak temperature-gradients, called *layers*, are from 2 to 5 m. thick, while the *sheets* separating them are only some 10 to 30 cm. thick. Individual sheets and layers have been traced over distances of several kilometres, and for periods of several hours, but the lamination's ultimate extent (and persistence) may be very much greater. This observation leads to our first tentative conclusion, that heat transported down through the thermocline must pass through the sheets and layers rather than around them.

The threefold aim of the micro-oceanographic investigation is now clearly seen. It must (1) describe the structure and energy balance of the turbulence field inside the laminated thermocline, (2) explain how this turbulence effects the vertical transport of heat and momentum through the laminae, and (3) explain why the thermocline *is* divided into sheets and layers.

The Measurement of Shear

In order to understand the generation of turbulence inside the thermocline it is necessary to know both the density gradient and the velocity shear. The former is obtained from the temperature and salinity soundings described in the previous section. It has not proved possible to

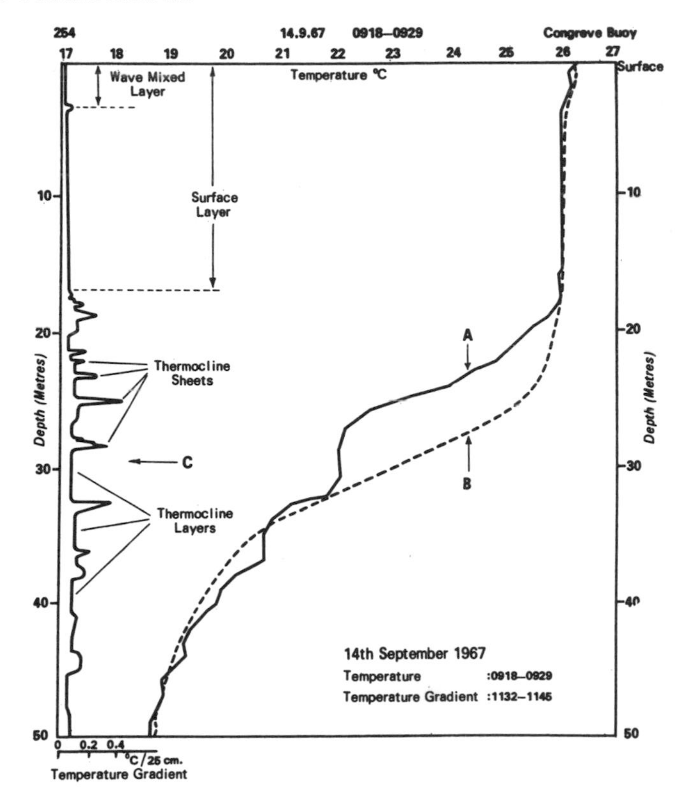

Fig. 100
Typical temperature microstructure sounding in the seasonal thermocline off Malta.

incorporate current meters into the probe, so shear is measured *in situ* by filming the distortion of dye streaks. This method has the great advantage that the shear can be measured at identically the same time and place as the turbulence measurements are made. The need for simultaneous shear measurements is far more important than simultaneous temperature-gradient since the former fluctuates rapidly in response to internal waves while the latter changes only slowly over the surface of a sheet.

The most convenient technique for laying a shear streak is to drop a tiny pellet of congealed fluorescein through the layer under study. Pellets in the form of a flat disc 3 mm. diameter and 1·5 mm. thick are particularly useful as they give a regular Karman vortex wake whose dimensions are reproducible and so may be reliably used, as a 'built-in' scale. It will be noted that the only disturbance to the existing flow caused by the pellet's passage through the sea is in the formation of this Karman vortex wake, whose individual vortices rapidly lose their own motion and then proceed to follow the ambient flow. The pellets are sealed in waterproof polythene strips (see Fig. 101a) until needed. Three sizes, each with the same aspect ratio, are used: the smallest, described above, give the most regular wake, but this only lasts for about 5 m., the largest (6 mm. diameter by 2·3 mm. thick) can lay a streak through the whole thermocline (see Fig. 101c). As the fall speeds of these pellets are comparable with the differences in horizontal velocity encountered along any streak, their fall path is often quite complex, so the velocity profile cannot be obtained from a single photograph. Instead the mean shear across any given layer is obtained from measurements of the relative displacement of the Karman vortices lying at the top and bottom of the chosen layer and identified in successive frames of a timed sequence of still photographs or on a motion film. The author uses a 16-mm. Telford underwater motion camera fitted with a temperature compensated electronic time-lapse control giving 1, 2, or 4 frames per second. Often the vertical shear in the thermocline is two-dimensional, so it becomes necessary to take stereoscopic pairs; for example, by linking together two Nikonos 35-mm. still cameras fitted with corrected Underwater Nikkor 28-mm. lens. A good introduction to the interpretation of stereo pairs is given by Atkinson and Newton in *Photography for the Scientist* (Engel 1968): an additional

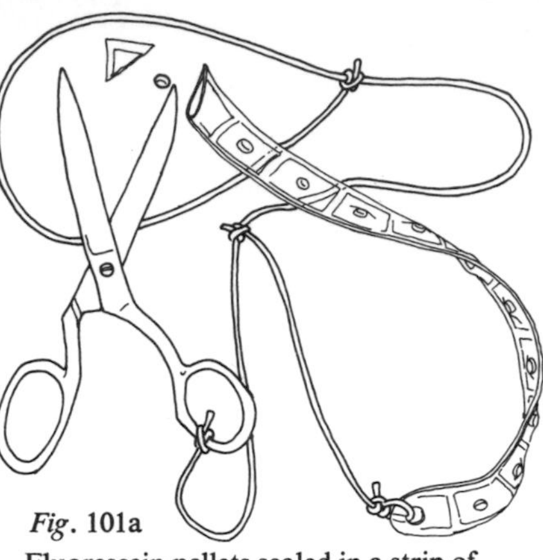

Fig. 101a
Fluorescein pellets sealed in a strip of polythene are cut open with scissors and dropped to give a vertical dye streak through the sea.

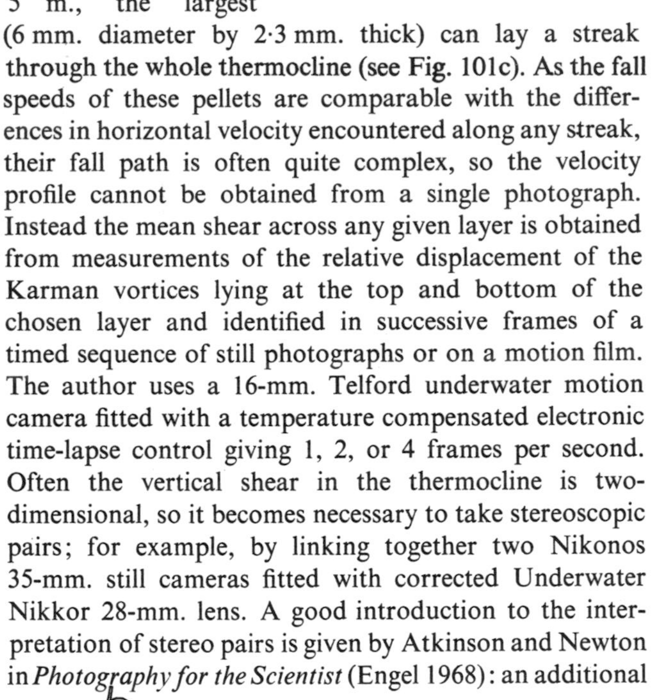

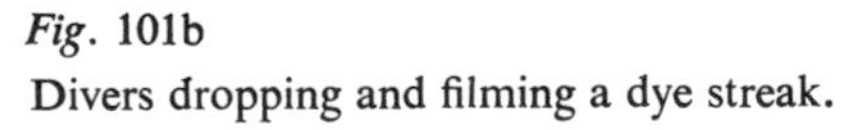

Fig. 101b
Divers dropping and filming a dye streak.

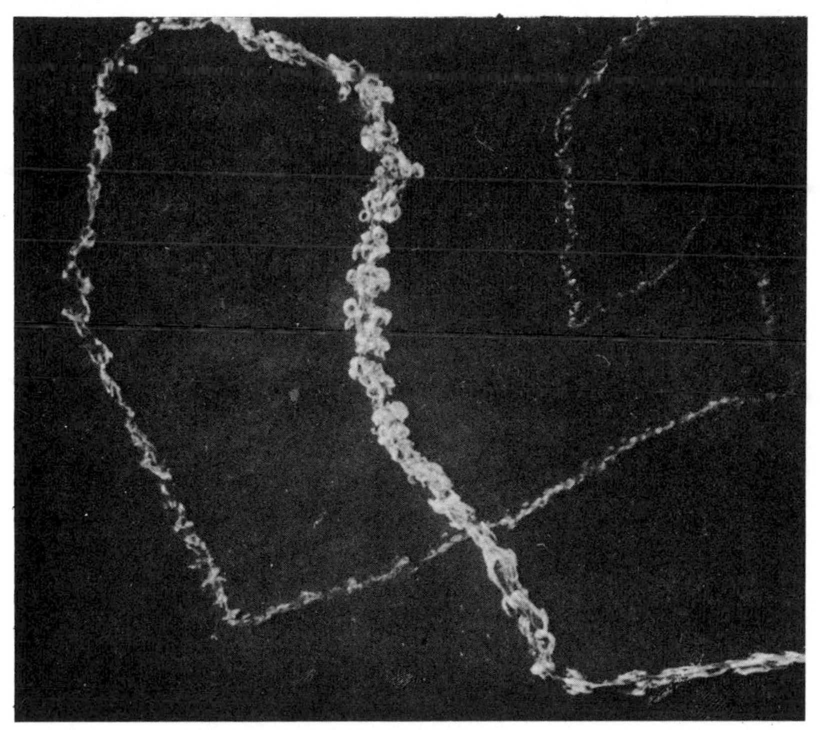

Fig. 101c

A typical dye streak viewed from above. The eddy structure (Karman vortices) in the streak is due to the wake of the falling pellet; since the streak is not rapidly diffused through the water we deduce that the ambient flow is laminar (i.e. non turbulent).

numerical correction will be required if uncorrected lenses are used.

The underwater experiment proceeds as follows. After identifying the area of interest by dropping a trial pellet, the camera-man positions himself at the chosen level and signals to his assistant, who is floating some 3–4 m. above and upstream, to release a second pellet (Fig. 101b). As the pellet begins to fall the assistant increases buoyancy by letting air into his Fenzy life-jacket; this allows him to rise gently away from the dye streak without disturbing it by turbulence from his fins. Whenever possible the assistant positions himself above the sheet overlying the layer being filmed; this sheet isolates his movements from the subject. As the camera-man films the dye streak (keeping the sun behind the camera to increase contrast), the assistant lowers the scale into the camera's field of view. Provided the scale moves only slowly relative to the dye streak it is possible to approach within half a metre of it without provoking any unnatural disturbance. However, if this does happen, it occurs on a scale comparable with the thickness of the limbs of the scale and this disturbance is easily recognized since it is rather small compared with any naturally occurring instability in the thermocline.

The results of these shear measurements are not yet completely understood, but the following essential features have been identified. The shear across the *sheets* detected on the soundings is considerably greater than the shear across the intervening *layers*. In all cases the peak shear inside the sheets lies very close to the peak in temperature-gradient, the latter being measured simultaneously with a special thermistor probe positioned by the diver. Usually the transition from strong to weak shear occurs within a centimetre and the shear/temperature-gradient inside the sheet is approximately linear. The value of the velocity difference across the sheet fluctuates rapidly with space and time. The short-term fluctuations, with periods of a few minutes, can be associated with the single-sheet internal waves to be described later; while slower changes are believed to reflect changes in the relative drift of the adjacent layers due to external stresses arising from thermocline waves (i.e. long internal waves which raise and lower the whole thermocline), tidal motions or geostrophic currents.

The magnitude of the shear found in a *layer* is usually between $\frac{1}{4}$ and $\frac{1}{10}$ that found in the adjacent *sheets*, but since the layer is much thicker the total velocity difference across it may be several times greater than that across the bounding sheets. However, the shear in a layer can seldom be described in terms of the simple linear shear model found useful for the sheets; its magnitude fluctuates in an irregular manner through the layer's depth. The source of these fluctuations, which typically have scales of half a metre, is still unclear although it seems unlikely that such fluctuations could arise from external stresses, which are more likely to affect the layer as a whole. The solution may prove to be associated with sporadic injection of momentum from the boundaries by the aperture conduction process to be described below. Some of the fluctuating shear in the layers may be attributed to the single sheet waves and this wave shear may play a part in triggering the observed sporadic transition to turbulence in the layers.

Internal Waves in the Thermocline

Of the wide variety of internal waves found in the ocean, one particular class exists only in the summer thermocline. These *thermocline waves*, which raise and lower and distort the whole thermocline have been described theoretically in detail by, to quote two recent texts, Phillips (1966) and Krauss (1966). The main feature of these progressive gravity waves is that they are trapped in the 30 m. thick thermocline where the static stability is sufficiently high to give a period of natural oscillation, the Brunt-Väisälä period, shorter than the internal wave period.

i.e.
$$T_{\text{wave}} \geqslant 2\pi \left(\frac{g}{\rho}\frac{dp}{dz}\right)^{-\frac{1}{2}}$$

where g = gravitational accelerate;
ρ = local sea-water density;
$\frac{d\rho}{dz}$ = local density gradient.

Exploration of the thermocline with thermistor chains towed behind research ships (e.g. Lafond 1962), or by long series of frequent bathythermograph soundings from weather ships (e.g. Tully and Giovando 1963) have revealed complex undulations and distortions on the thermocline with a spectrum of periods with peak energy near the semi-diurnal tidal period and the inertial period, but extending down to periods of less than an hour. It is generally believed that these thermocline movements can, in principle at least, be explained in terms of combinations of thermocline waves of the kind described above, but attempts to carry out the appropriate analysis (e.g. Charnock 1965) have often revealed a bewildering complexity. It seems that vertical motions on the thermocline are made even more complex than surface waves. The question that concerns us here is whether our new knowledge concerning the internal lamination of the thermocline enables us to explain some of the difficulties encountered in earlier analysis. Furthermore, since our concern is with micro-oceanography, we hope to throw some light on the processes by which thermocline waves lose energy due to turbulence.

One of the first discoveries made by divers in the thermocline was the existence of internal waves on the sheets. These waves are revealed by staining the sheet with fluorescein dye, using the technique illustrated in Fig. 102, although the strong gradients of optical refractive index and turbidity found at the sheets generally makes them just visible to the diver. The periods of these waves (a few minutes) are rather longer than the Brunt-Väisälä period (typically 100 seconds) of the sheet on which they travel. Their wavelengths lie in the range 3 to 10 m. and those detected in the photographs are remarkably steep, often 1 part in 10. Three additional features identify this new class of internal waves: they propagate along a single sheet, they have broad crests and they are often found in long coherent trains. To a first approximation the properties of these 'single-sheet waves' are described by the classical theory for sinusoidal internal gravity waves on an interface between two infinite uniform layers (Lamb 1932). The wave shear across the sheet is greatest at the crest and trough where it is given by Phillips (1966):

$$\left[\frac{\Delta U}{h}\text{ wave}\right]_{\text{crest, trough}} = \left(\frac{N^2}{n^2} - 1\right)^{\frac{1}{2}} nka,$$

where N = Brunt-Väisälä frequency;
n = wave frequency;
k = wave number;
a = wave amplitude.

This gives, for example, on a 10 cm. thick, $\frac{1}{4}$°C sheet, bearing a wave of 5 m. length and 50 cm. height, a velocity shear $(\Delta U/h) = 0{\cdot}15$ s.$^{-1}$ at the crest and trough. The vertical disturbance, η, is given by

$$\eta(z, x, t) = \eta_0 \exp\left[-kz + i(wt + kx)\right]$$

so the maximum disturbance to the neighbouring sheets is thus

$$\eta_0(H) = \eta_0 \exp(kH),$$

where H = layer thickness.

If, for example, the layer thickness is 4 m., the disturbance produced by a 50 cm. high, 8 m. long wave will be only about 3 cm. The name, 'single-sheet' wave is, thus, quite valid. The wave will slowly lose energy to the neighbouring sheets at a rate depending upon $\eta_0(H)$ and upon the closeness of match of the phase speeds of the wave of length λ on the neighbouring sheets to that on the central sheet.

The mechanism for the generation of these waves is unclear, although it seems plausible to suggest that the current shear acts across the sheet in a manner analogous to the wind in surface wave generation. If, as it is supposed, the shear across the sheet comes primarily from the rather long thermocline waves, then it will persist for several hours before reversing; this gives a maximum time for the waves to grow to the observed steepness.

The observation that single-sheet waves often occur in clearly recognizable, coherent trains may be interpreted in several ways. Firstly, it implies that we seldom see a complex interaction between several wave trains from different sources, such as we usually find on the sea surface. This may result either from some physical barrier to their transmission, such as a break in the sheet; in which case it may be possible to deduce the likely dimensions of individual sheets. Alternatively they may lose energy so rapidly that they never travel very far from the source. The mechanism by which they lose energy is discussed in Woods (1970) where it is concluded the waves may reach an equilibrium between growth and decay. Another possibility is that the generating process excites one dominant wavelength on any given sheet, though this explanation seems unlikely in view of the discovery (to be described below) that Kelvin-Helmholtz instability occurs on a scale far smaller than the wavelengths of single sheet waves. Furthermore, the latter's time period is far greater than the Brunt-Väisälä period on the sheets.

To conclude, there is as yet no clear description of this interesting new class of internal waves, but it seems likely that the evidence gained from underwater photographs will lead to an adequate model and that such a model will have important implications for the microstructure of the thermocline.

Kelvin-Helmholtz Instability

One of the first conclusions drawn from the dye studies was that a large part of the thermocline exhibits laminar flow; turbulence occurs sporadically within clearly defined regions. It seems, therefore, that the thermocline should be treated as though it were in a state of sporadic transition from laminar to turbulent flow, rather like the flow through a pipe at the critical Reynolds number.

In the layers, turbulence occurs as discrete spots reminiscent of those described by Klebanoff, Tidstrom, and Sargent (1962) for transitional flow over a solid boundary, while turbulence on the sheets takes the form of a raft-like patch up to a hundred square metres in area, but only two to four times thicker than the sheet they envelop. In this section we shall describe the processes responsible for the initiation, growth, and decay of these patches of sheet turbulence.

The first problem is to explain why certain regions of a sheet become turbulent while the great majority continues to exhibit laminar flow. The criterion for stability in a sheared density interface was given by Kelvin (1887) and Helmholtz (1868); Rosenhead (1932) later analysed the rolling up of the interface once it had become unstable due to excessive shear and, recently, Miles and Howard (1964) have given two relations; the first linking the wavelength of the fastest growing instability with the thickness of the interfacial zone and the second linking the growth rate of the instability with its wavelength and the Richardson number, *Ri*, of the interface. This theory has stimulated experimental investigations in laboratory tanks (Thorpe 1968) and has led to a new interpretation of billow clouds by Ludlam (1967), who confirmed that Miles-Howard relation for K-H rolls growing on atmospheric inversions.

It has long been suspected (e.g. Defant 1960) that Kelvin-Helmholtz instability is to be found in the ocean and several workers have analysed short-lived temperature inversions found on bathythermograph soundings in terms of localized shear instability. The boundary between the Atlantic water flowing in through the straits of Gibraltar over the outflowing Mediterranean water is a particularly well-known site (Defant 1948) and has received considerable attention in recent years. However the evidence of soundings is indirect and could not be compared satisfactorily with Thorpe's dye photographs or Ludlam's billow cloud photographs. The filling of this gap in our knowledge of the ocean microstructure has been one of the successes of underwater research, with the result that the most vivid support for the Miles-Howard-Rosenhead theory now comes not from the laboratory, but from the ocean.

The experimental technique follows that illustrated in Fig. 102; a selected sheet is coloured with fluorescein dye from a continuous

Fig. 102a An array of dye packets moored at a depth that coincides with a thermocline sheet. The dye flows in the sheet giving a thin carpet of yellow dye.

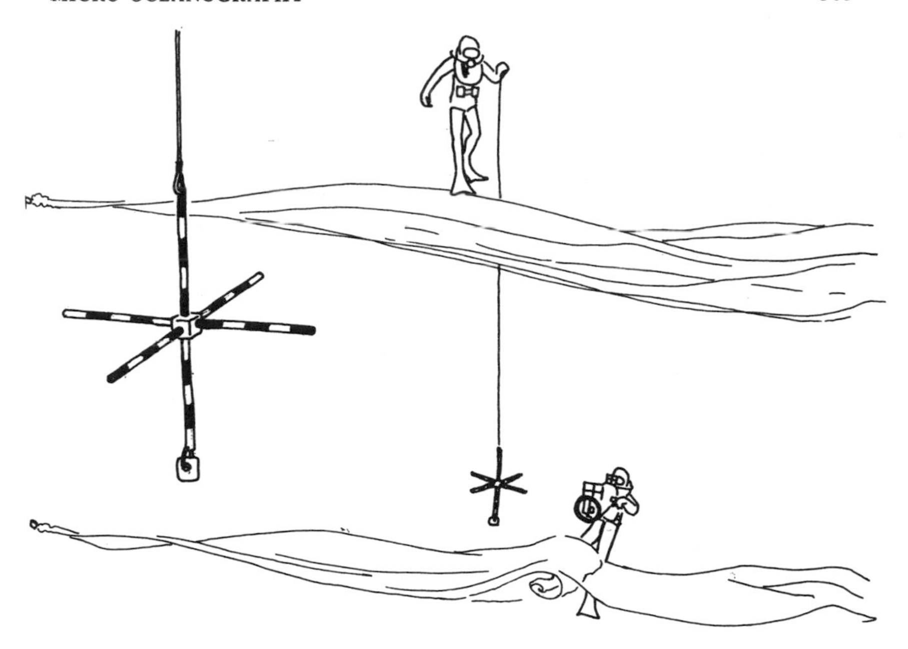

Fig. 102b

Filming the motions of dye in a sheet. The scale which has 10 cm. markings is lowered into the camera's field of view by a second diver who keeps well above the next higher sheet, which isolates his movement for the area being filmed.

multiple source moored at a fixed level. Over the great majority of the sheet's surface the dye forms long smooth filaments, confirming that it is in laminar flow. Simultaneous measurements of the temperature gradient and shear profiles through the sheet (using the techniques described previously) reveal that its Richardson number is greater than $\frac{1}{4}$, as predicted by Miles and Howard. Here the static stability is sufficiently strong to prevent any small disturbances from extracting energy from the shear; the sheet is stable.

Occasionally, however, and over a limited area the velocity difference across the sheet exceeds the critical value $(\Delta U)_c$ for which $Ri = \frac{1}{4}$

$$(\Delta U)_c = [2g\alpha\Delta Th]^{\frac{1}{2}}$$

For a $\frac{1}{4}$°C sheet, 10 cm. thick, the critical velocity is about 3 cm./s. Steady velocity differences of between 1 and 2 cm./s. are common across the stable, laminar flow parts of the sheets, so an additional local velocity difference of about the same value is needed for K-H

Fig. 102c

Kelvin-Helmholtz billows generated by the shear on the crest of a single sheet wave that can be seen on the *horizon* of the dye sheet.

instability. The only source of so large and yet so localized a shear is the single-sheet wave described in the last section, where it was shown that a 5-m. long, 50-cm. high wave produces a velocity difference of about ± 1·5 cm./s. across the sheet at its crest and trough. A careful examination of the photographs confirms that patches of sheet turbulence do indeed originate on the crests of single-sheet waves and subsequent analysis has shown that the Richardson number does fall below ¼ in these regions. The wavelength and growth rate of the K-H wavelets also follows the prediction of Miles and Howard and their subsequent rolling up takes the form predicted by Rosenhead. A detailed account of this analysis has been published elsewhere (Woods 1968d).

These observations confirm earlier theoretical studies and experiments in other media; in the remainder of this section we shall discuss two quite original results from the underwater investigation. The first of these concerns the detailed structure of the sheet during the early stages of K-H instability before the wavelets break into rolls. The first sign of instability, even before the wavelets have detectable asymmetry, is given by a periodic thinning of the sheet between the growing crests; this gives bands of reduced dye intensity (due to the shorter optical path length of dyed water) running parallel to the wave crests, as shown in Fig. 102c. By the time the rolls are fully grown the sheet between them is so thin that no dye can be detected there, although the rolls themselves are brilliantly stained. Thorpe has since reported a similar redistribution of the interfacial fluid during K-H instability in his tank experiments. This observation has led quite unexpectedly to a reinterpretation of the appearance of billow and noctilucent clouds. Kelvin-Helmholtz instability occasionally manifests itself in the former, but invariably in the latter as a striation on an existing thin cloud sheet. It had previously been assumed that these bands represented regions of the cloud which are raised or lowered by the K-H wavelets so causing condensation or evaporation respectively of the cloud particles. Thus the bright bands were considered to be regions of temporary uplift, and the dark bands regions of depression. However, the underwater studies have led us to an alternative explanation in which condensation-evaporation plays no part, the bands being formed by alternate thickening and thinning of the cloud sheet. Since this is an *inevitable* consequence of K-H instability it is preferred to the former explanation which needs rather special humidity conditions (Woods 1968b).

The other novel result from the investigation of K-H instability in the thermocline concerns the evolution of the patch of turbulence after the rolls have been formed. It is not practicable to extend observations to this stage in laboratory tanks where wall effects lead to increasing errors, or in the atmosphere where the release of latent heat in billow clouds introduces buoyancy forces which modify the air motion. These objections do not apply in the thermocline where the evolution follows the stages shown schematically in Fig. 103. Secondary rolls develop between and on the backs of the initial rolls again by K-H instability, but these are smaller since they form on parts of the sheet that has been thinned by the primary instability. Further instabilities follow until the region becomes continuously turbulent with eddies on all scales down to a few millimetres diameter. The smaller eddies become irregularly orientated and constitute a cascade although, since the Reynolds number of the original rolls is only about 1,000, this cannot accurately be called an *equilibrium* inertial cascade such as was described by Kolmogoroff (1941). The patch of turbulence floats on the sheet like a raft, extracting energy from the mean shear across itself and losing energy to viscosity (almost exclusively through eddies smaller than 1 cm.

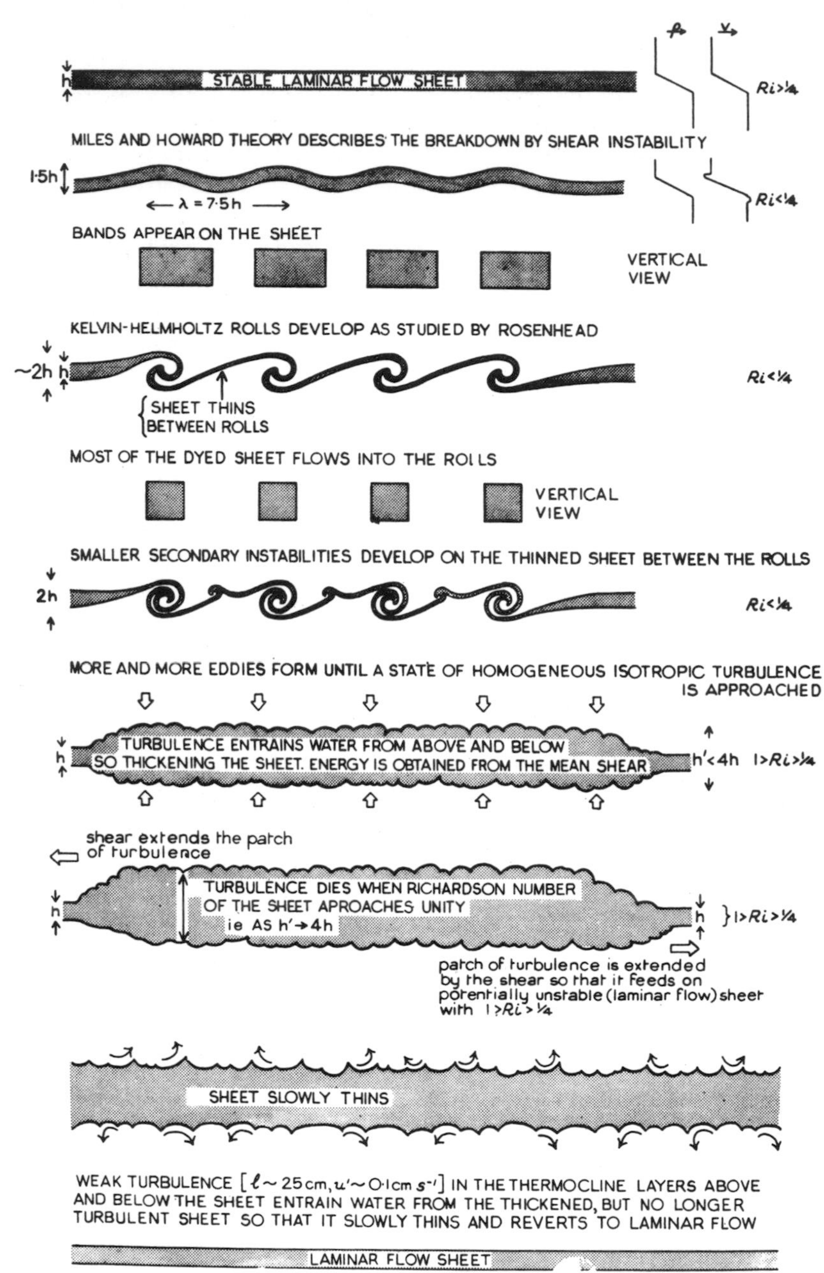

Fig. 103 Evolution of a patch of turbulence generated on a thermocline sheet by Kelvin-Helmholtz instability. The time from the start of instability to the termination of turbulent motions is approximately 10 minutes. The subsequent thinning of the thermocline sheet to its original thickness takes much longer. (From Woods 1969b.)

across), by working against gravity (i.e. by transporting heat down through the patch) and by extending the thickness of the patch through entrainment. These processes will be examined in the next sections; it suffices at this stage to indicate that the turbulence thickens until the shear is no longer able to supply the energy losses. The dye studies show that the turbulence finally decays when the patch of turbulence has reached three to four times the thickness of the unbroken sheet.

As the single-sheet wave whose shear provoked the instability advances along the sheet at a phase speed of a few centimetres per second it continues to break up the sheet on its crest until eventually the wave enters a region where the combined wave and drift shears no longer exceed the critical value. The wave then travels along the sheet without losing any more energy to *sheet* turbulence, although we shall see later (pp. 308–9) that it may be losing some energy to turbulence in the adjacent *layers*.

At its edges a patch of sheet turbulence feeds on the potentially unstable, laminar-flow sheet, which generally has a low Richardson number well below unity (but greater than $\frac{1}{4}$ since here the sheet has not spontaneously become unstable). The spreading of a patch of turbulence in this manner may readily be demonstrated by creating a patch of artificial turbulence as shown in Fig. 104. The original patch of turbulence shown in the photographs occupied an area of about 4 sq. m., but ultimately spread over an area 3 m. wide by over 30 m. long. The existence of such ribbons of turbulence had previously been the subject of conjecture by Phillips (1966), who called them *blini*. Grant, Moilliet, and Vogel (1968) have detected similar patches of turbulence by means of a hot film flowmeter mounted on the prow of a submarine on level flight through the thermocline.

Turbulence in the Layers

It is not practicable to explore the structure of layer turbulence by laying a horizontal dye sheet as we do in the thermocline *sheets* because the weak density gradients found in the *layers* are insufficient to suppress the turbulent wake of the dye packets used as sources and consequently the dye dilutes rather rapidly. So the only reliable evidence comes from dye streaks formed at low Reynolds number by falling pellets. These are difficult to use in a manner that permits easy, yet unambiguous interpretation, but, after two seasons' investigation it is possible to propose the following, rather tentative description of the turbulent motions inside the thermocline layers.

The first conclusion from the dye streak investigation is that the *layers* are also basically in laminar flow with sporadic localized patches of turbulence; like the *sheets* they are in a state of sporadic transition from laminar to turbulent flow. However, since (by definition) there are no extensive temperature steps in thermocline *layers*, it is unlikely that regions of layer turbulence will lie in sharply defined horizontal patches of clearly measurable thickness. On the contrary, we expect the turbulence to occur rather irregularly through the layers at sites determined more by the fluctuating shears than by the rather weaker, more slowly varying fluctuations in temperature gradient. Variations of both shear and temperature gradient may occur inside the layers by virtue of the slow diffusion of momentum and heat injected sporadically by the aperture conduction process described in Woods (1968a), but, as in the case of sheet turbulence, the extra shear needed to trigger off any particular instability probably comes from single-sheet waves.

The detailed structure of the individual regions of layer turbulence has not yet been fully established, but the evidence so far available suggests that breakdown occurs in isolated 'spots' in much the same way that transition has been found to occur near the critical Reynolds number in a neutrally stratified boundary layer over a solid surface (see, for example, the review by Kline *et al.* 1967). One point is clear, the diameter of the eddy formed by the initial breakdown lies, like the K-H rolls on a sheet, within a narrow range of sizes from about 15 to 30 cm. This 'primary eddy' lies in a vertical plane, but it soon breaks down to form smaller 'secondary eddies' which eventually form a cascade of scales with the smaller motions appearing to take no preferred orientation. During the five minutes or so that they persist, the turbulent motions have velocities of the order of 0·1 cm./s. and extend down to a viscous limit at about 2–3 mm. If, as in Woods (1968a), we accept the use of similarity analysis at such low Reynolds numbers, then the effective

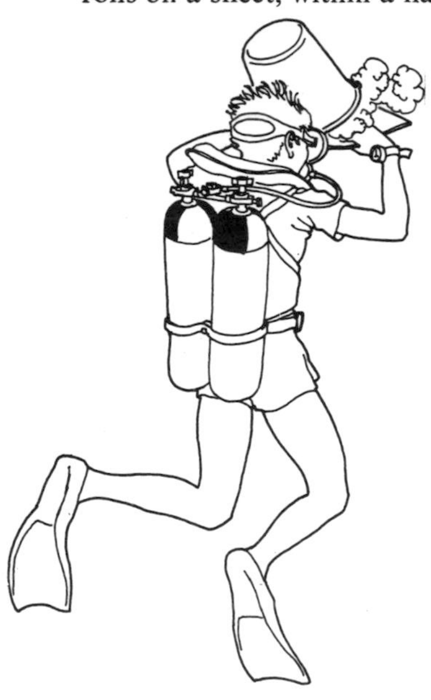

Fig. 104a

A diver opening a bucket of dyed fresh water below a thermocline sheet.

Fig. 104b The rising, buoyant dyed water entrains the surrounding sea water.

Fig. 104c The patch of artificial turbulence propagates across the surface of the sheet.

mean transport coefficient $K = 3$ cm.^{2}/s. and the viscous dissipation rate is of the order of 3×10^{-3} erg/gs. The heat budget in the subsequent analysis leads to the suggestion that these spots of turbulence may occupy on average about $\frac{1}{10}$ of the volume of the thermocline layers. In this case they will be extracting energy at a rate of some 10 erg per cm^2 cross-section of the whole thermocline (assumed to be 30 m. thick). Comparing this figure with the corresponding estimate of $\frac{1}{4}$ erg/cm.^{2} cross-section for the sheet turbulence (based on 10 sheets, 5 per cent covered by turbulence at 16 milli erg/gs.) we conclude that most of the turbulent dissipation in the thermocline comes from layer turbulence (although sheet turbulence is most effective in causing the decay of single-sheet waves).

Turbulent Entrainment in the Thermocline

Since the thermocline is in sporadic transition from laminar to turbulent flow, there are strong gradients of turbulence intensity present and turbulent entrainment inevitably follows. Phillips (1966) has shown that the turbulent entrainment velocity across an interface between laminar flow water on one side and water with turbulent fluctuations U' on the other is given by

$$U_e = \left(\frac{\nu U'}{l}\right)^{\frac{1}{2}}$$

where ν = viscosity of water;

l = diameter of eddy with water velocity U'.

The patches of sheet turbulence and the spots of layer turbulence both exhibit entrainment. In the case of the sheets this leads to a thickening of the patch of turbulence at a rate given by Phillips's entrainment velocity, which, for $l = 30$ cm., $U' = 0{\cdot}2$ cm./s., and $\nu = 10^{-2}$ cm.^{2}/s., is 1/000 cm./s. This compares reasonably with the estimated time $t = 300$ s. taken by a patch of turbulence to double its thickness, achieved by a mean entrainment velocity $U_e = 1/30$ cm./s. Entrainment at a comparable rate increases the diameter of the turbulent spots formed in the layers, although the effect here is masked by the growth of the turbulence as it feeds on the surrounding potentially unstable water (i.e. where $\frac{1}{4} < Ri < 1$) as occurs at the edges of the patches of sheet turbulence. Laboratory experiments by Turner (1968) have recently given an insight into the details of the entrainment process and our observations provide strong evidence in support of his physical model for an entraining interface.

Turbulent Transport through the Thermocline

If we are to understand the structure and development of the thermocline we must measure the heat, salt, and momentum fluxes due to the mixing elements described above. This means that we must develop an operational technique for detecting the spots of turbulence and making the necessary measurements of each one encountered. It turns out that probably the best way to do this is by an extension of the dye streak photography method used during the initial exploratory stages of our investigation. Frame by frame analysis of motion film of these dye streaks can yield the concentration, diameter (l) and overturning speed (u) of the billows, which start each spot of turbulence. We have shown (Woods 1970) that these parameters l and u are sufficient to let us calculate the effective vertical transport coefficients of heat, salt, and momentum. Thus

$$K_H = c_1 lu$$
$$K_S = c_2 lu$$
$$K_M = c_3 lu$$

where the coefficients c_1, c_2, and c_3 are likely to be constants if the Reynolds number is large ($\mathrm{Re} = lu/\nu \simeq 1{,}000$). But at small Reynolds numbers, such as we encounter in the thermocline, the coefficients are likely to depend on the Reynolds number of the billows. They must be determined by experiment. Meanwhile, we can reduce the number of unknowns by noticing that all the thermocline mixing occurs at $\frac{1}{4} < \mathrm{Ri} < 1$, say $\mathrm{Ri} = \frac{1}{2}$. Ellison (1957) has shown that for boundary layer turbulence at $\mathrm{Ri} = \frac{1}{2}$,

$$\frac{K_H}{K_M} = \frac{c_1}{c_3} = \frac{1}{5}$$

Ellison's theory leads to the further conclusion that only 10 per cent of the energy in the spot of turbulence is used to transport heat downwards, the remaining 90 per cent being lost to viscosity. This allows us to predict the rate at which the turbulence is extracting energy from the currents and waves in the thermocline. In a sheet, for example, this turns out to be (Woods 1970):

$$E = \frac{20\alpha}{s}\, ghQ \ \mathrm{erg/cm.^2\ s.}$$

where Q = heat flux through the sheet (cal./cm.^{2} s.);
h = sheet thickness (cm.);
α = coefficient of value expansion of sea-water;
s = specific heat of sea-water;
g = acceleration due to gravity.

The interest in this formula lies in the fact that, although it was derived from a knowledge of internal mixing gained by divers, the only unknowns Q and h can be measured by microstructure soundings made from a boat without the need to dive.

An instrument has now been made to sink freely through the ocean filming dye released from an arm below the camera (Woods 1969a). This device is being used to measure profiles of Richardson number and to collect the statistics on the concentration of spots of turbulence and their diameter and overturning speed from the sea surface to a depth of 1 km. It is hoped that the heat, salt, and momentum fluxes calculated from these data will help to explain why the thermocline is divided up into layers.

Micro-oceanography in Ocean Forecasting

The justification for extending our investigation of any natural phenomenon to a finer scale is that by doing so we become able to explain fluctuations of behaviour whose average properties are described adequately by coarser studies. In the atmosphere, the meteorologist can make useful forecasts of the broad pressure distribution of a large area using a three-layer model, but has to use a ten-level model with mountains and seas if he wishes to extend his forecast to include rainfall distributions. Similarly, the oceanographer is able to describe the broad distributions of current, temperature, and salinity and their seasonal variations by analysing water samples made at standard depths hundreds of metres apart, but these methods are quite inadequate to answer such questions as 'how does the sea temperature distribution change under a storm?' In order to do this we must construct new models involving the relevant microstructure features of the upper ocean. The role of micro-oceanography is to establish what structure is important in controlling variability in the oceans and to construct physical models that will meet the individual requirements of potential users of oceanic forecasting.

At present there are three main customers for oceanic forecasts, the Navy, the fishing industry, and the meteorologist. The military interest in oceanic variability arises from the use of echo-sounding apparatus (SONAR) in anti-submarine warfare. The speed of sound in sea-water varies with temperature, salinity, and pressure and a sound wave is refracted as it travels through water in which these properties vary. The effective range of a ship's sonar may be calculated by plotting the sound waves' curved path as they propagate through the sea, whose local temperature profile is measured by soundings with a Spilhaus bathythermograph. One consequence of the thermocline study described in this chapter will be the need to modify the detailed ray tracing procedure to take account of the laminae and to calculate the transmission loss due to scattering by patches of turbulence. Perhaps the resulting improvement in range prediction may have operational significance only in rather special circumstances, but the second naval application of micro-oceanography, in forecasting, is likely to have wider application. The aim of naval ocean forecasting is to predict the sea temperature profile from (i) the climatological mean profile, and (ii) the immediate past, the present and forecast weather for the chosen area. The procedure is to calculate the heat exchange at the surface and the rate at which this heat is carried down into the thermocline and by comparing these estimates with climatological mean values, to predict the probable temperature profile from day to day. The heat transport processes discussed on page 310 are an essential part of this process and will have to be included in any satisfactory forecasting model.

The fishing industry also needs to know the temperature profile of the sea, since fish movements are quite sensitive to temperature. However, as most fishing is carried out in shallow coastal waters where the weather can cause large changes in sea temperature by advection, it is less easy to demonstrate that the additional changes brought about by vertical conduction will have significance in fish behaviour. One case where the need for the micro-oceanographical method *has* been demonstrated is in tuna fishing; these pelagic fish are caught in the thermocline, at the 17°C isotherm. Traditionally fishermen have caught tuna by towing an inclined line carrying hooks which cover a broad depth range, but better catches can be achieved if the hooks are concentrated at the 17°C level, though at some risk of missing it altogether. In order to guarantee success the fisherman must fit a thermometer amongst his hooks, but failing this the local fisheries officer should be provided with some simple rules for forecasting the depth of the chosen isotherm.

A third application of micro-oceanography lies in the forecasting of the sea surface temperature (SST) from a knowledge of the existing weather. Over short periods of, say, from one to two days changes in the sea surface temperature may be neglected in forecasting the broad development of a given meteorological situation. However, over longer periods a persistent SST anomaly may significantly alter the weather and such features become particularly important to the meteorologist attempting to produce the 'long-term' forecast for a month or even a season ahead. Measurements of SST are obtained from ships' water intakes or from spot samples taken by bucket. In the winter the sea is practically isothermal so these two methods give the same result even though they sample water at significantly different depths, and the reported temperature is the actual SST seen by the overlying weather (i.e. no amount of additional wind mixing can alter the SST). In the summer, however, there are strong temperature gradients near the surface, so the 'bucket' temperature may be significantly higher than that measured at the ship's intake and these may both be lower than the actual SST (which may be measured by airborne infra-red radiation thermometers, though not on a routine operational basis). Furthermore, a change in the weather may alter the surface mixing rate (and perhaps the internal mixing rate, through the generation of internal waves) so changing the ocean temperature as seen by the weather. As a result the long-range forecaster is limited at present to using only the winter SST monthly anomaly patterns, which are reliable to the required accuracy. Eventually it may become possible to develop an accurate model that uses micro-oceanographic principles for the prediction of SST in summer.

Conclusion

The theme of this chapter has been the diver's use of dye tracers to explore the microstructure of the upper layers of the ocean. In the study of the Maltese thermocline, which occupied the greater part of the chapter, the evidence derived from a detailed analysis of dye tracer films has been sufficiently precise to support a novel model for vertical heat transport through the upper ocean. The unexpected discovery that a large proportion of the thermocline exhibits laminar flow explains why the tracers disperse so slowly and hence their effectiveness in revealing such relatively slow motions as a new class of (single-sheet) internal waves and the Kelvin-Helmholtz rolls that grow on their crests.

Our interpretation of the photographs of tracers in the sea owes much to the results of other workers who studied stratified flows in laboratory tanks. The underwater studies have the advantage of a larger scale and the elimination of unwanted boundaries: their greatest disadvantage is that the motions being studied occur naturally and so are not subject to the same control as can be applied in laboratory experiments. In certain circumstances, however, the diver can control the underwater environment; one example being the creation of an artificial patch of turbulence on a thermocline sheet in order to study its thickening by entrainment (see p. 310). To sum up, divers and dye tracers make an excellent combination for the initial exploration of small-scale flow patterns in a weakly turbulent sea; however, the diver must eventually give way to precision instruments tailor-made to measure accurately the flow field he has established in broad outline. In the early stages of development of these instruments it may prove useful to have divers position the sensors against dye tracers, but the ultimate aim should be to develop a model of the structure that permits a complete interpretation from the records of remote sensors. The interval between the diving scientist's original discovery and his final replacement by instruments is probably shorter in micro-oceanography than in any other discipline; the complete cycle took about five years for the Maltese thermocline, but may take much longer in the less tractable biological sciences.

Acknowledgement

This chapter is published by permission of the Director-General of the Meteorological Office.

References

Atkinson, K. B. and I. Newton (1968), 'Photogrammetry.' In *Photography for the Scientist*. Edited by C. E. Engel. London: Academic Press.
Banner, A. H. (1955), 'Note on a visible thermocline.' *Science*, **121,** 402.
Bowden, K. F. (1962a), 'Turbulence.' In *The Sea*. Edited by M. N. Hill. New York: Interscience.
Bowden, K. F. (1962b), 'Measurements of turbulence near the sea bed in a tidal current.' *J. Geophys. Res.*, **67,** 3181–6.
Bowden, K. F. (1965), 'Oceanography and Marine Biology.' In *Annual Review*, Vol. 2. Edited by H. Barnes. London: Allen & Unwin.
Brooke-Benjamin, T. (1968), 'Gravity currents and related phenomena.' *J. Fluid Mech.*, **31,** 209–48.
Charnock, H. (1965), 'Preliminary study of the directional spectrum of short-period internal waves.' *Proc. Second U.S. Navy Symposium on Military Oceanography*, 177–8.
Defant, A. (1948), 'Uber interne Gezeitenwellen und ihre Stabilitatsbedingungen.' *Arch. Met. Geophys.*, **1,** 52.
Defant, A. (1960), *Physical Oceanography*. London: Pergamon.
Ellison, T. H. (1957), 'Turbulent transport of heat and momentum from an infinite rough plane.' *J. Fluid Mech.*, **2,** 456–66.
Ellison, T. H. and J. S. Turner (1960), 'Mixing of dense fluid in a turbulent pipe flow.' *J. Fluid Mech.*, **8,** 514–44.
Engel, C. E. (Ed.) (1968), *Photography for the Scientist*. London: Academic Press.
Grant, H. L., R. W. Stewart, and A. Moilliet (1962), 'Turbulence spectra from a tidal channel.' *J. Fluid Mech.*, **12,** 241–68.
Grant, H. L., A. Moilliet, and W. M. Vogel (1968), 'Observations of turbulence in and above the thermocline.' *J. Fluid Mech.*, **34,** 443–8.

Helmholtz, H. (1868), *Uber discontinuirliche Flussigkeits—Bewegungen*. Akad. Wiss, Berlin, Monatsber, 215.

Kelvin, Lord (1887), 'Rectilinear motion of a viscous fluid between parallel plates.' In *Mathematical and Physical Papers*, Vol. 4, 321–30. London and New York: Cambridge University Press.

Klebanoff, P. S., K. D. Tidstrom, and L. M. Sargent (1962), 'The three-dimensional nature of boundary layer instability.' *J. Fluid Mech.*, **12,** 1–34.

Kline, S. J., W. C. Reynolds, F. A. Schraub, and P. W. Runstadler (1967), 'The structure of turbulent boundary layers.' *J. Fluid Mech.*, **30,** 741–7.

Kolmogoroff, A. N. (1941a), 'The local structure of turbulence in incompressible viscous fluid for very large Reynolds numbers.' *C.R. Acad. Sci. URSS.*, **30,** 301.

Kolmogoroff, A. N. (1941b), 'On degeneration of isotropic turbulence in an incompressible viscous liquid.' *C.R. Acad. Sci. URSS.*, **31,** 538.

Kraus, E. B. and C. Rooth (1961), 'Temperature and steady state vertical heat flux in the ocean surface layers.' *Tellus*, **13,** 231–8.

Krauss, W. (1966), *Methoden und Ergenbnisse der Theoretischen Ozeanographie*, II. Berlin: Interne Wellen Gerbruder Bornstraeger.

Lafond, E. C. (1962), 'Internal waves.' In *The Sea*. Edited by M. N. Hill. New York: Interscience.

Lamb, H. (1932), *Hydrodynamics*. London: Cambridge University Press.

Limbaugh, C. and A. B. Rechnitzer (1955), 'Visual detection of temperature–density discontinuities in water by diving.' *Science*, **121,** 395.

Ludlam, F. H. (1967), 'Characteristics of billow clouds and their relation to clear-air turbulence.' *Quart. J. R. Met. Soc.*, **93,** 419–35.

Miles, J. W. (1965), 'A note on the interaction between surface waves and wind profiles.' *J. Fluid Mech.*, **22,** 823–7.

Miles, J. W. and L. N. Howard (1964), 'Note on a heterogeneous shear flow.' *J. Fluid Mech.*, **20,** 331–6.

Munk, W. H. and E. R. Anderson (1948), 'Notes on a theory of the thermocline.' *J. Mar. Res.*, **7,** 276–95.

Pasquill, F. (1962), *Atmospheric Diffusion*. London: Van Nostrand.

Phillips, N. (1966), 'Large scale eddy motion in the Western Atlantic.' *J. Geophys. Res.*, **71,** 3883–91.

Phillips, O. M. (1966), *Dynamics of the Upper Ocean*. Cambridge University Press.

Rosenhead, L. (1932), 'The formation of vortices from a surface of discontinuity.' *Proc. Roy. Soc.*, **A134,** 170–92.

Schmitz, H. P. (1962a), 'A relation between the vectors of stress, wind and current at water surfaces and between the shearing stress and velocities at solid boundaries.' *Deut. Hydrograph Z.*, **15,** 23–36.

Schmitz, H. P. (1962b), 'On the interpretation of profiles of vertical velocity near the bottom in ocean and atmosphere as well as of wind stress over water surfaces.' *Deut. Hydrograph Z.*, **15,** 46–72.

Stewart, R. W. and H. L. Grant (1962), 'Determination of the rate of dissipation of turbulent energy near the sea surface in the presence of waves.' *J. Geophys. Res.*, **67,** 3177–80.

Sutton, O. G. (1953), *Micrometeorology*. London: McGraw-Hill.

Sverdrup, H. U., M. W. Johnson, and R. H. Fleming (1942), *The Oceans*. New Jersey: Prentice-Hall.

Thorpe, S. A. (1968), 'A method of producing a shear flow in stratified fluid.' *J. Fluid Mech.*, **32,** 693–704.

Tully, J. P. and L. F. Giovando (1963), 'Seasonal temperatures in the Sub-Arctic Pacific.' In *Marine Distributions*. Edited by M. J. Dunbar. Univ. of Toronto Press.

Turner, J. S. (1968), 'Effect of molecular diffusion on entrainment across a density interface.' *J. Fluid Mech.*, **33,** 639–56.

Woods, J. D. (1968a), 'An investigation of some physical processes associated with the vertical flow of heat through the upper ocean.' *Met. Mag.*, **97,** 65–72.

Woods, J. D. (1968b), 'On the formation of certain billow clouds.' *Quart. J. R. Met. Soc.*, **94,** 209–10.

Woods, J. D. (1968c), 'CAT under water.' *Weather*, **23,** 224–35.

Woods, J. D. (1968d), 'Wave-induced shear instability in the summer thermocline.' *J. Fluid Mech.*, **32,** 791.

Woods, J. D. (1969a), 'On designing a probe to measure ocean microstructure.' *Underwater Science and Technology J.*, **1**(1), 6.

Woods, J. D. (1969b), 'On Richardson's number as a criterion for transition from laminar-turbulent-laminar flow in the ocean and atmosphere.' *Radio Science*, **4** (12).

Woods, J. D. and R. L. Wiley (1971), 'Billow turbulence and ocean microstructure.' (In preparation.)

Zhukov, L. A., A. V. Mayer, and G. R. Rekhtzamer (1964), 'Use of underwater photo and movie survey for investigation of turbulence in the sea.' *Mat. 11 Konf. Probl.* 'Vzaimodeystviye Atmos Gidros Severn Chasti Atlantisch Okeana,' 151–5.

Author Index

General Index